AN INTRODUCTION TO MOLECULAR ELECTRONICS

edited by

Michael C. Petty, Martin R. Bryce and David Bloor

Centre for Molecular Electronics, University of Durham, UK

Oxford University Press
New York
1995

First published by Edward Arnold in 1995

Copyright © 1995 by Edward Arnold

First published in the United States by Oxford University Press, Inc.,
198 Madison Avenue, New York, New York 10016

Oxford is a registered trademark of Oxford University Press

ISBN: 0-19-521156-1
1 2 3 4 5 6 7 8 9
Printed in Great Britain

Contents

Contributors   xi
Preface   xiii

1   Molecular Electronics: Science and Technology for Today and Tomorrow   **1**
    D Bloor
    1.1   Introduction   1
    1.2   Molecular Materials for Electronics   2
           (Overview. Current applications. New opportunities)
    1.3   Molecular Scale Electronics   13
           (Background. Techniques for molecular scale science. Materials for molecular scale electronics)
    1.4   Conclusions   24
    References   25

2   Theory   **29**
    R W Munn
    2.1   Introduction   29
    2.2   Molecular Properties   30
    2.3   Molecular Arrangement   33
    2.4   Molecular Interactions   37
    2.5   Material Properties   41
    2.6   Conclusions   44
    References   45

3   Piezoelectric and Pyroelectric Materials   **47**
D K Das-Gupta
  3.1   Introduction   47
  3.2   Basic Concepts   48
         (Piezoelectric effect. Pyroelectric effect)
  3.3   Piezo- and Pyroelectricity in Polymers   60
         (Structural forms of poly(vinylidene fluoride). Copoly-
         mers of vinylidene fluoride. Other polymers)
  References   70

4   Molecular Magnets   **72**
R J Bushby and J-L Paillaud
  4.1   Introduction   72
  4.2   Basic Concepts   73
  4.3   Magnets Based on Transition Metal Complexes   73
  4.4   Organic Ferromagnets   85
  4.5   Conclusions   89
  References   91

5   Organics for Nonlinear Optics   **92**
G H Cross
  5.1   Introduction   92
  5.2   Basic Concepts   93
  5.3   Molecular Nonlinear Optics   98
  5.4   Linear and First Nonlinear Polarizabilities, α and β 99
         (p-NA, a model NLO molecule. Design for enhanced β)
  5.5   Second Nonlinear Polarizability, γ 103
         (Free electron models. Semiconductor models. Quantum
         chemical models. Non polymeric materials)
  5.6   Macroscopic Assemblies   109
         (Poled polymers)
  5.7   Conclusions   110
  References   111

6   Photochromism   **112**
P J Martin
  6.1   Introduction   112
  6.2   Basic Concepts   112
  6.3   Data Storage and Other Applications   114
  6.4   Classes of Photochromic Materials   117
         (Hydrogen tautomerism. Dissociation. Dimerization. *Cis-
         trans* isomerization. Cyclization. Charge-transfer)
  6.5   Conclusions   139
  References   139

7   Physics of Conductive Polymers   **142**
A P Monkman
  7.1   Introduction   142

7.2   Basic Concepts   142
7.3   Polyacetylene   146
7.4   Solitons   151
7.5   Doping   153
7.6   Polarons, Bipolarons and Charged Solitons   155
7.7   Chain Alignment   159
7.8   Photoinduced Absorption   160
7.9   Polyaniline   162
7.10 Device Applications   163
7.11 Conclusions   165
References   166

8   Conductive Charge-Transfer Complexes   168
M R Bryce
8.1   Introduction   168
8.2   Basic Concepts   170
8.3   TCNQ and TTF Systems   171
8.4   Metal-dithiolate Systems   178
8.5   Metallomacrocycles   179
8.6   Fullerenes   181
8.7   Conclusions   183
References   183

9   Liquid Crystals and Devices   185
D Lacey
9.1   Introduction   185
9.2   Basic Concepts   186
9.3   The Mesophases   188
      (Smectic phase. Nematic phase. Cholesteric phase (chiral
      nematic). Phase sequences)
9.4   Identification of Phases   196
9.5   Liquid Crystal Polymers   198
9.6   Applications of Liquid Crystals   200
      (Orientational elasticity. Anisotropic properties. Twisted
      nematic display. Supertwisted nematic display. Active
      matrix display. Ferroelectric display)
9.7   Conclusions   217
References   218

10   Langmuir–Blodgett Films   220
T Richardson
10.1 Introduction   220
10.2 Basic Concepts   222
      (Formation of insoluble monolayers. Langmuir–Blodgett
      deposition)
10.3 Characterization of LB Assemblies   228
      (General methods. Structural characterization tech-
      niques)

10.4   Fundamental and Applied Research   234
       (Fundamental research. Applied research)
10.5   Conclusions   241
References   242

11   Organic Molecular Beam Epitaxy   **243**
M Hara and H Sasabe
11.1   Introduction   243
11.2   Basic Concepts   246
11.3   *In situ* Observation of Film Growth—RHEED   249
11.4   Characterization of Assembled Molecules—STM   253
11.5   Prospects for OMBE   254
11.6   Conclusions   258
References   259

12   Scanning Tunnelling Microscopy   **261**
J P Rabe
12.1   Introduction   261
12.2   Basic Concepts   262
12.3   Solid Surfaces   264
       (Organic conductors. Noble metals. Layered materials)
12.4   Molecular Adsorbates under Ultra-High Vacuum   269
12.5   Molecules at Solid-Fluid Interfaces   269
       (Molecular structure. Molecular dynamics. Electronic
       properties.
12.6   Prospects for Molecular Manipulation   274
References   276

13   The Biological Membrane   **279**
J B C Findlay
13.1   Introduction   279
13.2   Structure   279
       (Lipid bilayers. Proteins. Carbohydrate)
13.3   Transmembrane Signalling   284
13.4   Material Transport   285
       (Nonchannel systems. Channel systems. Electron trans-
       port system)
13.5   Conclusions   291
References   294

14   Biosensors   **295**
M Aizawa
14.1   Introduction   295
14.2   Molecular Information Transduction in Biological Sys-   296
       tems
14.3   Design Principles of Biosensors   296
       (Molecular recognition and signal transduction. Two
       types of biosensor. Biocatalytic sensors. Bioaffinity sen-
       sors)

14.4   Electrochemical Biosensors   302
       (Electrochemical sensing principle. Amperometric glu-
       cose sensors. Molecular interfacing for electron transfer
       of enzymes. Potentiometric enzyme sensors. Enzyme
       FETs. Potentiometric immunosensors. Amperometric
       immunosensors)
14.5   Optical Biosensors   310
       (Optical sensing principles. Optical enzyme sensors.
       Optical immunosensors)
14.6   Conclusions   313
References   313

15   Biomolecular Optoelectronics   315
       R R Birge and R B Gross
15.1   Introduction   315
15.2   Linear Optoelectronic Devices   316
       (The function and photochemistry of bacteriorhodopsin.
       Holographic optical recording. Spatial light modulators.
       Holographic associative memories)
15.3   Nonlinear Optoelectronic Devices   332
       (Two-photon properties. Higher-order polarizabilities.
       Second-order polarizability of bacteriorhodopsin. Third-
       order polarizability. Nonlinear optical volumetric
       memories based on bacteriorhodopsin. Technique of
       writing two-photon data. The read operation)
15.4   Conclusions   341
References   342

16   Molecular Electronic Logic and Architectures   345
       J R Barker
16.1   Introduction   345
16.2   Inorganic Electronic Systems   346
16.3   Molecular Electronic Systems   348
16.4   Guidelines from Biological and Chemical Systems   349
16.5   Complexity and Systems Specification   351
16.6   Physical Representations of Data and Logic   351
       (The charge packet model. Molecular representations.
       The granular limit)
16.7   Logic Possibilities   354
       (Restoring logic. Neural logic and neural networks.
       Conservative logic. High level logic. Cellular automata.
       Floating architectures)
16.8   Molecular Representations   359
       (Molecular wires and switches. Molecular memory and
       logic. Molecular matrices. Single electronic systems.
       Quantum cellular automata systems. Optoelectronic sys-
       tems. Quasi-mechanical systems: molecular networks.
       Artificial bio-neural networks. Fluid models)

x   Contents

16.9   Practical Issues   370
        (Functionality and emplacement. The input-output prob-
        lem. Hybrid routes)
16.10   The Long-Term View   371
        (Emergent computation and artificial life. Bio-molecular
        systems. The silicon factor)
16.11   Conclusions   373
References   374

Index of subjects   377
Index of materials   387

Contributors

Masuo Aizawa
Department of Bioengineering, Tokyo Institute of Technology, Nagatsuta, Midori-ku, Yokohama 227, Japan

John R. Barker
Nanoelectronics Research Centre, Department of Electronics and Electrical Engineering, The University, Glasgow G12 8QQ, UK

Robert R. Birge
Center for Molecular Electronics, Center for Science and Technology, Syracuse University, Syracuse, New York 13244-4100, USA

David Bloor
Department of Physics and Centre for Molecular Electronics, University of Durham, South Road, Durham DH1 3LE, UK

Martin R. Bryce
Department of Chemistry and Centre for Molecular Electronics, University of Durham, South Road, Durham DH1 3LE, UK

Richard J. Bushby
School of Chemistry, University of Leeds, Leeds LS2 9JT, UK

Graham H. Cross
Department of Physics and Centre for Molecular Electronics, University of Durham, South Road, Durham DH1 3LE, UK

Dilip K. Das-Gupta
School of Electronic Engineering Science and Computer Science, University of Wales, Bangor, Dean Street, Bangor, Gwynedd LL57, 1UT, UK

John B. C. Findlay
Department of Biochemistry, University of Leeds, Leeds LS2 9JT, UK

Richard B. Gross
Center for Molecular Electronics, Center for Science and Technology,
Syracuse University, Syracuse, New York 13244-4100, USA

Masahiko Hara
Frontier Research Program, RIKEN, 2-1 Hirosawa, Wako, Saitama
351-01, Japan

David Lacey
Department of Chemistry, The University, Hull HU6 7RX, UK

Philip J. Martin
Formerly of: Centre for Molecular Electronics, Cranfield Institute of
Technology, Cranfield, Bedford MK43 0AL, UK

Andrew P. Monkman
Department of Physics and Centre for Molecular Electronics, University of
Durham, South Road, Durham DH1 3LE, UK

Robert W. Munn
Department of Chemistry and Centre for Electronic Materials, University
of Manchester Institute of Science and Technology, PO Box 88,
Manchester M60 1QD, UK

Jean-Louis Paillaud
School of Chemistry, University of Leeds, Leeds LS2 9JT, UK

Jürgen P. Rabe
Institut für Physikalishe Chemie, J. Gutenberg Universität Mainz, Jakob
Welder Weg 11, D-55099 Mainz, Germany

Tim Richardson
Department of Physics and Centre for Molecular Materials, University of
Sheffield, Hounsfield Road, Sheffield S3 7RH, UK

Hiroyuki Sasabe
Frontier Research Program, RIKEN, 2-1 Hirosawa, Wako, Saitama
351-01, Japan

*Preface*_____

The subject of molecular electronics has evolved during the 1980s as scientists and technologists have become aware of the potential applications for organic materials. As a highly interdisciplinary field, progress has depended on successful interactions across the boundaries of traditional disciplines. Molecular electronics encompasses biology, chemistry, computing, electronics and physics, each subject with its own methodology and jargon. This can be daunting to the newcomer.

Here we have attempted to draw together some basic ideas behind molecular electronics and present them in a coherent manner. The book is based on a series of short courses given in the University of Durham during the summers of 1987 to 1991. Many chapters are contributed by the original lecturers. Each is self-contained, with fundamental principles, the key-references and a look towards the future. We hope that the resulting text will be readily understandable to both final year undergraduates and research students from a wide range of backgrounds.

Chapter 1 is an overview of molecular electronics, providing a formal definition and outlining the scope of the subject. The important role of theory is emphasized in Chapter 2. Chapters 3 to 5 concern the dielectric, magnetic and nonlinear optical properties of organic materials. Photochromism is discussed separately in Chapter 6. Conductive organics—polymers and charge-transfer complexes—are covered in Chapters 7 and 8. To date, the commercial success story of molecular electronics has been liquid crystals. These, and devices based on them, are the subject of Chapter 9. Chapters 10 and 11 describe two important methods to manipulate organic molecules in thin film form: the Langmuir–Blodgett technique and organic molecular beam epitaxy. Chapter 12 introduces the scanning tunnelling microscope. This is not only a means to observe organic materials at the molecular level, but also to manipulate the individual molecules. Chapter 13 examines the function of the biological membrane, while Chapters 14 and 15 focus on specific applications of biological materials, in sensors and optoelectronics. Finally, Chapter 16 concentrates on systems architectures, highlighting implications for computational systems based on organic materials.

MICHAEL PETTY
MARTIN BRYCE
DAVID BLOOR

Durham

1

Molecular Electronics: Science and Technology for Today and Tomorrow

D Bloor

1.1 Introduction

Molecular electronic devices are both a reality and a possible basis for a future generation of electronic systems. The best known of molecular electronic devices currently incorporated in products is the liquid crystal display. The hope for the future is that molecules, either singly or as aggregates of nanometre dimensions, can be utilized to provide the elementary active units of electronic systems with extremely high component density. These two areas have been described as molecular materials for electronics and molecular scale electronics respectively.[1,2] Often molecular electronics has been taken as exclusively dealing with the prospects for molecular scale electronics. However, new technologies rarely spring up spontaneously. They may involve significant changes of direction but usually evolve from less sophisticated use of similar materials and properties. The 'cats whisker' diode in early radio receivers was an antecedent of today's integrated circuits. Furthermore the use of the macroscopic electrical and optical properties of molecular materials is facilitated by a thorough understanding of microscopic molecular scale models. The latter are essential if new molecules with either improved or specifically targeted properties are to be developed. Thus inevitably the development of molecular materials for electronics leads to a better understanding of microscopic properties and the potential that exists for molecular scale electronics. The two branches of molecular electronics are strongly inter-related and cannot be separated if the subject is to progress.

One consequence of this is that molecular electronics is inevitably a broad and interdisciplinary topic. The materials available for study include synthetic low molecular weight molecules, and macromolecules and their natural counterparts. The preparation of samples for study involves the chemist, biologist and physicist, i.e. synthesis, extraction and manipulation respectively. The investigation of the properties of these materials involves the same three disciplines bringing to bear their different expertise and perspectives. The practical utilization of materials of interest requires input from physicists, engineers and computational scientists. The former two groups are involved in the fabrication of devices and the evaluation of their performance, while the latter can provide new concepts for devices and systems which employ the unique properties of molecular materials to best effect. The outcome of the applied science is often a requirement for materials with more closely specified properties to enable improvements in device performance to be achieved. While such a scenario can be considered appropriate for many areas of materials development, that described above embraces more disciplines and skills than are generally required.

While the current uses of molecular materials in electronics are clearly defined, which materials will be used and how they will be used in molecular scale electronics remains to be determined as our knowledge of molecular scale science grows. Because of this and the cross-disciplinary nature of molecular electronics, no single book can hope to give a comprehensive account of molecular electronics. In the following chapters selected topics within molecular electronics will be described. This introductory chapter attempts to provide an overview and alert the reader to some other aspects of molecular electronics not considered in more detail later.

1.2 Molecular Materials for Electronics

Overview The technological utilization of materials depends on an overall set of properties that are either unique or superior to those of other materials. These properties include those essential for a particular device or application and those required for the manufacturing process. The demands of the latter often mean that it may not be possible to achieve optimum values for the former properties in a practical device. In most instances a compromise between the demands of performance and processing is inevitable. Outstanding properties, e.g. non-linear optical coefficients, conductivity, are of little use if the material is not stable under conditions appropriate for both processing and use.

Some reactivity with the normal ambient can be tolerated, most semiconductors currently in use are not inert, however, extreme sensitivity to air exposure makes processing difficult and costly. Normal operation of electronic devices involves exposure to temperature of 60–100°C over extended periods, while processing often results in brief excursions to temperatures in excess of 200°C. If the material is intrinsically unstable under any of these conditions it is obviously of little use. The chemical and thermal stability of organic compounds can be adversely affected by reactive impurities. The preparation of highly pure materials is therefore of utmost importance. This is also true in basic scientific investigations, where the aim is to understand the relationship between properties and structure, from the micro- to the macroscopic scales. While the environmental conditions to which samples are exposed are less severe than those encountered in processing and use, there is little point in making measurements where the results are dictated by the presence of impurities and defects. Achieving high levels of purity is not easy. Hence there is a need for thorough characterization of materials, whether they are for pure or applied research. A detailed knowledge of sample composition and structure provides the basis for distinguishing between artifacts that adversely affect properties and those whose presence is immaterial. This point cannot be emphasized too strongly, since the development of molecular electronics has been hampered by spurious results obtained on poorly characterized materials.

The need to produce high quality materials with tightly specified properties is one of the factors which accounts for the time required for the progression from the discovery of an interesting property to the marketing of a product based on it. Good examples of this are liquid crystals and conductive polymers. The existence of liquid crystalline phases has been known since early this century. The potential of liquid crystals as materials for novel displays was realized when the first patents for this application were filed in the 1930s. However, the first materials with the properties required to realize displays functioning at room temperature were not produced until the 1960s. Impurities were finally reduced to an acceptable level and commercial displays marketed in the 1970s.

Some of the materials currently classified as conductive polymers were known in the last century, e.g. oligo- and polyanilines. The development of chemical and electrochemical syntheses in the 1950s and 1960s attracted little attention but laid the basis for subsequent developments. The discovery that polyacetylene could be turned into a metallic conductor by the action of strong oxidizing and reducing agents in the 1970s sparked a huge upsurge in research activity. All kinds of technological application were promised to be just round the corner. It has taken more than a decade to find solutions to the problems of processability and stability. Commercial products are just coming onto the market in the form of polymer powders, blends and coatings. These are primarily aiming at the simpler applications of electromagnetic screening and elimination of static charging. Batteries utilizing polyaniline have also been put on sale. More demanding applications to active electronic devices are still some way off in the future. Here the continual development of existing technology presents a moving target, and though conductive polymers do offer some unique properties how best to translate these into a clear advantage in device technology remains a challenge.

The introduction of new materials into high technology devices is a slow process requiring many years of underpinning scientific effort. Even when materials with the desired properties are produced it will probably require about ten years to realize a marketable product employing them. The aim of molecular electronics is to provide the scientific basis which can lead to commercialization. In this area, molecular materials for electronics, this is on a near to mid-term time scale; five years for well established materials and ten years or more for newer materials. Molecular scale electronics is a much longer term activity with little chance of making a commercial impact well into the next century.

Current Applications   Liquid crystals have been mentioned already and are the prime example of how properties unique to a class of molecular materials have led to significant application.[3-5] Because of this liquid crystals will be considered at some length here, despite the more detailed account to be found in Chapter 9.

The liquid crystalline state is intermediate between the high order of crystalline solids, where the long range of molecules on lattice sites is only weakly perturbed by thermal motion, and the disorder of liquids, where only short range order exists and molecules are free to translate and rotate. In liquid crystals the inter-molecular interactions limit the freedom for molecular reorientation, so that the degree of order is intermediate between that of crystalline solid and isotropic melt. In nematic liquid crystals there is a tendency for molecules to align in a common direction, the directrix. However, this alignment is not perfect and there is no translational order, Fig. 1.1(a). In smectic liquid crystals there is also on average a common orientation but in addition there is a tendency for the molecules to cluster into separate planar sheets. However, there is little positional order within each sheet, Fig. 1.1(b). There are numerous smectic liquid crystalline phases depending on the orientation of the directrix with respect to the layers and from its variation from layer to layer. A special case is the class of cholesteric liquid crystals; the directrix lies in the plane of the ordered layers but rotates from layer to layer to trace out a helix.

These structures have usually been associated with linear molecules such as the cyanobiphenyl and the fluorinated biphenyl shown in Fig. 1.2. Which phases occur depend on the intermolecular interactions and hence on the detail of the molecular structure. The lateral substitution of the fluorines favours the occurrence of a tilted smectic phase rather than the nematic phase of the cyanobiphenyl compound. More recently flatter molecules have been shown to occur in discotic phases in which the rotating molecules arrange in stacks much like a stack of plates, Fig. 1.1(c). Understanding how the different liquid crystal phases arise is important because it enables us to understand how interactions between molecules can produce structure over intermediate (meso) and macroscopic regions.

As we will see later the idea that structures with complex large scale features can be obtained simply by the correct choice of intermolecular interactions, so-called self-assembly, is an important concept for molecular scale electronics. This is a prime example of how the need to tailor materials for devices, e.g. liquid crystal displays, can be helpful in the longer term scientific endeavour.

The use of liquid crystals in displays depends on the fact that the structure of liquid-crystal phases is sensitive to external influences. The fact that the pitch of the helix in cholesteric liquid crystals depends on

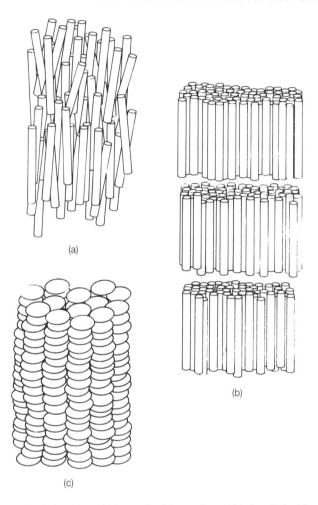

(a)

(b)

(c)

Fig. 1.1 Generalized structures of (a) nematic, (b) smectic and (c) discotic liquid crystals.

temperature has been used in thermal sensing. Since the pitch of the helix is comparable to optical wavelengths cholesteric liquid crystals reflect strongly at particular wavelengths. As the pitch varies with temperature so does the colour of reflected light giving a simple visual temperature indication. In the displays most familiar to us all an electric field is used to rotate the directrix of a nematic liquid crystal from the plane of the display to the normal to the plane. On its own this effect would not produce a display, however, the directrix is found to align with surfaces of glass or polymer which have been rubbed to produce a microtextured surface. Though the exact mechanism of this effect is still not fully understood it

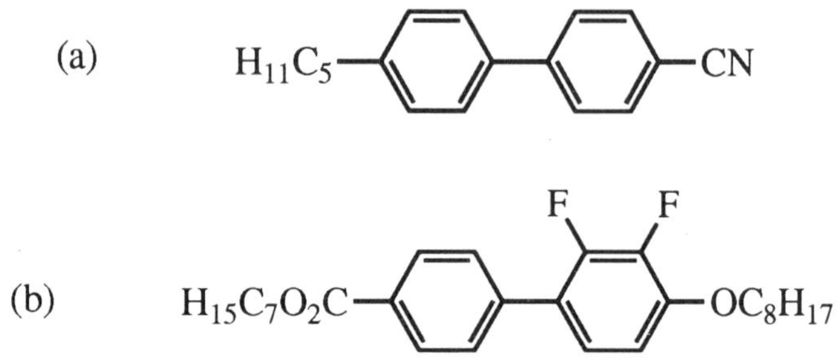

(a) $H_{11}C_5$—⟨benzene⟩—⟨benzene⟩—CN

(b) $H_{15}C_7O_2C$—⟨benzene⟩—⟨benzene with F, F⟩—OC_8H_{17}

Fig. 1.2 Molecular structure of (a) cyanobiphenyl and (b) fluorinated biphenyl.

allows a simple device structure. A liquid crystal layer is held between a transparent top electrode and a back electrode. The rubbed surfaces in contact with the liquid crystal layer are arranged at right angles. With no field applied the directrix rotates through 90° due to the surface orientation imposed by the rubbed surfaces. The plane of polarization of light traversing this structure will rotate in sympathy with the rotation of the directrix. Thus both transmitted and reflected light will have its plane of polarization rotated by the liquid crystal layer. With an electric field applied the molecules reorient parallel with the normal. The plane of polarization of light travelling through the film is now unaffected. Thus use of either a polarizer and analyser in transmission or a single polarizer in reflection gives a light/dark contrast between the two states of the film, cf. Chapter 9, Section 9.6.

The technology is currently being used in watches, calculators, computer displays and televisions. However, there is a finite switching speed which causes problems where fast switching is required. This problem is being addressed by the use of ferroelectric liquid crystals. The fluorinated biphenyl, Fig. 1.2(b), has an electric dipolar moment perpendicular to the long axis of the molecule. In the tilted smectic liquid crystal phase these dipoles order parallel to one another to give a ferro-electric phase. Reorientation of the overall electric moment can occur rapidly by a collective motion, in which the directrix swings around a cone defined by the tilt angle. The use of these materials in displays and fast electro-optical devices is under investigation, cf. Chapter 9, Section 9.6.

This account of liquid crystal properties and applications illustrates many aspects of the interplay of pure and applied science. Had there not been the drive towards producing a practical display, then far fewer compounds would have been studied and our understanding of the interplay of molecular structure and the order which has resulted would be far less complete. Many of the unusual features displayed by liquid crystals would probably have been a novelty rather than the subject of a major research activity. It should also be noted that complete scientific understanding is not needed for an application to be successful. The empirical use of rubbed surfaces to align liquid crystals is central to current technology but not precisely understood. The development of less sophisticated uses should not be ignored, temperature indication with cholesteric liquid crystals has found a range of uses from thermal mapping for medical diagnostics to executive desk top thermometers. It lacks the sophistication of displays but has been commercially successful. The simpler applications should not be overlooked in efforts to produce materials for high technology.

In comparison organic semiconductors have had a chequered history. They were the subject of intense study in the 1950s and 1960s in parallel with the work on which current silicon integrated circuits are based. The success of silicon resulted in a loss of interest in its organic counterparts. Progress has been achieved[6-7] but the published literature does not reflect their technological importance. This is because the prime use of organic semiconductors is in electrophotographic reproduction (Xerography). In Xerography a photosensitive drum charged electrostatically is discharged in regions exposed to illumination. The charge is transferred to the copy paper and toner is picked up by the electrostatic action. Organic semiconductors provide a combination of high dark resistance, photoconductivity and carrier mobility ideal for use as the charge storage layer in a Xerographic copier.[8] Combinations of organic semiconductors and polymeric matrices are now widely used commercially. Often several layers are used with different additives to control photoconductive response, charge injection and carrier mobility to improve speed and sensitivity of the Xerographic process. The toners used to render the image visible are mixtures which incorporate dyes and organic semiconductors. The commercial sensitivity of this application means that much of the important research on new materials is not published. However, the general problem of charge transport in disordered organic materials has been extensively discussed in the open literature since it is a basic topic applicable to a wide range of organic semiconductors.[9]

Organic materials possess good piezo- and pyroelectric coefficients which have resulted in their technological use, cf. Chapter 3. Some use has been made of organic materials in infra-red detectors, e.g. triglycine sulphate, however the optimum pyroelectric response has been shown to occur in films 10–100 nm thick.[10] Such thin films can be prepared by the Langmuir–Blodgett method, cf. Chapter 10. However, the difficulty encountered in preparing optimized materials has not allowed this possibility to be realized yet. In contrast extensive use has been made of the piezoelectric properties of the ferroelectric polymer, polyvinylidene fluoride, and the related copolymers,[11] cf. Chapter 3, Section 3.3. Poled polymer films are available commercially and the polymer properties, notably the low density, provide an excellent acoustic match to water. This has made it useful for sonar transducers and ultrasonic transducers for medical diagnosis. The use of these polymers in microphones has been investigated, where the ability to produce a free standing piezo-electric film has been put to good use in noise cancelling microphones.

Though there has been a trend to replace many materials of natural origin with synthetic alternatives in traditional, nonelectronic uses there are areas where it has not yet been possible to reproduce the particular properties displayed by natural materials. The use of enzymes in biosensors with a high specificity for a particular metabolite is a case in point,[12] and Chapter 14. The metabolic processes in living systems depend on redox reactions involving a range of small molecules. The redox processes are carried out by specific enzymes which have evolved to recognize a particular metabolite and effect a particular redox reaction. Abstraction of the enzyme from its natural environment and immobilization of the enzyme at an electrode provides a system in which the redox reaction can give rise to electron transfer either to or from the electrode. The current observed with such a system is related to the concentration of the specific metabolite in the solution in contact with the electrode.

Interest in such sensors has been driven by the fact that many metabolites are common across different living systems and their concentration is symptomatic of the well being of the system. Biosensors are, therefore, potentially of considerable value in medical diagnosis and monitoring. The prime example is the glucose sensor based on glucose oxidase, which is a valuable monitor of diabetes. The redox cycle of GOD at an electrode is shown in Fig. 1.3.

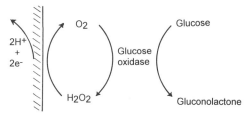

Fig. 1.3   The redox cycle of glucose oxidase at an electrode surface at a potential of 320 mV.

Though biological macromolecules are often envisaged as rather fragile molecules which degrade if not kept in a living system this is not always the case. Enzymes either adsorbed onto electrode surfaces or incorporated into a conductive polymer matrix can be quite robust. Glucose oxidase based sensors can be stored over reasonable periods and, though currently used as a one off test, considerable effort is being expended to develop the methodology to provide devices capable of *in vivo* monitoring over extended periods.

Since the potential markets for medical sensors are large and the glucose oxidase sensor has been successful, it is not surprising that there is a high level of research into other systems which could be utilized as biosensors. Many other mechanisms and detection methods are being investigated, e.g. specific adsorption of antibodies and antigens detected by methods sensitive to the binding of submonolayer quantities of materials. These include optical methods, cf. Chapter 14, Section 14.5, surface acoustic waves and quartz crystal microbalances.

Biosensors provide an excellent example of how the highly developed, specific function of biological materials can be utilized in practical devices.

The same processes can be considered for use in molecular scale devices as discussed later, Section 1.3. A further example of use as an active material for electro-optical devices is considered in the next section.

New Opportunities In addition to these current uses of molecular materials for electronics, the progress achieved over the last decade in the science of conductive polymers has created new opportunities. Many conjugated polymers such as those shown in Fig. 1.4 are intrinsic semicon-

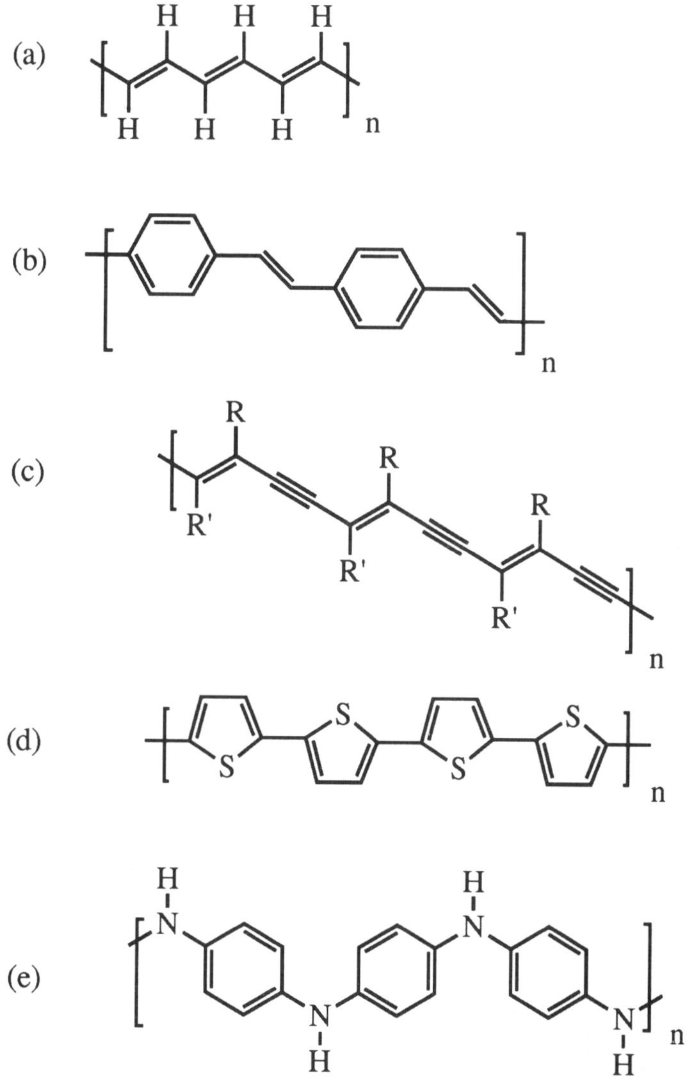

Fig. 1.4 Idealized structures of conjugated polymers, (a) trans polyacetylene, (b) polyparaphenylenevinylene, (c) polydiacetylene, (d) polythiophene and (e) polyaniline.

ductors which can be made conductive by either oxidation or reduction. The structural relaxations, which occur in these polymers in the vicinity of added charge carriers, have created much excitement and a degree of controversy. A wide range of applications have been considered. In the metallic state conductive polymers have high work functions which result in efficient charge injection into conventional semiconductors. The difficulty encountered in producing pure well characterized samples of conductive polymers and their reactivity in air, particularly for polyacetylene, has hindered progress. However, a variety of different approaches have enabled cleaner materials to be prepared either *in situ*[13] or in soluble, processable forms.[14] These have enabled Si:polyacetylene Schottky diodes to be prepared at low temperatures. Such devices display ideal photovoltaic characteristics, due to the absence of surface states generated in the Si by damage resulting from the high temperature depositions of normal metals.[15] Polyacetylene with conductivity greater than 10^5 Scm^{-1} has been produced and intensively studied.[16] The use of precursor polymers has enabled spin coating to be used to fabricate field effect transistors [FETs] and light emitting diodes [LEDs] with active semiconducting polymer layers.[17,18] The former were initially limited by the low carrier mobility in the amorphous polymer; however, recent studies of simple oligomeric analogues has shown performance comparable with amorphous silicon devices.[19] The latter offer the prospect of inexpensive large area electroluminescent displays. Many other materials in addition to conductive polymers have been investigated in this context.[20] While device efficiency can be reasonable, the device lifetimes achieved in the research laboratory are not yet sufficient for commercial exploitation. These advances are encouraging but considerable work in both the pure and applied arenas is needed before they make the transition from research laboratory to commercial device, cf. Chapter 7.

Another class of organic materials, which has been the focus of recent attention, is that of nonlinear optical materials,[21] cf. Chapter 5. The increasing use of optical communication systems has resulted in a demand for improved electro-optical materials, i.e. materials with large non-linear optical coefficients. The two classes of materials studied, first those with a large quadratic term in the response to applied electric fields (second-order nonlinearity) and secondly those with a large third-order (electric field cubed) response, provide an interesting contrast. The second-order nonlinearity of polar organic molecules, cf. Fig. 1.5, can be predicted using relatively simple quantum mechanical calculations involving the ground and first excited electronic states. This had enabled many potentially active molecules to be designed prior to synthesis. The resulting practical realization has resulted in significant increases in the maximum second-order nonlinear optical coefficients. The combination of these active species into polymers allied with the processability of the products offers reasonable prospects for practical devices.

In contrast the prediction of third-order nonlinear properties requires complex multi-level, multi-excitation calculations. Beyond the simple prediction of the activity of linear systems with highly polarizable elec-

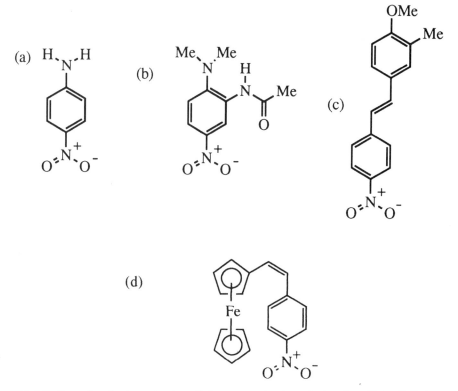

Fig. 1.5 Examples of polar molecules with large second order optical nonlinearities, with their common abbreviations, (a) mNA, (b) DAN, (c) MMONS and (d) ferrocenyl nitrophenylethylene.

trons, there have been few concrete guidelines to aid synthesis. Little progress has been made in increasing third-order nonlinear coefficients over the last decade. Despite this, optical image correlation on a picosecond timescale has been demonstrated in polyacetylene thin films.[22] A simple device has been demonstrated with an effective data rate of 7×10^{16} pixels per second. In a different class of conjugated polymers, the polydiacetylenes, Fig. 1.4, fundamental studies have shown that there is a narrow wavelength region below the absorption edge where the nonlinear absorption, i.e. the light induced absorption which can render a material useless for practical applications, becomes zero. In some materials the linear absorption is also low at this point but the intensity dependence of the refractive index is large.[22] Thus, these materials offer prospects for applications in optical, logic and dynamic holography.[23] However, if large scale use of the third-order nonlinearity is to emerge it seems that a breakthrough in either theory or experiment is required.

One possibility for such a breakthrough is provided by a natural material, bacteriorhodopsin.[24] The perceived fragility of natural electro-optically active materials has inhibited their study and exploitation, however bacteriorhodopsins have proved to be an exception. Bacterior-hodopsin containing membranes are produced by organisms which are evolved to utilize either available oxygen or light as an energy source. These membranes are produced whenever the available oxygen becomes insufficient and the photoreaction system is required to sustain the organism. The membrane can be extracted and sedimented as a film which displays a complex photocycle, cf. Chapter 15, Section 15.2. By the use of either low temperatures or modified material (either chemically or geneti-cally) a photochromic behaviour of potential use in optical data storage, dynamic holography, spatial light modulators, etc. has been obtained. Such films also display useful second- and third-order optical nonlinearities and a fast photoresponse which has been used in photodetectors with 20 psec rise time. Rapid progress with applications has been reported,[25,26] cf. Chapter 15. This is a timely reminder that, as well as providing model systems with which to study molecular scale electronic processes, naturally derived materials must not be overlooked as candidates for conventional electronic and opto-electronic applications.

Another topic which is of relatively recent origin is that of molecular magnetic materials, cf. Chapter 4. One challenge in this area has been the preparation of a purely organic ferromagnet. Some of the early claims of success have been discounted due to ferromagnetic impurities. Examples of ferromagnetism at low temperatures are now well established, and a recent claim for room temperature ferromagnetism in a copolymer of polyaniline[27] suggests that rapid progress will be achieved. A number of interesting metal-organic materials have been synthesized. Of these com-pounds those possessing a co-operative transition between two spin states of the metal centres display large changes in their absorption spectra. This effect shows a large thermal hysteresis and can be used to produce displays.[28]

The study of conductive organic charge-transfer salts has been a fruitful field for the study of the interplay of electron–electron and electron–lattice interactions in determining overall physical properties, e.g. insulating (antiferromagnetic), semiconducting, conducting and superconducting. This study has been made possible by the synthesis of tailored molecular electron donors and acceptors, cf. Chapter 8. These materials have been incorporated in electrolytic capacitors as a protective layer, i.e. when local breakdown occurs the current flowing destroys the conductive CT salt and prevents further current leakage. Despite this these materials remain primarily of interest for pure research exploring the details of their structure-property relationships.

Another topic which has considerable potential for application is photo-chromism, the reversible photoinduced colour change observed in many organic compounds, cf. Chapter 6.

As our knowledge of the properties of molecular materials grows there will be further opportunities for application in devices. There is a clear need for a continuing dialogue between those in pure and applied research and between academics and industrialists. Without this, potential areas for

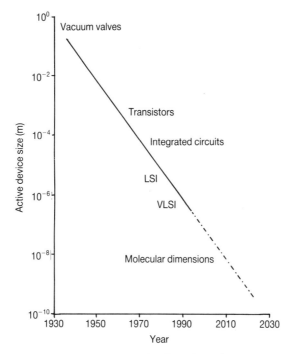

Fig. 1.6   Historical and projected reduction in the dimensions of active semiconductor devices.

application will be missed. The future of molecular electronics depends on such opportunities not only being spotted but also being converted into commercially viable devices by vigorous collaboration between the disciplines and academic and industrial sectors. An increased awareness of the value of molecular materials and the realization that they meet the requirements of purity and robustness needed for successful exploitation in devices is important, if the much longer range potential of molecular scale electronics is to stand any chance of being realized.

1.3   Molecular Scale Electronics

Background   There have been two conceptual bases for the possibility of molecular scale electronics. The first is the continual reduction in the size of an elementary, active electronic device over most of the twentieth century, Fig. 1.6. If this trend is extrapolated linearly then somewhen early in the next century the molecular scale is reached. Such a plot disguises the fact that major changes in technology have been necessary to enable the past reduction in size to be achieved. These are from vacuum to solid state devices, from discrete to integrated devices and from micron to submicron integration. Thus it is probable that a further reduction to the molecular scale will require a comparable technical advance. In this context one possibility is the emergence of molecular scale electronics based on molecules.

An alternative view is that of complexity. As silicon device technology has advanced so has the complexity of the devices produced. Indeed many of the devices now being fabricated would have been considered as systems until quite recently. Such complexity creates problems of device yield and testing, and is potentially a factor limiting the development of silicon based technology. Contemporary advances in biology have moved in the opposite direction; starting from studies of whole animals and plants attention has focused on to the component parts down to the molecular level. Now the structure and processes within individual biomacromolecules and small subassemblies of molecules are being investigated. The complexity of the subassemblies is comparable with that of highly integrated silicon devices. Hence there is the possibility of synergistic, interdisciplinary study which may lead to molecular scale electronics based on either naturally derived materials or their synthetic analogues.

The latter approach is attractive, as living systems can be considered to be an existence theorem for molecular electronics based on molecular scale processes. However, it must be treated with caution as there are numerous examples of technology which, while based on the same physical principles, have an embodiment which contrast sharply with that found in nature. The aerodynamics of the flight of birds and aeroplanes has a common basis but the mechanisms adopted are totally different. Thus, while there is much that can be learnt from nature, this may only extend to providing principles which can be realized by totally synthetic means.

Early attempts to explore the possibility of molecular scale electronics failed because of the lack of methods which enabled experiments to be conducted at the molecular scale. This was true of efforts to make a technological leap from the first integrated circuits to the molecular scale.[29] It was also true in the 1970s when the discovery of metallic conductivity in doped polyacetylene seemed to offer the prospect of a molecular wire. This prompted much speculation in the early 1980s as to the possible structures of molecular systems that would provide functionality similar to that offered by silicon integrated circuits.[30–32] Though these models were based on known natural and synthetic materials they presented major synthetic difficulties and could not be investigated at the molecular level. This together with rather extravagant claims for the realization of biochips, i.e. self-organizing, self-repairing integrated systems based on biomacromolecules, created considerable scepticism about the scientific value of the area.[33]

The prospects for progress towards molecular scale electronics are now much improved, as scientific developments have provided the means to explore the properties of molecular systems at the molecular scale. The worldwide initiation of programmes, which address both molecular materials for electronics and the possibilities for molecular scale electronics and emphasize the interdependence of the areas, have focused serious scientific attention on the topic. The convergence of activities in several disciplines at the molecular scale has created a fertile ground for interdisciplinary

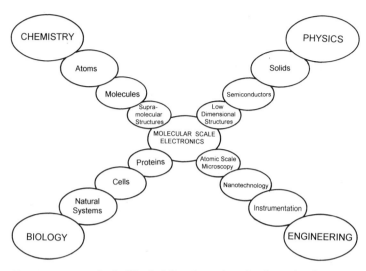

Fig. 1.7 The convergence of scientific disciplines towards molecular scale science.

research, Fig. 1.7. As noted earlier, within the biological sciences there is now an interest in understanding the structure and function of biomacromolecules in either isolation or small assemblies. Chemistry starting from the synthesis of simple compounds has developed methods for the production of complex molecular structures, and is beginning to explore how to control primary and secondary structure simultaneously. This will provide means of mimicking the self-assembly of biomacromolecules but with the possibility of totally new, synthetic molecular architectures. The physical sciences have developed methods to produce controlled superstructures with molecular dimension, e.g. by the Langmuir–Blodgett method for organics and molecular beam epitaxy for inorganic semiconductors. Allied with advances in the engineering of submicron dimension structures, nanostructures, this has focused attention on the properties of low dimensional structures, i.e. electrons confined in sheets, lines and dots. The advent of the scanning tunnelling microscope and related scanning probe microscopies has made it possible to interact directly with individual molecules either electronically, optically or mechanically. This, together with the development of spectroscopic techniques for studying single molecules in amorphous matrices, has provided the tools with which to carry out investigations at the molecular scale.

Now the barrier preventing progress in molecular scale science seems finally to have been removed. It is, therefore, appropriate to press forward in an effort to lay a scientific basis with which the potential for molecular scale electronic devices can be properly assessed.

Techniques for Molecular Scale Science   Given that there are now methods which enable measurements to be made at the molecular scale there is still the question: what next? Should we study either specially synthesized molecules designed to provide different, detectable states of either conformation, charge transport or optical absorption; or should we utilize naturally derived materials likely to behave in a similar manner? What methods should we use to study them? How can we manipulate such materials to provide supra-molecular structures, which will inevitably be required if a molecular electronics system is to be produced? In short, where do we begin and which direction do we take? Finding an answer is not simple since the questions posed above are not independent, e.g. manipulation may be coupled with either synthesis or scanning probe microscopy, the method of study will be related to the molecular switching function, etc. In the following some of the possibilities will be outlined.

The two principal classes of experimental probe seem likely to be scanning probe microscopes and single molecule microscopy.

Scanning tunnelling microscopy originated in the early 1980s and rapidly progressed to provide atomic resolution images of high quality semiconductor surfaces,[34] and Chapter 12. An electron tunnelling current is established between a sharp conductive tip and the conductive surface being studied, Fig. 1.8. The exponential dependence of the current on tip surface separation provides vertical resolution of atomic dimensions, and also ensures comparable lateral resolution since the current generally flows through a single atom on the tip, i.e. the one closest to the surface. The method was soon utilized to study conductive organic substrates, such as conductive organic charge-transfer salts and molecules adsorbed on conductive surfaces.[34,35] The images obtained have been difficult to interpret, first, because of the dynamical behaviour of monolayers adsorbed on

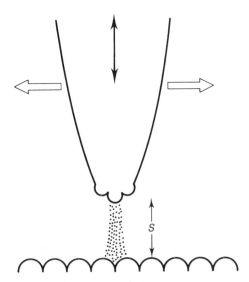

Fig. 1.8 Schematic view of a scanning tunnelling electron microscope, the vertical and lateral positions of the probe tip are controlled by piezoelectric transducers. The tunnelling current, which depends exponentially on the probe-surface separation, S, from the atom at the tip of the probe dominates (atoms in the probe are indicated only near the tip).

surfaces, and secondly because of the uncertainty concerning the mechanism giving rise to contrast in the STM images. Theoretical models are being developed[35] and as more materials are studied the role of the molecule-substrate interactions are becoming clearer.

Other scanning probe microscopies depend on the direct force interaction between the probe and the surface, the atomic force microscope (AFM),[37] and the coupling of evanescent optical fields between the surface and tip, near field optical microscopy (NFOM).[38] Closely related to the STM is the scanning electrochemical microscope,[39] although the resolution is inherently lower owing to the diffusion of the electrochemical species in the solution between the probe tip and surface. However, useful information about the electrochemical behaviour of substrates can be obtained down to 100 nm resolution. The AFM has been used to measure differences in frictional forces between molecular scale regions in two component Langmuir–Blodgett films where separation of the two components has occurred.[40] It has also been modified to determine the surface electrical potential.[41] Together with the determination of surface densities of states for carriers and current versus voltage characteristics with the STM there are now a wide variety of tools for probing local surface properties.

The STM has been used to modify surfaces, dissociate individual molecules and move individual atoms to form structures of an atomic scale. It has opened up the possibility of fabricating structures at the sub-nanometre scale.[42] However, this raises questions of how stable such structures are and what mechanical forces arise when the tips in AFMs and STMs are brought close to surfaces both in vacuum and liquid media.[43]

A number of other methods may be utilized to prepare organic molecular materials as samples with nanometre dimensions. The Langmuir–Blodgett film method[10] and Chapter 10, is a long established method with which to manipulate monolayers of amphiphilic molecules and build up layered supra-molecular structures. It has the disadvantages of access to a restricted set of molecules, i.e. those that form monolayers at the air–water interface, and a somewhat disordered structure within the layers, since highly crystalline monolayers are difficult to transfer from the water surface to a solid substrate. Despite this it has been an important tool in studying monolayers and supramolecular layered structures. Alternatively ultra-high vacuum deposition techniques have been used,[44] here solvent free films are obtained but again the choice of molecule-substrate combinations which give high quality monolayers appears to be limited. The ultimate aim of developing a flexible organic MBE method for molecular scale structured multilayers is still to be realized. Electrochemical methods have been devised to produce multilayer films of conductive polymers to be prepared with layer thicknesses as small as 10 nm.[45]

Conventional nanolithography is capable of producing lateral dimensions down to a few nanometres in size.[46] Methods are being developed to impress patterns into organic materials.[47] Thus, it will be possible to fabricate nanostructures in organic materials either directly or by adsorption of molecules onto a patterned substrate. Such regular structures can be produced directly using biomacromolecules and this has been used to fabricate nanostructures by transferring the structure from the biological master to an inorganic copy.[48]

The spectroscopic technique of 'hole burning' has been studied since the 1970s. In this method either a photophysical or a photochemical process leads to a change in the electronic structure, and hence in the optical absorption spectrum of a molecule in a polymer matrix. Since the molecule is just one member of an ensemble which gives rise to an inhomogeneously broadened absorption band, the change reduces the absorption at a specific wavelength within the band, i.e. 'burns' a hole in the absorption band profile, Fig. 1.9.[49] Extension of the method to the wings of the absorption band, where the density of site specific absorption lines is low, has enabled the hole burning,[50] fluorescence[51] and ESR[52] of single molecules to be studied. These studies are already turning up unexpected results when isolated molecules are probed compared with the situation where an ensemble is involved. While the applications normally considered for this

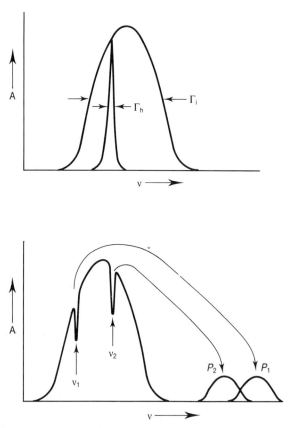

Fig. 1.9   (a) Spectral profile of an inhomogeneously broadened absorption bond of width Γ_i is composed of numerous homogeneous bonds of width Γ_h but different centre frequencies. (b) Hole-burning caused absorption intensity to be removed from bands at ν_1 and ν_2 and transferred to product bands, P_1 and P_2.

effect are optical data storage and dynamic holography it can also in principle be applied as a molecular based neural network.[53] It offers an alternative means of studying the properties of isolated molecules in matrices where there are a range of sites available to the active molecule.

Materials for Molecular Scale Electronics Where does one start in selecting materials for study at the molecular scale? There are a number of possibilities; first, one can attempt to prepare a material to implement one of the many postulated molecular electronic devices. Secondly, one may prepare a molecule designed to have a switching function, e.g. a photochromic molecule, etc. Thirdly, one may utilize a naturally derived material with a specific functionality, which may be of use for molecular scale electronics.

The first of these is probably the most difficult. Consider for example the proposal of Aviram and Ratner for a molecular rectifier made in 1974.[54] A monolayer of molecules with a strong electron accepting group at one end and a strong electron donor at the other is sandwiched between two metal electrodes, Fig. 1.10. With a voltage applied in one direction electrons can tunnel easily from the metal into the energy levels of the donor and acceptor and out into the other metal electrode. For a reverse bias no easy path is available and the effective resistance of the film is much higher. The problems of synthesizing a model molecule that would form a densely packed, stable monolayer and coating it with metal electrodes without damage were such that over a decade elapsed before they were solved. Finally using zwitterionic molecules and a magnesium evaporated electrode rectification was demonstrated in 1990, Fig. 1.11.[55] While the final embodiment has interesting properties it is unlikely to replace conventional rectifiers in electronic systems.

Thus the second approach may be preferable since the likelihood of finding novel properties, unique to molecular systems, is greater. Considerable effort has been expended in producing photochromic molecules for macroscopic applications, cf. Chapter 6. However, though there has been discussion of bistability in these and similar molecules[56] there has been little effort to investigate their behaviour at the molecular scale. Alternatively chromophores coupled to form molecules in which there can be excitation transfer have been suggested as potential molecular switches.[57,58] Such molecules can also be considered as synthetic analogues of the natural photosynthetic systems.[59,60] Again, though many studies have been reported of macroscopic samples in solution and glassy phases, study at the molecular scale has yet to be undertaken.

The ability of biomacromolecules to 'self-assemble' into higher order supra-molecular structures with specific functionalities has prompted chemists to explore synthetic methods of self-assembly. It has been argued that this will be a necessary prerequisite for the development of molecular electronic systems, since component by component assembly would severely limit the scale of systems. One starting point is to view complex fluids, such as liquid crystals, as simple systems which self-assemble.[61] It is possible to go one stage further, using molecular interactions to produce

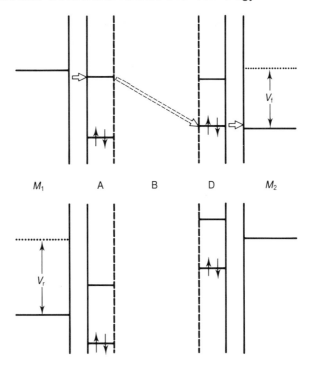

Fig. 1.10 Energy level schemes for an idealized monolayer molecular rectifier; for forward bias V_f (above) and reverse bias Vr (below), electron tunnelling path shown by open arrows. M_1, M_2 metal contacts, A accepting moiety, D donating moiety and B molecular bridge.

aggregates which display liquid crystalline properties.[62] Alternatively molecular self-assembly can be considered as a means of producing nanometre-scale structures in a reproducible manner without recourse to lithography.[63] A variety of approaches have been developed which provide molecules trapped in molecular cages,[64] interlocked ring molecules [65] and rings on linear molecular chains, Figs 1.12–1.14.[66,67] The methodology of fabricating purely carbon cages and tubules, i.e. fullerenes and graphitic nanotubes is now well established.[68,69] Numerous other molecular architectures have been considered, e.g. Kohnkenes.[70] This molecular engineering is still in its early stages but offers prospects of new and unusual materials for study at the molecular scale, e.g. the ring on chain molecular systems have been considered as a molecular abacus, cf. Chapter 16, Section 16.8.

The kinetics of chemical reactions may also have an important role to play. Molecular interactions can lead to the autocatalytic synthesis of simple molecules, i.e. self replication.[71,72] Furthermore large scale spatial and temporal organization can occur when chemical reactions are driven far from equilibrium in fluids[73] and on surfaces.[74]

The final class of materials which can be considered are those derived

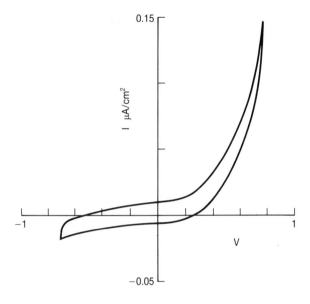

Fig. 1.11   Current voltage characteristic for the molecule shown below the plot (after Ashwell *et al.*[55])

from natural sources. Bacteriorhodopsin has been considered earlier in the context of macroscopic applications. The process involved in photoreaction centres and vision have been studied in considerable detail, e.g. Schoenlein *et al.*[75] These processes involve electron excitation and transfer and show that highly efficient electron transfer can occur in biomacromolecules.[76] Surface modification of these molecules enables the electron transfer to be directed to external electrodes. Thus, there is considerable scope for exploiting biological electron transfer systems in the context of molecular electronics.

Natural systems do not rely primarily on electron transport for the communication and manipulation of information, instead ions are utilized. The transport of ions through biological membranes, which is the basis of communication along and between nerves, is described in more detail in Chapter 13. It is a highly efficient process which can switch currents in

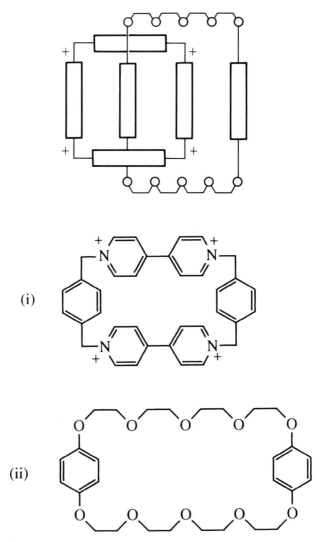

(i)

(ii)

Fig. 1.12 Chemically engineered molecules; a rotaxane formed from molecules (i) and (ii).

the range of 10^{-3} A, in a squid giant axon, to 10^{-12} A, in a single *trans* membrane ion-channel.[77]

The transport of ions through membranes can be affected by isolated natural ion channels, which have been extensively studied, and control can be realized by applied voltages[78,79] and chemicals, such as toxins and metabolites, Fig. 1.15.[80,81] Much simpler channels, which retain the ability to control the passage of ions, can be formed from specially synthesized

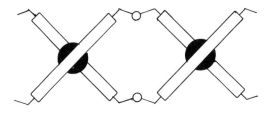

Fig. 1.13 Chemically engineered molecules; a double helix formed from bipyridyl units (rectangles) bonded to Cu (black circles), the repeat unit of the strand is shown below.

peptides. These have been studied primarily as models for natural ion channels but could be used to implement reproducible ion gates.[82] In this context ion channels offer two advantages compared to the conceptual models of electronic molecular switches. First, ions are massive particles and much less affected by quantum effects than electrons, and secondly ion channels have gain, i.e. many ions can flow through while the gate is open. Thus, ion channels have the possibility of fan-out from one gate to many,

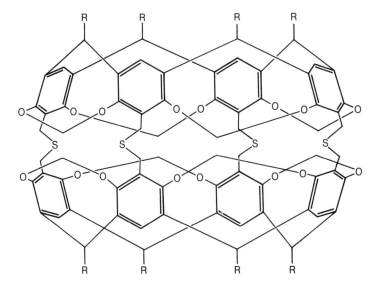

Fig. 1.14 Chemically engineered molecules; a carcerand, a molecular cage (R = $(CH_2)_4CH_3$).

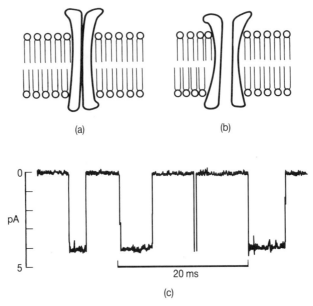

(a)   (b)

(c)

Fig. 1.15   A schematic model of a membrane ion channel (a) closed and (b) open; (c) typical signals from ion channels from a rat's muscle.

an important factor in the realization of silicon based integrated devices. Finally it should be noted that single molecules offer function as ion channels when incorporated in membranes and that light activated ion transport can be realized.[83]

Finally the ability of natural systems to move cells and subcellular structures should not be ignored in the context of molecular electronics. While nano-mechanics, i.e. the nanofabrication of machines, has attracted much interest the ability of biological systems to produce controlled motion, flow, etc. has attracted less attention. The crawling of cells[84] and the motion of proteins along the microtubules within nerve cells[85] are two examples from which we might learn. Growing interest in this topic[86] may open up new possibilities for molecular electronics.

1.4   Conclusions

Molecular electronics poses numerous challenges across a range of disciplines. Building bridges between the disciplines is a key step in initiating the research and development needed to make progress. Two factors are driving such collaborative research. First, in the area of molecular materials for electronics is the prospect of the commercialization of active optoelectronic and electronic devices based on organic materials. The second is the emergence of molecular scale science, where the problems relevant to one discipline can only be tackled by the application of techniques developed in another.

What does the future hold? Certainly liquid crystal devices will continue to be developed and new commercial applications will appear. It is to be hoped that these will be joined by nonlinear optical devices, based on the large nonlinearities displayed by organic molecules, and LEDs and other active electronic devices, where both polymers and low molecular weight materials may have a role to play. Biosensors will continue to be developed principally in the context of medical diagnostics. Other biological materials, e.g. bacteriorhodopsin, may open up other areas of application to naturally generated materials.

The best ways forward to molecular scale electronics have not yet been identified. Several possibilities have been considered above and others are discussed in Chapter 16. An interesting parallel development deriving from developments in nanolithography is the observation of single electron transport in nanostructures.[87] As metallic structures become smaller and their capacitance also decreases, the voltage induced by the presence of a effect have been proposed and it has been claimed that single electron transport has been observed at room temperature. Can such systems be realized using molecules? Indeed do they already exist in molecular conductors such as the metallic charge transfer salts and polymers? What can molecular electronics derive from these studies? These questions are considered in Chapter 16.

Silicon has come to dominate modern computing engines. As circuitry develops, the possibilities for artificial intelligence are being explored and the use of multiply connected models of natural neural networks considered. These are very simple models of natural systems but these ideas have been transferred back into biological science.[88] It has also stimulated the fabrication of silicon circuits which are more realistic analogues of real nerve cells[89] and the human visual receptor.[90] While these studies are based on silicon technology and are focused on gaining better insights into the function of natural systems, these results could provide new insights applicable to molecular scale electronics.

Molecular electronics is already playing a role in the rapid development of electronic and optoelectronic technology that increasingly affects all of us. It has the potential to become even more important in the future as new technologies emerge.

References

1. D. Bloor in *Fine Chemicals for the Electronics Industry II*, Eds. D. J. Ando and M. G. Pellatt, (Royal Soc. Chem., London, 1990), p. 265.
2. D. Bloor in *Chemistry of Advanced Materials*, Ed. C. N. R. Rao, (Blackwell, London, 1992), p. 295.
3. F. Funada in *Fine Chemicals for the Electronics Industry II*, Eds. D. J. Ando and M. G. Pellatt, (Royal Soc. Chem., London, 1990), p. 97.

4. I. C. Sage, *loc. cit.*, p. 114.
5. E. P. Laynes, *loc. cit.*, p. 130.
6. J. Simon and J. J. Andre, *Molecular Semiconductors*, Springer Verlag, Berlin, (1985).
7. *Conjugated Polymeric Materials: Opportunities in Electronics, Opto-electronics and Molecular Electronics*, Eds. J-L. Bredas and R. R. Chance, (Kluwer Academic, Dordrecht, 1990).
8. H. W. Gibson, *Polymer*, **25**, 3 (1984).
9. P. M. Borsenberger, L. Pautmeier and H. Bassler, *J. Chem. Phys.*, **94**, 5447 (1991) and references therein.
10. G. G. Roberts in *Molecular Electronics—Science and Technology*, Ed. A. Aviram, (United Eng. Trustees, N.Y., 1990), p. 309.
11. R. G. Kepler and R. A. Anderson, *Adv. in Phys.*, **41**, 1 (1992).
12. C. R. Lowe, N. C. Foulds, S. E. Evans and P. F. Y. Yontlin, in *Electron. Properties of Conjugated Polym. III*, Eds. H. Kuzmany, M. Mehring and S. Roth, (Springer Verlag, Berlin, 1989), p. 432.
13. W. J. Feast, *Phil. Trans. Roy. Soc.*, **A330**, 117 (1990).
14. J. Roncali, *Chem. Rev.*, **92**, 711 (1992).
15. E. J. Ginsberg, C. B. Gorman, R. H. Grubbs *et al.*, in *Conjugated Polymeric Materials*, Eds. J-L. Bredas and R. R. Chance, (Kluwer Acad., Dordrecht, 1990), p. 65.
16. J. Tsukamoto, *Adv. in Phys.*, **41**, 509 (1992).
17. J. H. Burroughes, C. A. Jones and R. H. Friend, *Nature*, **335**, 137 (1988).
18. P. L. Burn, A. B. Holmes, A. Kraft *et al.*, *Nature*, **356**, 47 (1992).
19. B. Xu, D. Fichou, G. Horowitz and F. Garnier, *Adv. Mater.*, **3**, 150 (1991).
20. Y. Hamada, T. Sano, M. Fujita, T. Fujii, Y. Nishio and K. Shibata, *Jpn. J. Appl. Phys.*, **32**, L511 (1993) and references therein.
21. D. Bloor in *Organic Materials for Non-linear Optics II*, Eds. R. A. Hann and D. Bloor, (Royal Soc. Chem., London, 1991), p. 1.
22. G. M. Proudley, P. D. Foote, L. M. Connors *et al.*, *Opt. Comput. Processing*, **2**, 139 (1992).
23. S. Molyneaux, A. K. Kar, B. S. Wherret, T. L. Axon and D. Bloor, *Opt. Lett.* **18**, 2093 (1993).
24. C. Brauchle, N. Hampp and D. Oesterhelt, *Adv. Mater.*, **3**, 420 (1991).
25. Q. W. Song, C. Zhang, R. B. Gross and R. Birge, *Opt. Lett.*, **18**, 775 (1993).
26. Q. W. Song, C. Zhang, R. Blumer, R. B. Gross, Z. Chen and R. R. Birge, *Opt. Letts.*, **18**, 1373 (1993).
27. S. Galaj and A. Le Mehaute, *Eur. Patent* 0545819 Al (1993).
28. O. Kahn, J. Krober and C. Jay, *Adv. Mater.*, **4**, 718 (1992).
29. E. Braun and S. MacDonald, *Revolution in Miniature*, Cambridge University Press, Cambridge, 1978.
30. A. Aviram, P. E. Seidan and M. Ratner, in *Molecular Electronic Devices*, Ed. F. L. Carter, (Marcel Dekker, New York, 1981), p. 5.
31. H. Sixl and P. Higelin, in *Proc. II Int. Workshop on Molecular Electronic Devices*, Ed. F. L. Carter, (Marcel Dekker, New York, 1985), p. 17.
32. J. J. Hopfield, J. N. Onuchic and D. N. Beratan, *Science*, **241**, 817 (1985).
33. D. Bloor, *Springer Ser. Sol. St. Sci.*, **107**, 437 (1992).
34. J. P. Rabe, *Adv. Mater.*, **1**, 13 (1989).
35. S. Buchholz and J. P. Rabe, *Angew. Chemie. Int. Ed. Engl.*, **31**, 189 (1992).
36. M. M. D. Ramos, *J. Phys. Condens. Matter*, **5**, 2843 (1993).
37. P. A. Christensen, *Chem. Soc. Rev.*, p. 197 (1992).
38. E. Betzig and J. K. Trautman, *Science*, **257**, 189 (1992).

39. A. J. Bard, F-R. F. Fan, D. T. Pierce, P. R. Unwin, D.O. Wipf and F. Zhou, *Science*, **254**, 68 (1991).
40. R. M. Overney, E. Meyer, J. Frommer *et al.*, *Nature*, **359**, 133 (1992).
41. M. Fujihira, H. Kawate and M. Yasukate, *Chem. Lett.*, p. 2223 (1992).
42. H. Rohrer, *Jpn. J. Appl. Phys.*, **32**, 1335 (1993).
43. U. Londman, W. D. Leutke, J. Ouyang and T. K. Yia, *Jpn. J. Appl. Phys.*, **32**, 1444 (1993).
44. M. Hara, P. E. Barrows, A. F. Garito and H. Sasabe, *Mol. Cryst. Liq. Cryst. Sect. B. Non-linear Optics*, **2**, 253 (1992).
45. M. Fujitsuka, R. Nakahara, T. Iyoda *et al.*, *Synth. Met.*, **53**, 1 (1992).
46. D. K. Ferry and R. O. Grandon, *Physics of Sub-micron Devices*, (Plenum, New York, 1992), Chap. 2; R. F. Pease, *Jap. J. Appl. Phys.*, **31**, Part 1, 4103 (1992).
47. R. Leuschner, H. Ahne, A. Hemmerschmidt *et al.* *Adv. Mater.*, **4**, 753 (1992).
48. K. Douglas, G. Devand and N. A. Clark, *Science*, **287**, 642 (1992).
49. J. Friedrich and D. Haarer, *Angew. Chemie.*, **96**, 96 (1984).
50. Th. Basche, W. P. Ambrose and W. E. Moerner, *J. Opt. Soc. Amer. B*, **9**, 829 (1992).
51. H. Talon, L. Fleury, J. Barnard and M. Orrit, *J. Opt. Soc. Amer. B*, **9**, 825 (1992).
52. J. Kohler, J. A. J. M. Disselhorst, M. C. J. M. Donkers, E. J. J. Groenen, J. Schmidt and W. E. Moerner, *Nature*, **363**, 242 (1993).
53. U. P. Wild, A. Rebane and A. Renn, *Adv. Mater.*, **3**, 453 (1991).
54. A. Aviram and M. A. Ratner, *Chem. Phys. Lett.*, **29**, 277 (1974).
55. G. J. Ashwell, J. R. Sambles, A-S. Martin, W. G. Parker and M. Szablewski, *J. Chem. Soc. Chem. Commun.*, 1374 (1990), A. S. Martin, J. R. Sambles and G. J. Ashwell, *Phys. Rev. Lett.*, **70**, 218 (1993).
56. U. Kolle, *Angew. Chem. Int. Ed. Engl.*, **30**, 956 (1991).
57. R. A. Bussell, A. P. de Silva, H. Q. N. Gunaratue, P. L. M. Lynch, G. E. M. Maguire and K. R. A. S. Sandanayake, *Chem. Soc. Rev.*, 187 (1992).
58. J. Jortner and M. Bixon, *Mol. Cryst. Liq. Cryst.*, **234**, 29 (1993).
59. M. K. Wasielewski, *Chem. Rev.*, **92**, 435 (1992).
60. D. Gust, T. A. Moore and A. L. Moore, *Acc. Chem. Res.*, **26**, 198 (1993).
61. P. G. de Gennes, *Science*, **256**, 495 (1992).
62. M. P. Taylor and J. Herzfeld, *J. Phys. Condens. Matter.*, **5**, 2651 (1993).
63. G. M. Whitesides, J. P. Mathias and C. T. Seto, *Science*, **254**, 1312 (1991).
64. D. J. Cram, *Nature*, **356**, 29 (1992).
65. D. Philp and J. F. Stoddart, *Synlett.*, 445 (1991).
66. P. R. Ashton, D. Philp, N. Spencer and J. F. Stoddart, *J. Chem. Soc., Chem. Commun.*, 1124 (1992).
67. G. Wenz and B. Keller, *Angew. Chem. Int. Ed. Engl.*, **31**, 197 (1992).
68. G. S. Hammond and V. J. Kuck, Eds., *Fullerenes, Synthesis, Properties and Chemistry of Large Carbon Clusters*, ACS Symp. Ser. No. 481 (ACS, Washington, 1992).
69. T. W. Ebbesen and P. M. Ajayen, *Nature*, **358**, 220 (1992).
70. J. P. Mathias and J. F. Stoddart, *Chem. Soc. Rev.*, 215 (1992).
71. A. Terfort and G. V. Kiedrowski, *Angew. Chem. Int. Ed. Engl.*, **31**, 654 (1992).
72. P. A. Bachmann, P. L. Luisi and J. Lanz, *Nature*, **357**, 57 (1992).
73. K. J. Lee, W. D. McCormick, Q. Ouyang and H. L. Swinney, *Science*, **261**, 192 (1993).
74. G. Ertl, *Science*, **254**, 1750 (1991).

75. R. W. Schoenlein, L. A. Peteanu, R. A. Mathies and C. V. Shank, *Science*, **254**, 412 (1991).
76. D. S. Wattke, M. J. Pjerrum, J. R. Winkler and H. B. Gray, *Science*, **256**, 1007 (1992).
77. A. Heller, *Acc. Chem. Res.*, **23**, 128 (1990).
78. E. Neher, *Science*, **256**, 498 (1992).
79. B. Sackmann, *Science*, **256**, 503 (1992).
80. N. A. Castle, D. G. Haylett and D. H. Jenkinson, *TINS*, **12**, 59 (1989).
81. B. Sommer and P. H. Seeburg, *TiPS*, **13**, 291 (1992).
82. K-S. Akerfeldt, J. D. Lear, Z. R. Wasserman, L. A. Chung and W. E. De Grado, *Acc. Chem. Res.*, **26**, 191 (1993).
83. T. Osa and J. Anzai in *Inclusion Aspects of Membrane Chemistry*, Eds. T. Osa and J. L. Atwood, (Kluwer Academic, Dordrecht, 1991), p. 157.
84. T. P. Stossel, *Science*, **260**, 1086 (1993).
85. R. B. Vallee, H. S. Shpetner and B. M. Paschal, *TINS*, **12**, 66 (1989).
86. D. H. Freeman, *Science*, **254**, 1308 (1991).
87. M. H. Devoret, D. Esteve and C. Urbina, *Nature*, **360**, 547 (1992).
88. R. J. Douglas and K. A. C. Martin, *TINS*, **14**, 286 (1991).
89. M. Mahowald and R. J. Douglas, *Nature*, **354**, 515 (1991).
90. M. A. Mahowald and C. Mead, *Sci. Amer.*, May 1991, p. 76.

2

Theory

R W Munn

2.1 Introduction

Molecular electronics arguably started in 1974 with a *theoretical* paper.[1] Aviram and Ratner proposed that a suitable molecule (such as that illustrated in Fig. 2.1) could act as an elementary electronic component, namely a rectifier. They used molecular orbital calculations to show that systems comprising an electron-donating fragment connected to an electron-accepting fragment by a sequence of carbon–carbon single bonds would allow electron transfer from donor to acceptor under the influence of a potential difference in one direction, but not in the reverse direction. Thus a molecule could mimic a semiconductor *p–n* junction. Since that time there has been extensive work aimed at synthesizing molecular rectifiers and studying their electrical behaviour (by no means a simple task).[2]

This example illustrates an important role for theory in molecular

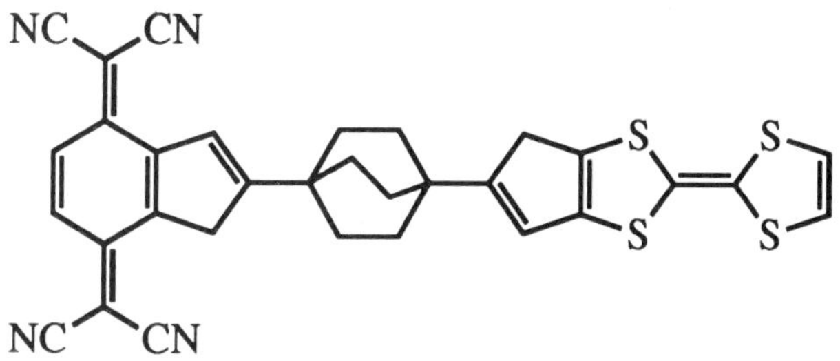

Fig. 2.1 A molecule designed to be a rectifier.

electronics. Theory allows us to explore ideas before we make the considerable commitment of undertaking experimental work. Such ideas need not be purely theoretical in origin: once a phenomenon is discovered, theory helps us to understand it, to predict where else it may occur, and to modify it to our advantage. In this way theory assists in making molecular electronics systematic; theory provides a framework that shows how the subject is organized.

There are two extreme views of theory, each not without some element of the truth. One is that theory is little more than common sense or the hypotheses any scientist uses, consciously or not, in planning research. Scientists already use it, like Monsieur Jourdain in Molière's *Le Bourgeois Gentilhomme*, who discovered that he had been speaking prose for more than forty years without knowing it. The other extreme view is that theory is impractical and useless, the opposite of common sense. Hence scientists need not bother with it: 'Theory is fine, but it can't stop things existing', as Charcot the nineteenth-century neurologist observed.

The truth lies in between. Theory goes beyond common sense, and indeed may correctly predict what common sense would deny—Einstein described common sense as 'a deposit of prejudice laid down in the mind prior to the age of eighteen'. Theory therefore accumulates a body of techniques and concepts that require some expertise in their manipulation. Sometimes its exponents may become too wrapped up in its mysteries, but ultimately it must submit to the test of experiment.

This chapter is concerned to describe theoretical techniques useful in molecular electronics. It asks what techniques have been used, how useful they are, and how accessible they are to the non-expert user. For instance one might want to know whether a standard computer package is readily available.

Molecular electronics brings together many conventional disciplines—chemistry, physics, electronics, computing, biology—each with its own body of theory. Hence one needs some basis for organizing those aspects of theory particularly relevant to molecular electronics. Molecular electronics necessarily uses molecular materials, in which the molecules largely retain their separate identities. As a result, the molecular properties, the molecular arrangement, and the molecular interactions together give rise to the molecular properties.[3] These will be considered in turn.

2.2 Molecular Properties

The theory of molecular properties is the province of quantum chemistry. Its techniques are well established and its practitioners are widespread. However, as its name suggests, it is best developed where it addresses the concerns of chemistry (some of which nevertheless remain rather intractable), and these do not always coincide with the concerns of molecular electronics. Most obviously, quantum chemistry usually deals with isolated molecules or perhaps two or three interacting small molecules rather than with the properties of molecules within the large assemblies typical of molecular electronics. We return to the consequences of this later, but for now consider an isolated molecule.

A molecule consists of a set of electrons i at positions r_i and nuclei N at positions R_N. Its states α are characterized by wavefunctions which depend on all these positions, $\Psi_\alpha(\{r_i\},\{R_N\})$, and its properties depend on these states. The states are solutions of the time-independent Schrödinger equation, but exact solutions cannot usually be obtained. Much of quantum chemistry therefore concerns the judicious choice of approximations that give useful results for moderate computational effort while retaining recognizably chemical features.[4]

First and best of the approximations is the Born–Oppenheimer or adiabatic approximation. Since nuclei are much heavier than electrons, then for most purposes one can regard the electrons as responding instantaneously or adiabatically to movements of the nuclei. As a result one can factorize the wavefunction into electronic and vibrational parts:

$$\Psi_\alpha(\{r_i\},\{R_N\}) = \psi_\alpha^{\mathrm{e}}(\{r_i\};\{R_N\})\psi_\alpha^{\mathrm{v}}(\{R_N\}). \tag{2.1}$$

The electronic wavefunction ψ_α^{e} refers to a particular fixed configuration of the nuclei, $\{R_N\}$. The energy varies as this configuration varies, being a minimum at the equilibrium structure. The variation of energy with configuration corresponds to the potential in which the nuclei move, which determines the vibrational wavefunction ψ_α^{v}. From now on we shall consider only the electronic wavefunction and will drop its superscript e; we shall also not specify explicitly its dependence on the nuclear configuration or the state α to which it refers.

One now seeks to represent the electronic wavefunction ψ as a composite of chemically significant functions. The valence-bond method builds up ψ from functions representing electron-pair bonds between atoms. Although this method is conceptually attractive and is currently enjoying increased popularity, it is not usually applied in molecular electronics. Normally, the molecular orbital (MO) method is the main tool. It builds up ψ from functions representing electron pairs in orbitals extending over the whole molecule. Thus if Φ_a denotes molecular orbital a we have

$$\psi = \prod_a \Phi_a. \tag{2.2}$$

It is also usual to construct Φ_a as a linear combination of atomic orbitals (LCAO) ϕ_{Ak}, where A denotes the atom and k the orbital on that atom; one expects that the properties of a molecule should be directly related to those of its constituent atoms. Thus

$$\Phi_a = \sum_{Ak} c_{Ak}^a \phi_{Ak}. \tag{2.3}$$

Here c_{Ak}^a are the molecular orbital coefficients that ultimately determine ψ.

The form (2.3) is particularly convenient for applying the variation theorem to determine ψ. The theorem says that any approximate form of ψ gives a higher energy for the ground state than the true form. Hence, of a set of approximate forms with different coefficients, the one giving the lowest energy is taken to be the best approximation (within that set) to the true ψ. Different versions of molecular-orbital theory then entail different prescriptions for constructing the orbitals and filling them with electrons.

Among the different versions, there are two broad classes. In *ab initio* methods, all calculations are carried out exactly. This does not mean that the methods are exact, but that for the chosen basis set of orbitals and the chosen method of constructing the wavefunction, the necessary integrals are evaluated exactly. The basis set may be minimal, with one atomic orbital per valence electron of either spin, or bigger and more flexible and hence more greedy of computer time. Electrons are assumed to move in the field of the nuclei and the average field of the other electrons in simple MO theory using the self-consistent field (SCF) method. This ignores correlation, which can be treated by a variety of methods, again at a cost in complexity and time. Many methods are available for mainframe computers, for example the GAUSSIAN and GAMESS packages.

Semi-empirical methods use experimental data (e.g. ionization energies) to avoid evaluating some integrals, and may use simple formulae to give others. Often these approaches are calibrated relative to *ab initio* methods on the one hand and experiment on the other. The commonest such methods are the all-valence-electron neglect of differential overlap (NDO) methods such as CNDO, INDO, and MINDO, where the acronym indicates that certain classes of integral are set to zero. This greatly reduces the amount of computation, to the extent that INDO calculations can now be performed on a PC with a maths co-processor. Again, standard packages are available. However, the parameters used in semi-empirical methods are chosen to optimize agreement with experiment, and hence depend on the experimental quantity chosen. As a result, semi-empirical methods are usually good for determining trends in values but not for absolute values. There are also π-electron-only theories, of which the simplest are Hückel theory and an NDO relative is Pariser–Parr–Pople (PPP) theory. For a reasonable compromise between effort and reliability, CNDO is perhaps the best semi-empirical treatment, although its more sophisticated cousins are now increasingly used as computing power increases.

So much for the methods: what sort of information do they provide? In using the variational theorem, all methods typically provide the molecular electronic energy. By minimizing the energy with respect to bond lengths and angles, calculations yield the equilibrium molecular structure. By giving changes in the energy with respect to deviations from this structure, they also yield vibrational and torsional frequencies and hence corresponding spectroscopic information. Energies of excited electronic states can be calculated, giving electronic spectra. For a given structure, one obtains the charge distribution, and thence the molecular multipole moments. One may also deduce molecular response coefficients such as the polarizability and nonlinear hyperpolarizabilities (see Chapter 5).

An example of the value of quantum-chemical methods is provided by calculations of the first hyperpolarizability β to assess the potential of molecules for applications in nonlinear optics: high β promises high nonlinearity.[5] These calculations use a variant of CNDO (referred to as CNDOVSB) parameterized to yield agreement with optical absorption maxima and dipole moment measurements for a set of compounds, but not with any nonlinear property. Once devised and implemented, this method allows one to calculate β rapidly for many molecules. Comparison with a test set of compounds confirms that the results are at least semi-quantitatively correct. Thus if the method predicts a low β for a molecule, one may be confident that it is not worth synthesizing out of all the many possibilities of different substituents, different conjugation paths, and so on, and thus the calculations provide a preliminary screening. (They also provide input to calculations of material properties, as will appear later.)

A particular case concerns the 2-pyrazolines, illustrated in Fig. 2.2.

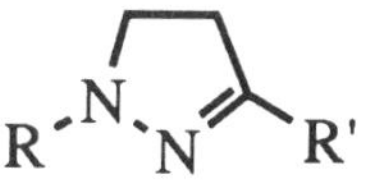

Fig. 2.2   Formula for 2-pyrazolines.

According to the usual design criteria, one of the substituents R and R′ should contain an electron-donating group and one an electron-accepting group—but which should contain which? Calculations for 4-methoxyphenyl as the donor and 4-nitrophenyl as the acceptor show that the static hyperpolarizability is four times larger with the donor as R′ than with the donor as R.[6] At a typical laser frequency both hyperpolarizabilities increase, but the ratio increases to about seventeen because the frequency is near a resonance. Clearly all this would take considerable time and effort to deduce experimentally.

2.3   Molecular Arrangement

Molecular properties often have a directional character. This may be a vector property such as dipole moment or a tensor property such as hyperpolarizability, or simply the molecular shape. Thus in order to understand how the material properties arise from the molecular properties, we need to know how the molecules are arranged.

To predict or interpret molecular arrangement, theory needs an intermolecular potential as basic input. (Note that the potential is therefore regarded as a special molecular interaction distinct from those to be considered in the next section.) Although such potentials can be calculated by quantum-chemical techniques, the results are reliable only for small or highly symmetrical molecules of little interest in molecular electronics. Hence empirical or semi-empirical potentials are used in which the algebraic form and some numerical values may be suggested by theoretical calculations, but some values are obtained by fits to experimental data.

One popular form of potential is the Lennard–Jones 6:12 potential, in which the energy at separation r is given by

$$W = A/r^{12} - B/r^6. \tag{2.4}$$

Here the term in $1/r^6$ is the leading term in the rigorous attractive potential between two neutral nonpolar species; higher terms in $1/r^8$, $1/r^{10}$ ... are usually neglected. The term in $1/r^{12}$ represents the repulsive potential, for which no general form is known, and so the power 12 is taken as algebraically convenient and approximately correct in its steepness. Theory suggests that a better approximation to the repulsive potential may however be an exponential, leading to the often-used exp-6 potential:

$$W = A \exp(-r/\rho) - B/r^6. \tag{2.5}$$

Here ρ represents the range of the repulsive potential.

Clearly the energy of a pair of non-spherical molecules cannot depend simply on one distance r. For example, two elongated molecules with centres separated by a distance a little less than their length will have a much higher energy when pointing along the line of centres then when pointing across it, as illustrated in Fig. 2.3. A convenient way of incorpor-

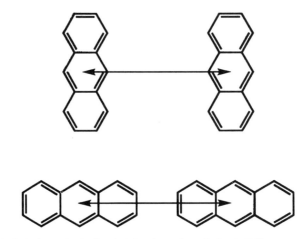

Fig. 2.3   The closest approach of molecules depends on their orientation.

ating this orientation dependence is to use a potential of the form (2.4) or (2.5) between every pair of atoms taken one from each molecule. Such an atom–atom potential requires a different set of parameters for each distinct type of atom pair, e.g. C–C, H–H and C–H for hydrocarbons. For large molecules, atom–atom potentials lead to a heavy computational effort, and instead potentials may be parametrized for groups such as the CH_2 group in alkanes and polymers.

All these potentials represent the quantum-mechanical repulsion and dispersion forces. For weakly polar molecules this may be adequate for the total potential, but in general one expects an electrostatic contribution from permanent and induced multipole moments. For example, atom–atom potentials devised for at best weakly polar aromatic hydrocarbons are unlikely to be satisfactory for the strongly polar azulene molecule which is superficially similar to them: see Fig. 2.4. One can then superimpose the

Fig. 2.4 Azulene.

electrostatic contribution on the others, or augment the atom–atom potentials by a Coulombic term C/r. Finally, one may discard the repulsion and dispersion terms to leave a purely electrostatic potential based on atomic charges derived from quantum-chemical calculations. For complex molecules one would need such calculations to deduce the multipole moments, and reliable electrostatic potentials require moments distributed over the molecules, and so one might as well cut out several intermediate steps. Certainly for biological problems such as enzyme-substrate interactions the electrostatic potentials offer an approach that is both tractable and effective.

In general the arrangement of the molecules will depend on their internal flexibility as well as their interactions. This can be modelled by adding potentials for bond stretching, bond angle deformation, torsional angle deformation, and internal non-bonded interactions. Such contributions are particularly important in modelling polymers and other systems composed of elongated molecules such as Langmuir–Blodgett films and liquid crystals.

Given the potentials, one needs a method for determining the molecular arrangement they imply. The simple direct approach is energy minimization: starting from a suitable arrangement one calculates the forces on the molecules, relaxes the system to a lower-energy arrangement, and repeats until a minimum is found. This is a standard problem with standard difficulties, notably that of finding a global minimum instead of the local minimum nearest the starting point. The method also has no way of including temperature and hence entropic effects that may be very important.

An alternative is the Monte-Carlo method, in which changes of arrangement are allowed if they satisfy an energy criterion governed by a random number generated at each step. This method is relatively straightforward but can become inefficient if few changes are accepted. However, new improved variants continue to be developed; one method of escaping from the local environment is simulated annealing, whereby one starts the system at a high temperature so that thermal fluctuations are large enough to surmount energy barriers, and then reduces the temperature in the lowest minimum thereby found.

Probably the best method is to use molecular dynamics. Here one solves the equations of motion for the system in one arrangement to calculate its arrangement a short time (typically 1 fs) later. As the arrangement develops in successive time-steps, one accumulates information on the configurations of the system and their probabilities and energies. This provides information on thermodynamic and structural properties. Techniques are also available to derive non-equilibrium properties by molecular dynamics, which the Monto-Carlo method cannot usually do directly.

Unfortunately, implementing molecular dynamics, and to a lesser extent the Monte-Carlo method, for a reasonable number of molecules in a realistic representation is very computer-intensive. Runs may take days on a dedicated superminicomputer. Hence for the interested enquirer, rather than the dedicated enthusiast, the most practical approach may be to use a standard molecular modelling program (energy minimization) of the sort commonly used for modelling enzyme-substrate interactions and the like. These are readily available on powerful workstations and increasingly on the more powerful PCs.

From various methods, then, one can obtain information on the molecular arrangement. This may first of all be the short-range molecular packing. This depends principally on the molecular shape, i.e. on the short-range repulsive potential, and may be predicted with reasonable accuracy. Next one may obtain information on the long-range order (if any). This depends on the longer-range attractive potential, though molecular shape is still important, and is usually harder to get exactly right. For example, one might regard a 1° error in one angle of a crystal unit cell as entirely acceptable, although it could mean the difference between one crystal symmetry and another. However, such an error usually has few practical consequences: contributions that are rigorously zero under the higher symmetry are likely to remain small when there is a small lowering of the symmetry. Hence getting the packing right is likely to suffice in order to predict whether a molecule will produce an effect of the desired size. For example, naphthalene and anthracene have monoclinic structures with equivalent molecules while tetracene and pentacene have triclinic ones with inequivalent molecules, but their properties are substantially the same. Another example is provided by *m*-nitrophenol, which exists in two slightly different forms.[7] One form is monoclinic and centrosymmetric, while the other is orthorhombic and non-centrosymmetric, but the molecular packing and the crystal properties are very similar, so that it would be difficult (and perhaps not really necessary) to predict the existence of the two forms. Only in special cases are the details likely to be particularly significant, e.g. in specifically collective phenomena such as ferroelectricity in liquid crystals.

An example of the detailed information obtainable on molecular electronic materials by the molecular dynamics method is provided by work on Langmuir–Blodgett films comprising one and three monolayers interacting with a substrate.[8] It is found that the molecules tilt away from the vertical to the substrate by about 20° in the direction of their next-nearest neighbours, with a decrease of about 1° in going from the first layer to the second and from the second to the third. In the first layer the packing is very well ordered, but by the third there are considerable variations about the average which amount to a roughening of the interface between layers. This feature seems to reflect the decreased effect of the substrate, and may explain the difficulty of obtaining stable thick films of high optical quality. Insight into the stability is also given by displaying the molecular arrangement every so many time-steps (perhaps every 100). At lower coverages of the substrate, molecules can bend back on themselves and even turn over; such events occurring during deposition would presumably reduce the film compactness, order and stability.

2.4   Molecular Interactions

Apart from determining the molecular arrangement, the molecular interactions are of significance in two respects: they modify the molecular properties in the condensed-phase environment, and they contribute directly to the material properties. In either case, it is convenient to divide the interactions into two broad classes. Classical electromagnetic interactions are generally well understood and tractable, while quantum-mechanical interactions are less well understood and less tractable (but often essential). We have already seen this dichotomy appearing in the intermolecular potentials.

Electromagnetic interactions of interest in molecular electronics usually involve electric or optical phenomena; magnetic interactions are of more specialized interest, for example in molecular ferromagnets. If magnetism is ignored except as an accompaniment to the electric field in a light wave, then electromagnetic interactions entail some kind of applied electric field. This may arise from charges external to the material, e.g. on a set of electrodes; from oscillating external charges generating electromagnetic waves; from excess charge carriers; or from the molecular permanent charge distributions, for instance expressed in their multipole moments. Theory tells us about the fields produced and about the response to the fields that results in modified fields.

Calculating applied electric fields is basically an application of Coulomb's law. A charge q produces at position r a potential

$$V(r) = q/4\pi\varepsilon_0 r \tag{2.6}$$

and hence an electric field

$$E(r) = -\nabla V(r) = -(q/4\pi\varepsilon_0)\nabla(1/r), \tag{2.7}$$

where ∇ is the gradient operator with components $(\partial/\partial x, \partial/\partial y, \partial/\partial z)$. Electric field gradients obviously follow from higher gradients of $1/r$. The nth moment of a distribution of charges q_i at positions r_i is defined by

$$M^{(n)} = \sum_i q_i r_i^n. \tag{2.8}$$

It gives rise to an electric field

$$E = (1/4\pi\varepsilon_0)M^{(n)} \cdot T^{(n+1)}, \tag{2.9}$$

where the centred dot denotes an n-fold scalar product and the multipole tensor $T^{(n)}$ is

$$T^{(n)} = \nabla^n (1/r). \tag{2.10}$$

Hence the various multipole tensors determine the fields and field gradients due to a charge distribution. In molecular electronics one would typically obtain the field at one molecule due to another and then sum over all other molecules to obtain the total field.

In the presence of materials it is necessary to consider electric fields further. The applied electric field causes the material to acquire induced electric moments via the molecular polarizabilities, and these moments themselves contribute to the total electric field. Two kinds of electric field need to be considered: the local field due to the applied field and the field of the induced moments of the other molecules, and the macroscopic field due to the applied field and the field of the induced moments of all the molecules averaged over a region large enough to eliminate microscopic variations (in a molecular crystal, it suffices to average over a unit cell). It is the local field that is responsible for polarizing molecules, and the macroscopic field that determines the potential difference across the sample, which is of course experimentally accessible. Dielectric theory is concerned with the relation between the polarization and the macroscopic field, and hence requires the relation between the local field and the macroscopic field.

This problem requires a self-consistent treatment. The local field has to be determined so as to induce a polarization that combines with the applied field to yield the initial local field. In molecular materials this problem is simplified because of the separation of molecular response and molecular interactions, which means that correlation between electrons forms part of the molecular response and need not be treated explicitly. Iterative numerical treatment is then straightforward. In molecular crystals the problem simplifies further because of the translational symmetry. This means that the coupled responses and interactions of all the molecules can be decoupled by Fourier transformation to yield algebraic results.[9]

For example, in the simplest case of a crystal with one molecule per unit cell subject to a uniform applied field E^0, the local field is given by

$$F = (1 - a \cdot t)^{-1} \cdot E^0. \tag{2.11}$$

Here a is the dimensionless reduced polarizability $\alpha/\varepsilon_0 v$, where α is the molecular polarizability and v the unit-cell volume, while t is the lattice dipole sum

$$t = (v/4\pi) \sum_{\ell(\neq 0)} T^{(2)}(r_\ell), \tag{2.12}$$

where ℓ labels the unit cells. Equation (2.11) constitutes an explicit calculation of the inverse dielectric function, which plays a central but often purely formal role in dielectric theory. In the present case it leads to an expression for the ordinary dielectric constant that includes the Lorentz–Lorenz or Clausius–Mossotti equation as a special case.

A more complex example is given by charge-transfer excitons. Photoconductivity in molecular materials takes place via intermediate states in which the photon transfers an electron from one molecule to the next to form an anion-cation pair. This pair is also called a charge-transfer exciton. Its energy consists of the Coulomb energy of interaction of the ions plus the energy of polarization of the surrounding molecules. This can be calculated for various separations of the two ions in different directions in a crystal, in order to determine the exciton states accessible to absorption spectroscopy and hence to clarify the detailed processes occurring during photoconductivity. The calculations also show that the energy behaves like the Coulomb energy in a dielectric continuum, provided the two ions are not adjacent. The energy is reduced from that *in vacuo* by an effective dielectic constant that depends on the direction between the ions but is related to the components of the dielectric tensor in an unexpected way, being smallest in directions where the components are largest.[10]

Quantum-mechanical interactions of interest in molecular electronics usually concern charge transfer. These involve the obvious interactions whereby an excess electron is affected by the nuclei on another so that there is a driving force for the transfer of charge from one molecule to the other. They also involve more subtle and complex contributions to the transfer of neutral excitations, which may entail correlated transfers of charge in opposite directions.

Such interactions present two kinds of difficulty. First of all, they involve aspects of electron exchange, electron correlation and orbital overlap, which are not very tractable even in isolated molecules. Furthermore, the interactions refer to intermediate distances. The distances between molecules in materials are neither the short distances for which bonding treatments have been developed, nor the longer distances for which intermolecular force treatments have been developed. Hence there is no great reservoir of theory to tap.

In the tight-binding approximation, the electronic states of materials are constructed from those of their constituents, broadened by the charge-transfer interactions. The approximation is well suited to molecular materials, and interactions were calculated many years ago to determine the excess carrier band structure of the aromatic hydrocarbon crystals such as naphthalene and anthracene. These crystals are experimentally tractable—readily purified, stable and with accessible electronic states—and the molecules are theoretically tractable: planar π-electron systems with some properties not greatly dependent on the particular treatment because of symmetry. At the time, the Pariser–Parr–Pople treatment was state of the art, and it sufficed to show the anisotropy of the energy bands as well as widths of the order to 0.1 eV (not as narrow as still often supposed for molecular systems).[11,12] In disordered materials such as polymers, variations in the charge-transfer interactions are significant in charge-carrier transport and trapping. These have been explored by calculations of the interaction between two carbazole rings in different orientations such as one might find in poly(vinylcarbazole) in which the carbazoles are pendant groups.[13]

As already indicated, the interactions not only contribute directly to the material properties but also modify the molecular response from that *in vacuo*. From one point of view, this statement could be seen as unnecessary. In an interacting system only material properties are measurable, and the molecular response in the material environment is not obtainable except through these properties. An alternative viewpoint is that already implicitly accepted: we seek to explain the material properties in terms of the molecular response and the molecular interactions (and of course the molecular arrangement), and it is a matter of convenience how we divide up the consequences of the interactions.

The theory of metals provides one example of a similar sort. Electrons interact strongly with one another and with the metal ion cores, but nearly-free electron theory proves successful because the electrons are 'dressed' by the interactions to become new particles that interact relatively weakly. Since strong interactions are usually difficult to treat, one often seeks to incorporate them in some such way. Electronic transport in molecular materials is often treated as involving weakly-interacting polarons (see Chapter 7) consisting of charge carriers plus the lattice polarization they cause.[11,12]

Another example occurs in molecular exciton theory. One may start from the free-molecule excitations and invoke all dipolar interactions to calculate the exciton band states. Alternatively one may incorporate some interactions to calculate effective molecular excitations which then couple further through the remaining interactions to yield the exciton bands. Here we take the latter arguably more pragmatic view of trying to keep the interactions simple at the cost of complicating the molecular response.

The main effect of electromagnetic interactions in a material is to subject the molecules to high permanent electric fields—of the order of GV m^{-1} in crystals of polar molecules.[14] The result is that the molecular response relevant to the material environment is that at the prevailing electric field. Such responses are not those usually calculated, although it is possible to devise schemes for doing so. Because the fields are high, the properties can be markedly different from those in zero field, with dipole moments increased by as much as 50 per cent.[15]

One important effect of quantum-mechanical interactions arises through modifying the environment of the molecular wavefunction. In a free molecule, the wavefunction can extend to infinity, at least in principle. In a molecular material, the molecular wavefunctions are constrained by those of the neighbouring molecules and so there is orbital confinement. This has a direct effect on the molecular energy and charge distribution. It also affects the molecular response. In the free molecule, an electron can be displaced to very large distances by an electric field, but in the molecular material the displacement is restricted by the orbital confinement (a crude analogy would be blowing up a single balloon as opposed to one balloon in a bunch of balloons). As a result, the polarizability is changed. These features have been successfully calculated by *ab initio* techniques for ionic crystals, where it is found that anions typically have polarizabilities that are rather sensitive to their environment, in contrast to the more compact cations.[16]

2.5 Material Properties

It now remains to put the molecular properties, arrangement and interactions together to provide a theory of the material properties. Two classes of property can be distinguished. One class exists without the need for any interactions between molecules, although of course the interactions do affect the properties. Optical properties such as the refractive index and nonlinear response fall into this class. Such properties exist even for a dilute gas, and those of the molecular material can be treated in the oriented-gas model.[17] The molecules are oriented in a definite way but otherwise behave like a gas because they do not interact. Deducing the properties of an oriented gas requires one to relate all molecular properties to a set of material axes, a procedure considered in detail below. The oriented-gas model is used to estimate the nonlinear optical behaviour of materials from the calculated molecular hyperpolarizability, frequently with local-field effects included approximately and occasionally included rigorously.[17]

The other class of properties exists solely because of the interactions between molecules. Transport properties such as charge-carrier mobility fall into this class. Here there are no simple general models—quantum-mechanical interactions are often difficult to treat anyway, and kinetic processes also present problems if one seeks to go beyond phenomenological theory.

An example of what can be achieved by theory in the latter class is provided by work on energy transfer. One may design molecules comprising two distinct chromophores (light-absorbing groups) separated by a bridging group with higher energy excitations. If one chromophore is excited, the other may luminesce, and this process can therefore serve to channel energy. How does the process depend on the states of the separate chromophores, on those of the bridging group, on the molecular geometry, on the temperature, and so on? This problem falls somewhere among the areas of molecular properties, molecular interactions and material properties. Detailed examination shows that the electron transfer rate is considerably enhanced when the bridge has energy levels degenerate with those of the donor and acceptor. The dependence on bridge length varies with the other parameters, and under some circumstances need not be monotonic, so that sometimes a longer bridge may give faster transfer.[18] Such conclusions offer useful guidance about synthetic targets.

Molecular electronics usually deals with directional properties and hence with materials that are anisotropic (different in different directions): crystals, Langmuir–Blodgett films, liquid crystals, polymers . . . Thus the directional properties of the material are important, and the question is how to relate these to the molecular properties and orientation. For scalar properties like energy or mass, the question does not apply. For other quantities, one needs to transfer the properties from a set of Cartesian axes $X\,Y\,Z$ fixed in the molecule to a set $x\,y\,z$ fixed in the material. The latter set may be determined by the axes of a crystal, the director in a liquid crystal, the draw direction in a polymer, and so on, and therefore can be referred to as laboratory axes.

For a vector like the dipole moment, the transformation is easy and familiar: we just resolve along the laboratory axes. Let the dipole moment be p, with components in the molecular axes $p^{\mathrm{M}} = (p_X^{\mathrm{M}}, p_Y^{\mathrm{M}}, p_Z^{\mathrm{M}})$. Then we want the components $p^{\mathrm{L}} = (p_X^{\mathrm{L}}, p_Y^{\mathrm{L}}, p_Z^{\mathrm{L}})$ in the laboratory axes. (Note that the vector p is the same in the two sets of axes; the superscript M or L merely serves as a reminder of the axes chosen to define the components of p.) We use α to denote one of $x\,y\,z$ and A to denote one of $X\,Y\,Z$, when we have

$$p_\alpha^{\mathrm{L}} = \sum_A a_{\alpha A} p_A^{\mathrm{M}}. \tag{2.13}$$

Here $a_{\alpha A}$ is the direction cosine between axes α and A. Often this equation is written without the summation sign as just

$$p_\alpha^{\mathrm{L}} = a_{\alpha A} p_A^{\mathrm{M}}, \tag{2.14}$$

using the Einstein convention that repeated subscripts are understood to be summed over all three possible values. Using subscripts α and A consistently would allow us to omit the labels L and M, but the labels provide an extra check.

For each laboratory axis there are three direction cosines, one for each molecular axis, and as there are three laboratory axes there are nine direction cosines in all, forming the 3×3 array $\{a_{\alpha A}\}$. The direction cosines are not all independent. Preserving the magnitude of p requires

$$a_{\alpha A} a_{\beta A} = 1 \quad \text{if } \alpha = \beta$$
$$\qquad\qquad = 0 \quad \text{if } \alpha \neq \beta, \tag{2.15}$$

and similarly

$$a_{\alpha A} a_{\alpha B} = 1 \quad \text{if } A = B$$
$$\qquad\qquad = 0 \quad \text{if } A \neq B. \tag{2.16}$$

This leaves three independent quantities, equivalent to the three Euler angles often used to denote orientation in space.

Other quantities have a more complicated dependence on direction. They may relate two or more vectors, e.g. the electric field gradient describing how the electric field vector depends on the position vector, or the polarizability describing how the dipole moment vector depends on the electric field vector. They may also represent a more complicated spatial arrangement, e.g. the electric quadrupole moment arising from the second moment of a charge distribution (the dipole moment being the first moment). Such quantities are tensors.[19]

In these cases, additional subscripts are required for the components of the tensors and for their transformations between sets of axes. For example, the transformation of the quadrupole moment $\boldsymbol{Q}$ is

$$Q^L_{\alpha\beta} = a_{\alpha A} a_{\beta B} Q^M_{AB}. \tag{2.17}$$

In general a tensor of rank n requires n direction cosines in its transformation, so that the quadrupole moment is a second-rank tensor. A scalar can be regarded as a zeroth-rank tensor, while a vector is a first-rank tensor.

The foregoing is rather abstract, so let us consider some specific examples. For simplicity, only molecules with axial symmetry will be treated. For such a molecule, with Z conventionally taken as the symmetry axis, the dipole moment has components in the molecular axes $\boldsymbol{p}^M = (0, 0, p)$, whence the components in the laboratory axes are $p^L_\alpha = a_{\alpha Z} p$. These are simple projections along the respective axes, as expected from an elementary treatment.

The intensity of optical absorption is governed by the Einstein coefficient[20]

$$B_i \propto |\boldsymbol{p}_i \cdot \boldsymbol{e}|^2. \tag{2.18}$$

Here $\boldsymbol{p}_i$ is the transition dipole moment of a molecule i and $\boldsymbol{e}$ is the polarization vector of the light (a unit vector). If the axis of molecule i makes an angle θ_i with $\boldsymbol{e}$, then

$$B_i \propto p^2 \cos^2 \theta_i. \tag{2.19}$$

In a nematic liquid crystal, the order parameter S is defined as the average

$$S = (3\langle \cos^2 \phi_i \rangle - 1)/2. \tag{2.20}$$

Here ϕ_i is the angle between the axis of molecule i and the director $\boldsymbol{n}$, a unit vector that characterizes the mean orientation (see Chapter 9). From these results we can derive the relative intensities of absorption or dichroic ratio for light polarized parallel or perpendicular to the director.

For light polarized parallel to $\boldsymbol{n}$, i.e. for $\boldsymbol{e} \| \boldsymbol{n}$, we have $\theta_i = \phi_i$. The mean Einstein coefficient in this case is

$$\langle B \rangle_\| \propto p^2 \langle \cos^2 \phi \rangle. \tag{2.21}$$

But we can solve for $\langle \cos^2 \phi \rangle$ in terms of S to obtain

$$\langle B \rangle_\| \propto p^2 (2S + 1)/3. \tag{2.22}$$

For light polarized perpendicular to $\boldsymbol{n}$, i.e. $\boldsymbol{e} \perp \boldsymbol{n}$, we have $\theta_i = \pi/2 - \phi_i$ so that $\cos \theta_i = \sin \phi_i$. Then we find

$$\langle B \rangle_\perp \propto p^2 \langle \sin^2 \phi \rangle \propto p^2 2(1 - S)/3, \tag{2.23}$$

so that the dichroic ratio is

$$\langle B \rangle_\perp / \langle B \rangle_\| = 2(1 - S)/(2S + 1). \tag{2.24}$$

Measurement of this ratio thus serves to determine S.

In molecular crystals, translational symmetry means that the bulk properties are determined by the unit-cell properties. Label the molecules in the unit cell k. The dipole moment of the unit cell in the laboratory axes (i.e. the crystal axes) is given by

$$p^L = \sum_k p^L(k), \tag{2.25}$$

so that from equation (2.13) we obtain

$$p_\alpha^L = \sum_k a_{\alpha A}(k)p_A^M. \tag{2.26}$$

Here $a(k)$ is the direction cosine matrix for molecule k and it has been assumed that the molecules are all equivalent and hence all have the same properties in the molecular axes, so that p_A^M is independent of k.

Now suppose that the unit cell has a centre of symmetry. Then the polar molecules are related in pairs such that, for example,

$$a_{\alpha A}(1) = -a_{\alpha A}(2), \tag{2.27}$$

whence it follows at once that $p^L = 0$. Similar arguments show why a centrosymmetric crystal has zero quadratic optical nonlinear response. The relevant susceptibility tensor $\chi^{(2)}$ is of third rank, and so transforming from molecular to crystal axes involves a product of three direction cosines. Between a pair of molecules related by a centre of symmetry, all three change sign, as does their product, and so $\chi^{(2)} = 0$.

One can also use this sort of approach to understand the polarization of optical spectra in crystals.[12] In crystals like naphthalene and anthracene, simple exciton theory shows that there are two Davydov components of the electronic spectrum with transition dipole moments

$$d^L(\pm) = d^L(1) \pm d^L(2), \tag{2.28}$$

where $d^L(1)$ is the transition dipole moment of molecules on sublattice 1. These can be expressed in terms of the transition dipole moments in the molecular axes to yield

$$d_\alpha^L(\pm) = [a_{\alpha A}(1) \pm a_{\alpha A}(2)]d_A^M, \tag{2.29}$$

where in these crystals the direction cosines satisfy

$$\begin{aligned} a_{\alpha A}(1) &= a_{\alpha A}(2) && \text{for } \alpha = a, c^* \\ a_{\alpha A}(1) &= -a_{\alpha A}(2) && \text{for } \alpha = b, \end{aligned} \tag{2.30}$$

c^* being the axis perpendicular to the ab plane. Thus the $(+)$ component has $d_b^L(+) = 0$ and is polarized perpendicular to b, i.e. in the ac plane, while the $(-)$ component has $d_a^L(-) = 0$ and $d_{c^*}^L(-) = 0$ and is polarized parallel to b, so that theory tells us the polarizations to be expected.

2.6 Conclusions

Theory offers a wide range of insights into molecular electronics. It provides much help in understanding and predicting molecular properties.

It is also useful for molecular arrangement, but this is in many ways more difficult and less accessible to the nonexpert. Molecular interactions are often subtle and not readily calculated by the novice, but theory does provide important concepts in this area. Material properties that exist even in the absence of interactions are easy to treat approximately in the oriented-gas model; the main labour is then transformations of axes, which have therefore been described in some detail here. Treating other material properties may require more investment of time.

In various of these areas, we may expect increasing availability of user-friendly interfaces to powerful PCs and workstations to enhance the role of modelling (if not of theory) in molecular electronics. As a long-term goal, one may seek to develop computer-aided design for molecular electronics. This is already a reality for isolated molecules,[21] and is now being developed in areas such as nonlinear optics.[8,22]

References

1. A. Aviram and M. A. Ratner, *Chem. Phys. Lett.*, **29**, 277 (1974).
2. R. M. Metzger and C. A. Panetta, in *Molecular Electronic Devices II*, Ed. F. L. Carter, (Dekker, New York, 1987).
3. R. W. Munn, in *Molecular Electronics: Materials and Methods*, Ed. P. I. Lazarev, (Kluwer, Dordrecht, 1991), p. 1.
4. D. B. Cook, *Structures and Approximations for Electrons in Molecules* (Ellis Horwood, Chichester, 1978).
5. V. J. Docherty, D. Pugh and J. O. Morley, *J. Chem. Soc. Faraday Trans. II*, **81**, 1179 (1985).
6. J. O. Morley and D. Pugh, in *Organic Materials for Nonlinear Optics*, Eds R. A. Hann and D. Bloor, RSC Special Publication 69 (Royal Society of Chemistry, London, 1989), p. 28.
7. G. Wójcik, J. Giermańska, Y. Marqueton and C. Ecolivet, *J. Raman Spectrosc.*, **22**, 375 (1991).
8. M. Bishop, J. H. R. Clarke, L. E. Davis et al. *Thin Solid Films*, **210–211**, 185 (1992).
9. R. W. Munn, *Mol. Phys.*, **64**, 1 (1988).
10. R. W. Munn, P. Petelenz and W. Siebrand, *Chem. Phys.*, **111**, 209 (1987).
11. E. A. Silinsh, *Organic Molecular Crystals: Their Electronic States* (Springer, Berlin, 1980).
12. M. Pope and C. E. Swenberg, *Electronic Processes in Organic Crystals* (Clarendon, Oxford, 1982).
13. J. H. Slowik and I. Chen, *J. Appl. Phys.*, **54**, 4467 (1983).
14. R. W. Munn and M. Hurst, *Chem. Phys.*, **147**, 35 (1990).
15. D. G. Bounds, A. Hinchliffe, R. W. Munn and R. J. Newham, *Chem. Phys. Lett.*, **28**, 600 (1974).
16. P. W. Fowler and N. C. Pyper, *Proc. Roy. Soc. Lond.*, **398A**, 377 (1985).
17. J. W. Rohleder and R. W. Munn, *Magnetism and Optics of Molecular Crystals* (Wiley, 1992).
18. J. R. Reimers and N. S. Hush, in *Molecular Electronics—Science and Technology*, Ed. A. Aviram (Engineering Foundation, New York, 1990), p. 339.

19. H. Jeffreys, *Cartesian Tensors* (Cambridge University Press, 1969); G. Temple, *Cartesian Tensors: An Introduction* (Methuen, London, 1960).
20. A. Hinchliffe and R. W. Munn, *Molecular Electromagnetism* (Wiley, 1985).
21. W. G. Richards, Ed., *Computer-Aided Molecular Design* (IBC Technical Services, London, 1989).
22. J. M. André, J. O. Morley and J. Zyss, in *Molecules in Physics, Chemistry and Biology*, Ed. J. Maruani (Reidel, Dordrecht, 1987), p. 615.

3

Piezoelectric and Pyroelectric Materials

D K Das-Gupta

3.1 Introduction

The efficient conversion of mechanical and thermal energy into electrical energy using piezoelectric and pyroelectric materials is of considerable importance in industrial, medical and other applications. High permittivity ferroelectric ceramic materials are extensively used at present for piezoelectric and pyroelectric transducers, high voltage generation by compressive stress, gas ignition, accelerometer, actuators, delay lines, wave filters, piezoelectric transformers, generation of sonic energy, infrared detection, radiometry, intruder alarm and thermal imaging.

The discovery by Kawai in 1969[1] that strong piezoelectricity can be induced in a semicrystalline poly(vinylidene fluoride) (PVDF) by an appropriate external electric field, and the subsequent observation of induced pyroelectricity in this material, were important milestones for polymeric electroactive sensors. PVDF is a ferroelectric polymer in which its dipoles can be permanently aligned by an electric field and the polarized film belongs to the point group 2 mm (for an uniaxially stretched film) or ∞ mm (for an unstretched film). These films are ultrasonic transducer materials as they have the longitudinal piezoelectric effect relevant to the thickness extensional anode. Piezoelectric polymer films are flexible and mechanically strong. In addition they have low acoustic impedance, comparable to those of biological materials and water, low dielectric constant and are quite stable in humid environment. These properties make PVDF an attractive material for hydrophone and ultrasonic imaging applications over a wide frequency range from 20 kHz to 10 GHz. The great potential of this flexible and strong ferroelectric film, which can be fabricated inexpensively in practically unlimited area, has immediately caught the imagination of both the scientific and technological communities. Since then various transducers, which are impossible or difficult to manufacture using conventional inorganic sensor materials, have been fabricated.

In the 1970s PVDF was the only potentially attractive ferroelectric polymer for piezo- and pyroelectric transducers. Recently, the copolymer of vinylidene fluoride—trifluoroethylene (VDF-TrFE)—has been found to provide a better electromechanical coupling coefficient, which has further encouraged the technological development of ultrasonic transducers.

To achieve this goal, extensive research work has been employed to study and determine the microscopic origin of piezo- and pyroelectricity in amorphous and semicrystalline polymers which have stable electroactive properties.

It is accepted that the piezoelectric (d-coefficients) and pyroelectric coefficients (p-coefficients) of PVDF are considerably smaller than those of ferroelectric ceramics. However, the dielectric constant of the polymer is quite small compared with the ferroelectric ceramics. As a result, the voltage coefficient per unit stress (g-coefficient) and the pyroelectric figure of merit are larger. In addition, the PVDF has a comparable acoustic impedance (acoustic velocity × density) to that of water which provides a good match for hydrophone and medical applications. Pyroelectric polymers are convenient for use in special detectors such as low-cost single element detectors and in two dimensional imaging devices. Fibre optic sensor technology is being developed with PVDF in which the operation is based on the interaction of light and propagating acoustic waves in birefringent fibres. It seems that these areas of research should lead to a number of new interesting and challenging developments.

Much effort is being expended to design new polymer structures which may provide improved piezoelectric, pyroelectric and dielectric properties. We have an extremely challenging time ahead to synthesize new and appropriate electroactive materials for microelectric sensor technology in the twenty-first century.

The present chapter provides a brief résumé on the basic concepts in piezo- and pyroelectricity and a survey of the piezoelectric and pyroelectric properties and their origin in PVDF and its copolymer VDF-TrFE.

3.2 Basic Concepts

Piezoelectricity and pyroelectricity refer to changes in the internal polarization of a dielectric material for small changes in stress and temperature, respectively. A material can be either piezoelectric, pyroelectric or ferroelectric if its crystalline symmetry is inherently asymmetric, i.e. if it lacks an inversion centre. All crystals are classified according to their symmetry. Crystals are classified into seven systems: triclinic (the least symmetrical), tetragonal, hexagonal, monoclinic, orthorhombic, trigonal and cubic. These systems are again subdivided into point groups (i.e. crystal classes) according to their symmetry with respect to a point. Of the 32 crystal

Table 3.1 *The 32 crystallographic point groups arranged by crystal systems*

Crystal system	Symbol		Pyro-electric	Piezo-electric	Centro-symmetric
	International	Schoen-flies			
Triclinic	1	C_1	✓	✓	
	$\bar{1}$	C_i			✓
Tetragonal	4	C_4	✓	✓	
	$\bar{4}$	S_4		✓	
	4/m	C_{4h}			✓
	422	D_4		✓	
	4mm	C_{4v}	✓	✓	
	$\bar{4}$2m	D_{2d}		✓	
	4/mmm	D_{4h}			✓
Hexagonal	6	C_6	✓	✓	
	$\bar{6}$	C_{3h}		✓	
	6/m	C_{6h}			✓
	622	D_6		✓	
	6mm	C_{6v}	✓	✓	
	$\bar{6}$m2	D_{3h}		✓	
	6/mmm	D_{6h}			✓
Monoclinic	2	C_2	✓	✓	
	m	C_s	✓	✓	
	2/m	C_{2h}			✓
Orthorhombic	222	D_2		✓	
	mm2	C_{2v}	✓	✓	
	mmm	D_{2h}			✓
Trigonal	3	C_3	✓	✓	
	$\bar{3}$	S_6			✓
	32	D_3		✓	
	3m	C_{3v}	✓	✓	
	$\bar{3}$m	D_{3d}			✓
Cubic	23	T		✓	
	m3	T_h			✓
	432	O			
	$\bar{4}$3m	T_d		✓	
	m3m	O_h			✓

A tick (✓) indicates that the point group is pyroelectric, piezoelectric, or centrosymmetric, as the case may be. (Lines and Glass, 1977).[1]

classes, eleven have a centre of symmetry and in one, a combination of symmetries effectively provides such a symmetry which endows them with no polar property. Thus only 20 classes can provide an asymmetric crystal structure and the materials belonging to these classes are piezoelectric. Ten of these 20 classes have a unique polar axis and they possess a spontaneous polarization (i.e. electric moment per unit volume) and are pyroelectric. Table 3.1 (see also Fig. 3.1) provides the details of the 32 crystallographic point groups arranged by crystal systems.[2] A restricted group of pyroelectrics have the further property of being ferroelectric. There is as yet no general basis for deciding whether a material will be ferroelectric. However, a crystal is regarded as ferroelectric when it has two or more orientational stages (in the absence of an electric field) which can be switched from one state to another by an electric field. These two

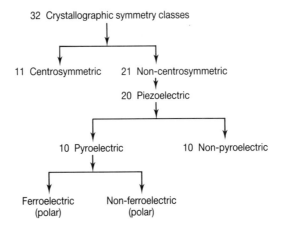

Fig. 3.1 Classification of symmetry relating piezoelectric, pyroelectric and ferroelectric materials.

orientational states have identical crystal structure but differ only in electric polarization vector at zero electric field. Thus there are no ferroelectrics which are not pyroelectric, just as there are no pyroelectrics which are not piezoelectric. However, the converse is not true, i.e. all piezoelectrics are not pyroelectric and all pyroelectrics are not ferroelectric as illustrated in Fig. 3.1.

Piezoelectric Effect   In piezoelectric materials an electric polarization develops in response to an applied mechanical stress. To a first approximation the induced polarization is proportional to the stress and this is the 'direct' piezoelectric effect. Similarly an electric field applied to such materials will cause the crystal to be strained and this is the 'converse' effect, in which the strain is directly proportional to the applied electric field.

Piezoelectric properties are described in terms of four parameters, D, E, X and x, where D is the dielectric displacement, E the electric field, X the stress and x the strain. If a stress X is applied to a material resulting in a strain, we may use the following simple one-dimensional relationship

$$X = Cx \text{ and } x = SX \tag{3.1}$$

where C is the elastic stiffness constant (stress per unit strain) and S the compliance (strain per unit stress). With a piezoelectric material the strain x also produces a polarization $P = Xd$, where d is a piezoelectric strain constant. Hence the dielectric displacement contains an extra term in the presence of a stress, i.e.

$$D = \varepsilon_0 \varepsilon E + Xd \tag{3.2}$$

where ε_0 is the permittivity of the free space and ε the dielectric constant of a meterial. Equation (3.2) represents the 'direct' piezoelectric effect. For the 'converse' effect, the corresponding relation in one-dimensional notation is

$$x = sX + Ed \tag{3.3}$$

In general, most crystalline materials are anisotropic so that the mechanical and electrical forces applied in one direction may produce components in other directions. Thus it is necessary to use tensor notation based on three mutually perpendicular directions. A stress is specified by a second rank tensor with nine components and the polarization, being a vector, is denoted by two components.[3] The components of a second rank tensor have two subscripts. Thus, in general, the stress X_{ij} coefficients can be conveniently expressed in a square array

$$X_{ij} = \begin{bmatrix} X_{11} & X_{12} & X_{13} \\ X_{21} & X_{22} & X_{23} \\ X_{31} & X_{32} & X_{33} \end{bmatrix} \tag{3.4}$$

This array within the square brackets symbolizes a tensor of the second rank and X_{11}, X_{12}, ... etc. are the components of the tensor (see Chapter 2). The first subscript of each component denotes the direction of the stress and the second subscript the direction of the normal to the plane at which the stress has been applied.[4] The shear stress X_{21} acts in the 2-direction on planes whose normals are in the 1-direction (see Fig. 3.2). A

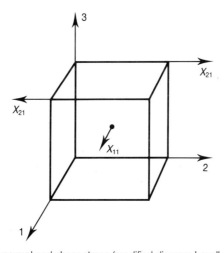

Fig. 3.2 Notation for normal and shear stress (modified diagram Lovell *et al.*, 1976).[3]

tensor is said to be symmetrical if $X_{ij} = X_{ji}$ for a body in equilibrium. The notation for shear stresses may be denoted by X_4, X_5 and X_6 where

$$\begin{aligned} X_4 &= X_{23} = X_{32} \\ X_5 &= X_{13} = X_{31} \\ X_6 &= X_{12} = X_{21} \end{aligned} \tag{3.5}$$

In this new (matrix) notation a single suffix running from 1 to 6 replaces the tensor notation thus

Tensor notation	11	22	33	23,32	31,13	12,21
Matrix notation	1	2	3	4	5	6

The matrix notation enjoys the advantage of superior compactness than the tensor notation.

Hence

$$X_{ij} = \begin{bmatrix} X_1 & X_6 & X_5 \\ X_6 & X_2 & X_4 \\ X_5 & X_4 & X_3 \end{bmatrix} \tag{3.6}$$

The relation between the polarization P and the stress X now becomes

$$P_i = d_{ij} X_j \tag{3.7}$$

where $i = 1, 2, 3$ and $j = 1, 2, \ldots, 6$. d_{ij} are the piezoelectric moduli, in which the first suffix represents the direction in which charge is generated and the second suffix indicates the direction of the applied stress.

The set of piezoelectric coefficients are properly defined by the following partial derivatives, where the subscripts indicate the variables held constant, T being the temperature.[5]

$$\left(\frac{\delta D}{\delta X}\right)_{E,T} = \left(\frac{\delta x}{\delta E}\right)_{X,T} = d \tag{3.8}$$

$$-\left(\frac{\delta E}{\delta X}\right)_{D,T} = \left(\frac{\delta x}{\delta D}\right)_{X,T} = g \tag{3.9}$$

$$\left(\frac{\delta D}{\delta x}\right)_{E,T} = -\left(\frac{\delta X}{\delta E}\right)_{x,T} = e \tag{3.10}$$

$$-\left(\frac{\delta E}{\delta x}\right)_{D,T} = \left(\frac{\delta X}{\delta D}\right)_{x,T} = h \tag{3.11}$$

These coefficients are not independent and it may be observed from equations (3.8) to (3.11) that

$$\frac{d}{g} = \varepsilon_0 \varepsilon^X \tag{3.12}$$

and

$$\frac{e}{h} = \varepsilon_0 \varepsilon^x \tag{3.13}$$

$$\frac{e}{d} = C^E \tag{3.14}$$

and

$$\frac{h}{g} = C^D \tag{3.15}$$

where C is the elastic stiffness constant (stress per unit strain). Following equations (3.12) and (3.13) we get

$$\frac{d_{ij}}{g_{ij}} = \left(\frac{\delta x_j}{\delta E_i}\right)_{X,T} \bigg/ \left(\frac{\delta x_j}{\delta D_i}\right)_{X,T} = \left(\frac{\delta D_i}{\delta E_i}\right) = \varepsilon_0 \varepsilon_{ii}^X \tag{3.16}$$

Thus

$$\frac{d_{31}}{g_{31}} = \varepsilon_0 \varepsilon_{33}^X \tag{3.17}$$

and similarly

$$\frac{e_{31}}{h_{31}} = \varepsilon_0 \varepsilon_{33}^x \tag{3.18}$$

$d(CN^{-1})$ and $g(VM^{-1})$ are known as the piezoelectric strain constants and $e(NV^{-1})$ and $h(NC^{-1})$ are the piezoelectric stress constants. The d- and the g-coefficients measure the performance of piezoelectric sensors in the receiver mode whereas the e- and h-coefficients are used to express the ability of the piezoelectric actuators in the transmitting mode. For hydrophone applications d_{33}-coefficients should be as high as possible, whereas for the bimorph actuators d_{31}-coefficient is desired to be high. For a hydrostatic stress the hydrostatic d_h-coefficient is given by

$$d_h = d_{31} + d_{32} + d_{33} \tag{3.19}$$

An important parameter characterizing a piezoelectric transducer is the electromechanical coupling coefficient k which is defined as follows

$$k^2 = \frac{\text{electrical (mechanical) energy converted}}{\text{into mechanical (electrical) energy}} \bigg/ \text{input electrical (mechanical) energy} \tag{3.20}$$

k^2 is of course less than 1. For most applications it is desirable to have a high value of k as it is a measure of the ability of a transducer to convert readily energy from one form to another.

The values of the piezoelectric parameters of loss-less ferroelectric ceramics may be derived from a study of their resonance behaviour, when subjected to a sinusoidally varying electric field. Figure 3.3 shows the equivalent circuit of a piezoelectric specimen close to its fundamental resonance and corresponding frequency response is illustrated in Fig. 3.4. The resonance and the anti-resonance frequencies are denoted by f_r and f_a, respectively, when the reactance X_e is zero. f_m and f_n are the frequencies at which the circuit impedance is minimum and maximum, respectively. f_s represents the frequency at which the reactance X_1 of the series arm is zero. It may be shown that for a thin disc of diameter d, electrodes on both surfaces and poled in the thickness direction, the planar coupling coefficient k_p is given by[6]

$$\frac{k_p^2}{1 - k_p^2} = f\left(J_0 J_1 v \frac{f_p - f_s}{f_s}\right) \tag{3.21}$$

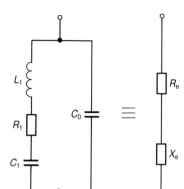

(a)   (b)

Fig. 3.3 (a) Equivalent circuit for a piezoelectric specimen vibrating close to resonance; (b) the equivalent series components of the impedance of (a). (Moulson and Herbert, 1990).[4,5]

where J_0, J_1 are Bessel functions and v the Poisson's ratio. Figure 3.5 shows the relationship between $(f_p - f_s)/f_s$ and k_p for the case $v = 0.3$. It should be mentioned that for the v-value within the range $0.27 < v < 0.35$ this curve is quite insensitive. Generally, for common piezoelectric ceramics v values lie within the range $0.28 < v < 0.32$. With the knowledge of k_p-value k_{31} may be determined thus

$$k_{31}^2 = \frac{1 - v}{2} k_p^2 \tag{3.22}$$

The compliance s_{11} may be obtained using the following expression

$$\frac{1}{S_{11}^E} = \frac{\pi d^2 f_s^2 (1 - v^2) \rho}{\eta_1^2} \tag{3.23}$$

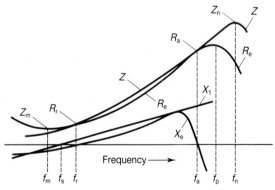

Fig. 3.4 The characteristic frequencies of the equivalent circuit exaggerating the differences between f_m, f_s and f_r and between f_a, f_p and f_n. (Moulson and Herbert, 1990.)[4,5]

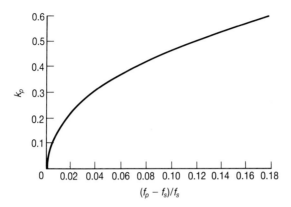

Fig. 3.5   The planar coupling coefficient k_p as a function of $(f_p - f_s)/f_s$. (Moulson and Herbert, 1990.)[4]

where ρ is the density of the material and η_1 (≈ 2) a root of the equation containing the Poisson ratio and the Bessel functions.[6] Furthermore

$$d_{31} = k_{31}(\varepsilon_{33}^X S_{11}^E)^{1/2} \tag{3.24}$$

and

$$g_{31} = \frac{d_{31}}{\varepsilon_{33}^X} \tag{3.25}$$

where the permittivity may be obtained from a measurement of the capacitance C, of the specimen at a frequency significantly below the resonance. It is given by

$$\varepsilon_{33}^X = CA/l \tag{3.26}$$

where A and l are the electrode area and the specimen thickness, respectively.

The mechanical Q factor Q_m can be obtained from the following expression if the magnitude of $|Z_m|$ is known[5]

$$\frac{1}{Q_m} = 2\pi f_s |Z_m|(C_0 + C)\frac{f_p^2 - f_s^2}{f_p^2} \tag{3.27}$$

where C_0 is the parallel capacitance of the equivalent circuit (see Fig. 3.3).

The expressions for the coupling coefficient and other piezoelectric parameters for loss-less ferroelectric ceramics are dependent on the geometry of the specimen. In general, the effective coupling coefficient is related to C_0, C_1, and the series and parallel resonant frequencies. Thus (see Fig. 3.4)

$$k_{\text{eff}} = \frac{C_1}{C_0 + C_1} = \frac{f_p^2 - f_s^2}{f_p^2} \simeq \frac{f_\infty^2 - f_r^2}{f_n^2}$$

$$\simeq \frac{f_n^2 - f_m^2}{f_n^2} \tag{3.28}$$

Table 3.2 *Typical values of some parameters of a few piezoelectric materials (Moulson and Herbert, 1990)[4]*

Parameters	Unit	BaTiO₃	PZT (wide range)	LiNbO₃ (single crystal)	LiTaO₃ (single crystal)	Modified PbTiO₃*
ε_{33}		1900	1200 to 2800	28	43	203
$\tan\delta$	$\times 10^3$	7	3 to 16	–	–	22
d_{31}	$\times 10^{-12}\,CN^{-1}$	−79	−119 to −234	−0.85	−3.0	−7.4
d_{33}	$\times 10^{-12}\,CN^{-1}$	190	268 to 480	6	5.7	47
k_{31}		0.21	0.33 to 0.39	0.02	0.07	0.052
k_{33}		0.49	0.68 to 0.72	0.17	0.14	0.35

* +5 mol% $Bi_{2/3}Zn_{1/3}Nb_{2/3}O_3$.

The d- and the g-coefficients may be determined from the k-coefficient which determines the transducer band width.[6] Typical values of important piezoelectric parameters of some materials are shown in Table 3.2.

Pyroelectric Effect The ancient discovery of pyroelectricity has been reviewed by Lang.[7] A pyroelectric material exhibits a spontaneous polarization, as in the case of a ferroelectric material, in the absence of an electric field. True (reversible) pyroelectricity originates from the temperature dependence of the spontaneous polarization. From thermodynamics, the pyroelectric coefficient p, may be expressed thus

$$p = \left(\frac{\delta P_s}{\delta T}\right)_{E,X} = -\frac{C_{E,X}}{T}\left(\frac{\delta T}{\delta E}\right)_{S,X} \tag{3.29}$$

where P_s is the spontaneous polarization, T the temperature, C the specific heat capacity, E the electric field and X the stress. The existence of a polar axis in a crystal allows the existence of a spontaneous electrical polarization. A required polar axis may also be introduced by the application of a high electric field.[8,9] This induced effect will be quite stable at ordinary temperatures and will be superimposed on any electroactivity of the material. The word 'polar' may refer to polar crystals, polar molecules or a polar direction.

The total dielectric displacement in a polar material on the application of an external electric field is given by

$$D = \varepsilon_0 \varepsilon E + P \tag{3.30}$$

where

$$P = P_s + P_{\text{induced}}$$

Assuming E to be constant

$$\frac{\delta D}{\delta T} = \varepsilon_0 E \frac{\delta \varepsilon}{\delta T} + \frac{\delta P}{\delta T} \tag{3.31}$$

or

$$p_g = \varepsilon_0 E \frac{\delta \varepsilon}{\delta T} + p \tag{3.32}$$

where p_g is known as a generalized pyroelectric constant and the true pyroelectric constant p is given by

$$p = \frac{\delta p}{\delta T} \tag{3.33}$$

As the temperature coefficient of the permittivity of ferroelectric materials can be quite considerable, particularly near their Curie temperatures, the contribution of the first term in equation (3.32) can be comparable in magnitude with that of the true pyroelectric coefficient p. Ferroelectric materials are likely to exhibit high pyroelectric coefficients just below their Curie temperatures. In this respect, high pyroelectric coefficients are observed with ferroelectrics possessing second-order transitions, such as TGS (triglycerine sulphate) which has a pyroelectric coefficient of $\approx 280 \times 10^{-6} C\ m^{-2} K^{-1}$ at 20°C which is well below its transition temperature of 49°C. Materials such as strontium modified barium titanate (BST), which have quite steep first-order transitions, are not suitable for pyroelectric sensors, because they show a hysteresis behaviour with their transition temperature on thermal cycling. It should, however, be emphasized that a contribution due to $E(\delta\varepsilon/\delta T)$ can appear for all dielectric materials. Furthermore, as the polar pyroelectric materials are also piezoelectric, a strain originating from a thermal expansion may also provide a secondary pyroelectric effect. This is, however, small in comparison with the primary pyroelectric effect. High permittivity ferroelectrics possessing desirable piezoelectric coefficients may not be the ideal ferroelectric materials for pyroelectric applications for the following reasons. First, for practical purposes it is of advantage to provide a good match to the low input capacitance of the field effect transistor (FET) which is coupled to the pyroelectric sensing material. Secondly, a high figure of merit ($\alpha p/\varepsilon_0 \varepsilon$) can be achieved with polar materials with low permittivity.

The pyroelectric coefficient has three components, defined by

$$\Delta P_i = p_i \Delta T \quad \text{with } i = 1, 2, 3 \tag{3.34}$$

where ΔP is the change in polarization arising from a corresponding change in temperature, ΔT. However, for practical applications, p_3 is of significance for cases when there is no net dipole moment along 1- and 2-directions.

The pyroelectric coefficient may be determined by a direct method.[10] For $E = 0$, i.e., for poled samples with shortened electrodes, we have from equation (3.31)

$$\left(\frac{\delta D}{\delta T}\right)_{E,X} = \left(\frac{\delta P}{\delta T}\right)_{E,X} = \left(\frac{\delta Q/A}{\delta T}\right)_{E,X} \simeq \frac{1}{A}\left(\frac{\delta Q}{\delta T}\right)_{E,X} \tag{3.35}$$

where Q is the liberated charge, and A the electrode area.

The pyroelectric coefficient p may be expressed thus

$$p = \frac{1}{A}\left(\frac{\delta Q}{\delta T}\right)_{E,X} = \frac{1}{A}\frac{\delta Q}{\delta t}\frac{\delta t}{\delta T} = \frac{1}{A}I\frac{\delta t}{\delta T} \tag{3.36}$$

where I is the current through the sample with the electroded 'short-circuited' and $\delta T/\delta t$ is the rate of rise of sample temperature. Thus equation (3.35) leads us to a determination in a change in Q rather than changes in Q/A. From equation (3.36) we get

$$\frac{I}{A} = J_{\text{pyro}} = p\frac{dT}{dt} \tag{3.37}$$

where J_{pyro} is the pyroelectric current density. Thus a plot of J_{pyro}/T provides directly the value of p-value[9] provided the rate of rise of temperature is known; and note that $J_{\text{pyro}} = 0$ for $dT/dt = 0$. Hence, any isothermal current cannot contribute to the pyroelectric current and such current, possibly due to space charge, should be eliminated.

The pyroelectric coefficient can also be determined by a dynamic method[11] in which a dielectric material is illuminated with a sinusoidally modulated thermal radiation. The pyroelectric response is given by

$$V = \frac{\alpha p}{\rho\omega C_p \varepsilon_0 \varepsilon A} \tag{3.38}$$

where V is the peak to peak open circuit voltage (per watt of incident radiation) across the pyroelectric material and α the absorption coefficient, ρ the density, ω the angular frequency and C_p the specific heat. This is an attractive way to measure the pyroelectric coefficient as it includes the material parameters α, ρ and C_p which are also involved in the assessment of the figure of merit $M(n)$. This is defined as follows[12]

$$M(n) = \frac{VA\omega\varepsilon_0}{\rho C_p \varepsilon^n l^{(1-n)}} \tag{3.39}$$

where l is the sample thickness and the exponent n has the value of $0, \frac{1}{2}$ or 1 depending on the noise source and resonant modes.[13] From equations (3.38) and (3.39) we get[13]

$$M(n) = \frac{VA\omega\varepsilon_0}{\varepsilon^{(n-1)} l^{(1-n)}} \tag{3.40}$$

Thus for constant values of A, ω and the incident of radiation, the figure of merit of pyroelectric materials can be compared by measuring their open circuit peak to peak voltage V, produced by the chopped incident radiation, the dielectric constant ε and the sample thickness.

The pyroelectric coefficient may also be determined by a dynamic method[14] using a step input of thermal radiation on a material and measuring the short circuited thermal current transient. This is given by

$$I(t) = \frac{F_0 A p(T)}{\rho C_p l}\left(\frac{1}{1-\theta}\right)[(\exp - t/\tau_T) - (\exp - t/\tau_E)] \tag{3.41}$$

where $I(t)$ is the pyroelectric current, F_0 the radiation power absorbed per unit area of the electroded material, θ the ratio of the electrical time

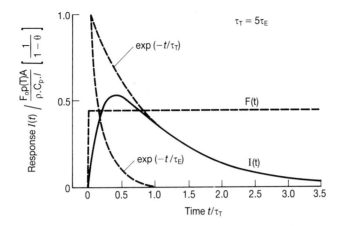

Fig. 3.6   Theoretical plot of the pyroelectric current response versus time (normalized).

constant τ_E and the thermal time constant (τ_T). Equation (3.41) represents the difference between two exponentials with the same initial value, one decaying with the thermal time constant τ_T and the other decaying with the electrical time constant τ_E. The two exponentials are shown in Fig. 3.6 by dash curves together with the dynamic pyroelectric response (solid curve) for the case $\tau_T = 5\tau_E$. Here we can also see that the peak response I_p which is reached in time t_p and it may be shown that

$$t_p = \tau_E \ln\left(\frac{\theta}{\theta-1}\right) \tag{3.42}$$

The peak current I_p is given by

$$I_p = \beta p(T) \tag{3.43}$$

where

$$\beta = \frac{F_0 A}{\rho C_p l}\,\theta^{\left(\frac{\theta}{1-\theta}\right)} \tag{3.44}$$

The initial slope K of $I(t)$ is,

$$K = dI(t)/dt \bigg| = \frac{F_0 p(T)A}{\rho C_p l}\frac{1}{\tau_E} \tag{3.45}$$

Equations (3.42), (3.43) and (3.45) may be used to analyse the nature and shape of the pyroelectric response, which originates from a dipolar mechanism, following a step input of thermal radiation.

It has been shown[15,16] that a low frequency temperature wave may be used as a direct dynamic measurement of the pyroelectric coefficient. These authors show that when a sample is heated with a small sinusoidal temperature wave, the pyroelectric current is directly proportional to the time derivative of temperature, whereas the nonpyroelectric current for

Table 3.3 *Pyroelectric materials and their properties*

Materials	Curie temperature T_c (°C)	Specific heat C_p (Jkg^{-1}K^{-1}) $\times 10^3$	Density ρ (Kgm^{-3}) $\times 10^3$	Dielectric constant ε	Pyroelectric coefficient p at 25°C (Cm^{-2}K^{-1}) $\times 10^{-4}$
TGS	49	1.5	1.7	43–50	2.8–3.5
LATGS	49.5	1.5	1.7	35	7.0
BaTiO$_3$	120	0.5	6.0	1900	2.0
LiTiO$_3$	618	0.43	7.45	43	1.7
LiNbO$_3$	1210	0.65	4.64	28	0.83
Sr$_{1-x}$Ba$_x$Nb$_2$O$_6$ (SBN) x = 0.27	≈47	≈0.4	5.2	8200	5.5
PZT 4	328	0.42	7.6	1300	2.7
PZT 5	365	0.40	7.75	2100	4.7
PLZT (8/65/35)	100		7.80	3800	4.0

small temperature intervals is either constant or proportional to the temperature. This particular technique is used to separate the pyroelectric current from the nonpyroelectric current if they are produced simultaneously in a pyroelectric material.

Table 3.3 gives a list of some of the important properties of a few useful pyroelectric materials with their properties as given by Jona and Shirane.[17]

3.3   Piezo- and Pyroelectricity in Polymers

There are several polymers in which piezo- and pyroelectricity can be induced by an external electric field. But the largest electroactive (piezo- and pyroelectricity) response observed so far has been from PVDF and some of its copolymers, such as vinylidene fluorine-trifluoroethylene (VDF-TrFE). PVDF is a semicrystalline polymer and its monomer unit $(CH_2 \cdot CF_2)_n$ has a dipole moment of 7.0×10^{-30} Cm perpendicular to the chain direction. If all the monomer dipoles were aligned along the chain direction, a maximum microscopic polarization of $100 \, mCm^{-2}$ can be obtained. However, semi-crystalline PVDF is approximately 50% crystalline and the observed polarization of $65 \, mCm^{-2}$ confirms its dipolar origin. It would thus appear that, if the dipole moment of the monomer unit in a polymer were increased, higher piezo- and pyroelectricity may be obtained due to the enhanced polarization. This is provided that the dipoles can be aligned by an appropriate electric field which does not cause an electrical breakdown of the dielectric, and that the dipoles maintain their alignment after the removal of the poling field. It should be noted, however, that it becomes difficult to rotate large dipoles.

In recent years copolymers of VDF-TrFE and vinylidene cyanide with a variety of comonomers have been introduced as new polymeric materials with strong piezo- and pyroelectric responses which are, however, not as high as those of ferroelectric ceramics. Ferroelectric thin films, on the

Table 3.4 *Comparison of piezo- and pyroelectric and other properties of a few polymers*

	Polyvinyl chloride	Nylon-11	PVDF	VDF-TrFE
Specific heat C_p $(Jkg^{-1}K^{-1}) \times 10^3$	0.93	1.6	1.3	
Density ρ $(kgm^{-3}) \times 10^3$	1.37	1.04	1.76	1.9
Dielectric constant (ε)	3.5	3.7	12	15–20
Coupling factor, k			0.16	0.3
Piezoelectric charge constant $d_{(ij)} (\times 10^{-12}CN^{-1})$	0.7	0.26	28	30
Piezoelectric voltage constant $g_{(kj)} (VmN^{-1}) \times 10^{-2}$	2.3	0.8	26.4	15
Pyroelectric coefficient, p $(Cm^{-2}K^{-1}) \times 10^{-6}$	0.1	0.5	4	4
Pyroelectric figure of merit $p/(\rho C_p \varepsilon_0 \varepsilon)$: $(Vm^2J) \times 10^{-4}$	253	917	1646	

other hand, enjoy higher mechanical strength in comparison with ceramics. There are several review papers on functional properties of PVDF.[18–26] A comparison of piezo- and pyroelectric and other properties of a few polymers is given in Table 3.4.

Structural Forms of Poly(vinylidene fluoride)   PVDF has at least four polymorphic phases,[23] designated as α-, β-, γ- and δ-phases (Fig. 3.7). The

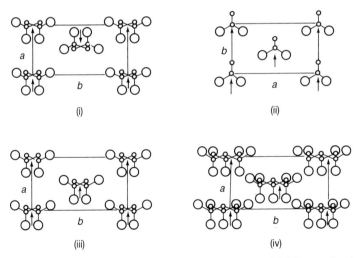

Fig. 3.7 Projection of C (small circles) and F (large circles) atoms of PVDF on to the *ab* planes of four of the five known crystal forms. Hs have been omitted. (i) α-phase (11), orthorhombic, a = 4.96 Å, b = 9.64 Å, c = 4.62 Å; (ii) β-phase (1), orthorhombic, a = 8.47 Å, b = 4.90 Å, c = 2.56 Å; (iii) δ-phase (IV), orthorhombic, a = 4.96 Å, b = 9.64 Å, c = 4.62 Å; (iv) γ-phase (III), monoclinic, a = 4.96 Å, b = 9.67 Å, c = 9.20 Å; angle β = 93°. Arrows indicate dipole moment perpendicular to chain axis. (Davis, 1988.)[37]

α-phase is the most common structure and the other three forms can be obtained from this parent phase by applications of mechanical stress, heat and electric field. The polymer chain in the α-phase has as TGTḠ conformation and the chain packs in an antipolar array resulting in no net dipole moment in the crystal.

When the α-phase of PVDF is mechanically deformed by stretching or rolling at temperatures below 100°C, β-phase of PVDF with all *trans* conformation is formed. The unit cell of β-phase has orthorhombic symmetry and the dipole moment of the polymer chains in the β-phase is 7.0×10^{-30} Cm which lies normal to the chain direction. The compressibility of the β-phase is significantly less than that of the α-phase and is anisotropic along the axis of the unit cell. The β-phase of PVDF is probably the most important of the four polymorphs in piezo- and pyroelectric applications. Although each crystallite of the β-phase PVDF has net dipole moment, because of the random orientation of the crystallites there is no net polarization until it is suitably poled by an external electric field.

The non polar α-phase of PVDF can also be converted to a polar phase by large electric fields[27–30] of the order of 1.2×10^8 Vm^{-1} at room temperature. This polar version of the α-phase is known as the δ-phase which has an orthorhombic structure with TGTḠ chain packing with the dipole vectors oriented in the same direction. At poling fields $\simeq 4.8 \times 10^8$ Vm^{-1} a further phase change occurs in which the δ-phase is converted to the β-phase.[30–32] The γ-phase chain conformation is similar to $T_3GT_3\bar{G}$ and the chains pack to form a monoclinic crystal.[24–37]

PVDF is a ferroelectric polymer containing polar crystals in which the direction of polarization can be reversed by an application of suitable electric field. Figure 3.8 shows[38] the nature of hysteresis phenomenon in PVDF at temperatures in the range of −100°C to 20°C; from which it may be observed that the hysteresis loop occurs even below its glass transition temperature provided the poling field is high (2×10^8 Vm^{-1}). The coercive field decreases significantly from 1.8×10^8 Vm^{-1} at −100°C to 0.25×10^8 Vm^{-1} at 20°C. These values are considerably higher than those of ferroelectric crystals (1×10^6 Vm^{-1} to 0.1×10^6 Vm^{-1}, respectively, for BaTiO$_3$). In ferroelectric crystals the residual polarization arises from the displacement of atoms, whereas in PVDF the dipoles are rotated together with the main chains, thus requiring high electric fields to promote rotation. The reversal of polarization may be achieved by a rotation of molecular dipoles in the crystalline region in discrete steps of 60° increments, with a maximum of three steps in a single reversal in β-phase PVDF which has a hexagonal symmetry around the C-axis.[39] Polarization reversal may also be explained with other models.[40–42]

The dielectric behaviour of PVDF is shown[43] in Fig. 3.9 where three relaxations α-, β- and γ-processes in the order of decreasing temperatures are observed. The α-process which occurs at 40°C is due to the molecular motion involving the change in dipole moment of the nonpolar α-form crystals of TGTḠ conformation. This relaxation process disappears on mechanical drawing of the α-form PVDF when β crystals are formed. The

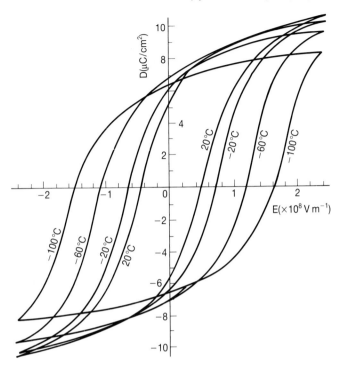

Fig. 3.8 Hysterisis loops of electric displacement versus electric fields for PVDF at various temperatures (after electric excitation at 20°C). (Furukawa et al., 1980.)[38]

β-relaxation process occurs at −40°C and is associated with the glass transition of the non crystalline molecules. On mechanical drawing the β-peak broadens and shifts slightly to a higher temperature. The origin of these changes is not yet understood. The γ-relaxation process at −100°C may arise from a local twist motion of the main chains in the crystalline/non crystalline regions.[43]

The components of the piezoelectric tensor for the poled PVDF are shown below where the units are in $\times 10^{-12}$ CN^{-1}

$$
d_{ij} = \begin{array}{c|cccccc}
 & \multicolumn{6}{c}{j} \\
i & 1 & 2 & 3 & 4 & 5 & 6 \\
\hline
1 & 0 & 0 & 0 & 0 & -27 & 0 \\
2 & 0 & 0 & 0 & -23 & 0 & 0 \\
3 & 21 & 1.5 & -32.5 & 0 & 0 & 0
\end{array} \qquad (3.46)
$$

It may be observed that d_{31}-coefficient is much larger than the d_{32} value and this is due to the difference in Poisson's ratio in the two directions. The 3-direction is the one where electrodes are deposited (i.e. thickness direction). By convention a tensile stress is positive, so that a positive stress in the 3-direction causes an increase in the film thickness and a corresponding decrease in the polarization and hence d_{33} is negative. Similarly, a

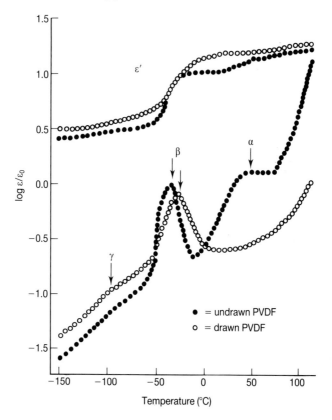

Fig. 3.9 Temperature spectra of complex dielectric constants for an undrawn form 11 PVDF, and drawn form 1 PVDF measured at 10 Hz. (Furukawa and Wang, 1988.)[43]

tension in either the 1- or 2-direction will cause a decrease in film thickness (through Poisson's ratio) and thus gives rise to positive d_{31}- and d_{32}-components. Figure 3.10 shows the nature of d_{31-}, d_{32}- and d_{33}-coefficients of PVDF as a function of remanent polarization.[43]

As polarization exists only in 3-direction, $p_1 = p_2 = 0$. An increase in temperature will produce an increase in volume, thus resulting in a decrease in the P_3 component which yields a negative pyroelectric P_3 coefficient. The pyroelectric tensor of poled PVDF is then

$$p_3 = \begin{bmatrix} 0 \\ 0 \\ -4 \end{bmatrix} \quad (3.47)$$

quite small compared with many ceramic materials, but the advantage of using thin polymer films is that a large change in temperature is obtained for a given input energy. However, the voltage output varies linearly with thickness. Figure 3.11 shows the plots of pyroelectric coefficient with temperature.[29] There are three basic models of the origin of piezo- and

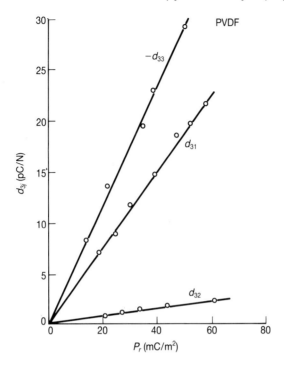

Fig. 3.10   Plots of d_{31}, d_{32} and d_{33} as functions of remanent polarization P_r for a uniaxially drawn PVDF. (Furukawa and Wang, 1988.)[43]

pyroelectricity in polymers. The model due to Broadhurst *et al.*[21,44,45] is based on field induced reorientation of crystalline dipoles, in which piezo- and pyroelectricity arise mostly from bulk dimensional changes due to changes in stress and temperature, respectively, and their effect on the

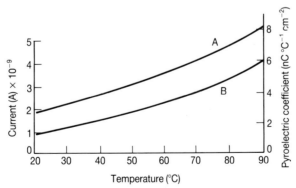

Fig. 3.11   Thermally stimulated current (reversible) during the second heating cycle and the pyroelectric coefficient for corona charged PVDF film of 25 μm thickness. Poling conditions for curves A and B are 10 kV and 5 kV (corona poling) respectively. (Das-Gupta and Doughty, 1978.)[29]

high frequency dielectric constant ε_∞ of the material. The frozen in polarization P_0, arising from aligned dipoles in a polar crystal is given by

$$P_0 = \left(\frac{\varepsilon_\infty + 2}{3}\right)\frac{N}{V}\mu_0\langle\cos\theta\rangle \qquad (3.48)$$

where N is the number of dipoles in a crystal volume V, ε_∞ the high frequency dielectric constant, μ_0 the dipole moment in vacuum and $\langle\cos\theta\rangle$ the average value of the cosine of the angle between the dipole and the polarization direction. In semicrystalline PVDF the crystals are assumed to be thin lamellae with the polarization vector being parallel to the large flat surface. For this case the vacuum dipole moment μ_0 becomes $(\varepsilon_c + 2)\mu_0/3$, where ε_c is the dielectric constant of the crystal due to the polarization of dipoles within the crystal by the fields of the neighbouring dipoles. For such a case the hydrostatic piezoelectric coefficient $d_{(\text{hydrostatic})}$ and the pyroelectric coefficients are reported to be[45]

$$d_{(\text{hydrostatic})} = P_0\left[\beta_c\frac{(\varepsilon_c - 1)}{3} + \frac{\beta_c\phi_0^2\gamma}{2} + \frac{\beta_s}{2}\right] \qquad (3.49)$$

$$P = -p_0\left[\alpha_c\frac{\varepsilon_c - 1}{3} + \frac{\alpha_c\phi_0^2}{2}\left[\gamma + (2T\alpha_c)^{-1} + \frac{\alpha_s}{2}\right]\right] \qquad (3.50)$$

and

$$P_0 = \phi\left[\left(\frac{\alpha_c + 2}{3}\right)\frac{N\mu_0}{V_c}\right]J_0(\phi_0)\langle\cos\theta\rangle \qquad (3.51)$$

where ϕ is the volume fraction of crystals, $J_0(\phi_0)$ the Bessel function arising from the harmonic oscillation of dipole fluctuations of average amplitude ϕ, γ a Grüneisen's constant arising from the change in the libration frequency with volume in the crystal, and α and β the volume thermal expansion and volume compressibility respectively. The subscripts c and s refer to crystal only and sample as a whole, respectively. P_0 is not calculated and needs to be obtained from the remanent polarization in a hysteresis loop. The first term within the brackets of equations (3.49) and (3.50) is due to the change in the dielectric constant of the crystal with temperature and constitutes electrostriction contribution. The second term arises from changes in the libration amplitude which produces changes in the net polarization and the third term is the contribution from the thickness changes in the polymer.[37] With $\phi = 16°$ and $\gamma = 5$, a numerical evaluation of each term[45] in equation (3.50) shows that the electrostriction, the dipole libration and dimensional changes of the sample contribute to 27%, 23%, and 50%, respectively.

The model due to Wada and Hayakawa[46] assumes that special polar crystals (rather than thin lamellae) with $\varepsilon_c = 3.3$ are located in a material consisting of amorphous medium with an average dielectric constant of ε_a. The overall polarization of the film is given by[24,47]

$$P = \phi\left(\frac{3\varepsilon_a}{2\varepsilon_a + \varepsilon_c}\right)P_s \qquad (3.52)$$

where P_s is the saturation polarization. There is no contribution from the libration amplitude in this model and the dimensional changes of the sample and the crystal contribute 47% and 53%, respectively. The calculated value of the pyroelectric coefficient ($p_3 = -2.7 \times 10^{-5}$ $Cm^{-2}K^{-1}$) is in reasonable agreement with the experimentally observed value. This model also provides a good agreement between the predicted and the experimentally observed values of the elastic and the piezoelectric constants of a single crystal of PVDF.[49]

The third model, due to Purvis and Taylor,[48] points out that the expression for the frozen-in polarization, given by equation (3.48) is strictly correct only for crystals in which a molecular dipole occupies a point of cubic symmetry; whereas β-PVDF crystals have an orthorhombic symmetry. These authors assume the dipoles to be point entities and incorporate the influence of base centred orthorhombic crystal structure and dependence of internal electric field on the lattice parameters. Their model also predicts a strong anisotropy for piezoelectric coefficients under uniaxial stress. Libration of dipoles, changes in chain conformation under stress and reversible crystallinity are not involved in this model, which also implies that the internal lattice field actually reduce the dipole moment of a single monomer ($-CH_2-CF_2-$). Despite these limitations, this model[48,50,51] provides an agreement between the expected and experimentally observed values of piezo- and pyroelectric coefficients of β-phase PVDF possessing an orthorhombic lattice. In this model[48] the pyroelectric coefficient arises from contributions of two sources, i.e. electrostriction (76%) and dimensional changes (24%) of the sample. It is difficult to assess the relative merits of these three models.

Thus our present knowledge of the origin of the piezo- and pyroelectricity in PVDF still remains inadequate, although it is generally agreed that a large portion of the piezo- and pyroelectric response may arise from the dimensional changes of the sample.

Copolymers of Vinylidene Fluoride   Vinylidene fluoride (VDF) copolymerizes readily with tetrafluoroethylene (VDF-TeFE) and trifluoroethylene (VDF-TrFE) providing copolymers with randomly distorted comonomer units in the molecular chains. In this process hydrogen atoms of the adjacent carbon atoms are replaced by large fluorine atoms, thus restricting the formation of TGTG polymer conformation and enhancing the all-*trans* conformation of the β-phase.[52] Hence the copolymers of VDF-TeFE and VDF-TrFE do not need to be mechanically stretched as they crystallize directly from the melt or solution into the analogous β-phase when the VDF content (x) is within the range of $0.6 \leqslant x \leqslant 0.82$.[53-56] VDF-TrFE copolymers exhibit a ferroelectric β-phase to a paraelectric phase transition and the Curie temperature increases with increasing VDF content.[57] The dielectric strength $\Delta\varepsilon$ is related to the remanent polarization P_r, thus,

$$P_r = \Delta\varepsilon E_p \tag{3.53}$$

where E_p is the poling field. For polar polymers $\Delta\varepsilon/\varepsilon_0$ is associated with free rotational dipolar motions. It has been shown that for VDF rich copolymers the dielectric anomaly ($\Delta\varepsilon/\varepsilon_0$) is small as VDF content is reduced, thus emphasizing that the ferroelectricity in VDF-TrFE copolymer is related with PVDF.[43]

It appears that in order to obtain a high piezoelectric response with VDF-TrFE (with 74 mol% VDF) it is necessary to anneal the copolymer near its ferroelectric to paraelectric phase transition temperature.[56] Such a thermal treatment increases the magnitude of the remanent polarization P_r, and the electromechanical coupling coefficient K_t. It has been possible to obtain 90% crystallinity with high VDF content in VDF-TrFE copolymer. For PVDF at $P_r \simeq 50$–60 mCm^{-2}, $K_t \simeq 0.2$ whereas for the copolymer, $K_t \simeq 0.3$ at $P_r \simeq 90 - 100$ mCm^{-2}. Figures 3.12 and 3.13 show[43] the piezoelectric and pyroelectric properties of VDF-TrFE and VDF-TeFE copolymers as a function of remanent polarization. Copolymers of VDF-TrFE with 70–80 mol% VDF can have very large remanent polarization (>100 mCm^{-2}) which provides corresponding higher d_{31} value.

Other Polymers   Piezo- and pyroelectricity have also been observed with copolymers of vinylidene cyanide (VDCN) and vinyl acetate (VCA) in spite of its amorphous structure.[58] The amorphous nature of the copolymer will make it optically more transparent, thus providing possible electro-optical applications which are not yet possible with PVDF because of its poor optical quality arising from its semicrystalline microstructure. Piezo-

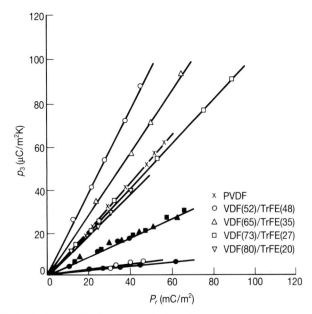

Fig. 3.12   Plots of e^a_{31} and e^a_{32} of undrawn (filled marks) and drawn (open marks) ferroelectric polymers as a function of P_r. (Furukawa, 1989.)[25]

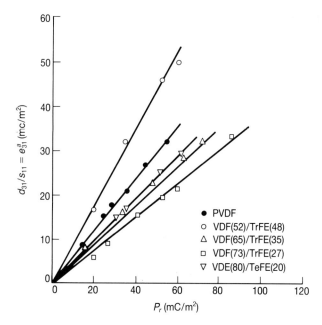

Fig. 3.13 Behaviour of the pyroelectric constant p_3 of ferroelectric polymers as a function of P_r. (Furukawa, 1989.)[25]

and pyroelectricity in poled thin films of aromatic poly(urea), oriented poly(trifluoroethylene), many biopolymers, synthetic polypeptides, oriented films of poly (γ-methyl-L-glutamate) and poly (γ-benzyl-L-glutamate) with α-helicoidal backbone structure have been observed. Pyroelectric effects in diacetylene and polydiacetylene have also been observed. However the electroactive responses of these polymers including nylon 11 are inferior to those of PVDF and its copolymer VDF-TrFE. The pyroelectric coefficient of VDCN-VAC, poly(urea) and poly(trifluoroethylene) at 20°C are 3×10^{-6} Cm^{-2}K^{-1}, 1×10^{-5} Cm^{-2}K^{-1} and 7×10^{-6} Cm^{-2}K^{-1}, respectively.

It is necessary to search for new piezo- and pyroelectric materials which are mechanically strong and have electro-active properties of ceramics (for micro- and nano scale) sensor applications.

In this respect, ceramics have been introduced in polymer matrices to produce composite materials with electroactive properties of ceramics and mechanical strength of polymers. At present, this is a growth area of research which may provide sensors suitable for microelectronics applications. An introduction of ferroelectric liquid crystal polymers of appropriate polar structure with large dipole moments in the main chain or side chain of commercially available synthetic polymers may be a fruitful approach to produce new sensor materials for the twenty-first century.

References

1. H. Kawai, *Japan J. Appl. Phys.*, **8**, 875 (1969).
2. M. E. Lines and A. M. Glass, *Principles and Applications of Ferroelectrics and Related Materials*, (Clarendon Press, Oxford, 1977), p. 608.
3. J. F. Nye, *Physical Properties of Crystals*, (Clarendon Press, Oxford, 1957), p. 110.
4. M. C. Lovell, A. J. Avery and M. W. Vernon, *Physical Properties of Materials*, (Van Nostrand Reinhold, London, 1976), p. 169.
5. A. J. Moulson and J. M. Herbert, *Electroceramics Materials Properties Applications*, (Chapman & Hall, 1990), p. 269.
6. IEEE Standards on Piezoelectricity, *ANSI/IEEE 176–1978*, 1978.
7. S. B. Lang, *Source Book of Pyroelectricity*, Eds I. Lefkowitz and G. W. Taylor, (Gordon and Breach, New York/London/Paris, 1974).
8. R. L. Zimmerman, C. Suchicital and E. Fukada, *J. Appl. Polym. Sci.*, **19**, 1373 (1975).
9. R. L. Zimmerman, *Biophys, J.*, **16**, 1341 (1976).
10. R. L. Byer and C. Roundy, *IEEE Trans. Sonics & Ultrasonics*, **SU-19**, 333 (1972).
11. E. H. Putley, The Pyroelectric Detection, *Semiconductors and Semimetals*, **5**, Infrared Detectors, Eds R. K. Willardson and A. C. Beer, (Academic Press, New York, 1970), p. 259.
12. R. G. F. Taylor and H. A. H. Boot, Pyroelectric Image Tubes, *Contemp. Phys.*, **14**, 55 (1973).
13. L. Garn and E. J. Sharp, *IEEE Trans on Parts, Hybrids, and Packaging*, **PHP10**, 208 (1974).
14. M. Simhony and A. Shaulov, *J. Appl. Phys.*, **42**, 37–41 (1971).
15. L. E. Garn and E. J. Sharp, *J. Appl. Phys.*, **53**, 8974 (1982).
16. E. J. Sharp and L. E. Garn, *J. Appl. Phys.*, **53**, 8980 (1982).
17. F. Jona and G. Shirane, *Ferroelectric Crystals*, (Pergamon Press, New York, 1962), p. 28.
18. Y. Wada and R. Hayakawa, *Japan J. Appl. Phys.*, **15**, 2041 (1976).
19. R. G. Kepler, *Ann. Rev. Phys. Chem.*, **29**, 497 (1978).
20. R. W. Whatmore, *Rep. Progr. Phys.*, **49**, 1335 (1986).
21. M. G. Broadhurst and G. T. Davies, *Topics in Applied Physics, Electrets*, Ed. G. M. Sessler, (Springer-Verlag, Berlin, 1980), p. 285.
22. D. K. Das-Gupta, *Ferroelectrics*, **33**, 76 (1981).
23. A. J. Lovinger, *Polyvinylidene Fluoride, Development in Crystalline Polymers—1*, Ed. D. C. Bassett, (Applied Science, London, 1981), p. 195.
24. Y. Wada, *Electronic Properties of Polymers*, Eds J. Mort and G. Pfister, (J. Wiley and Sons, New York, 1982), p. 109.
25. T. Furukawa, *IEEE Trans Electrical Insulation*, **24**, 375 (1989).
26. D. K. Das-Gupta, *Ferroelectrics*, **118**, 165 (1991).
27. D. K. Das-Gupta and K. Doughty, *Appl. Phys. Lett.*, **31**, 585 (1977).
28. D. Naegele, D. Y. Yoon and M. G. Broadhurst, *Macromolecules*, **11**, 1297 (1978).
29. D. K. Das-Gupta and K. Doughty, *J. Appl. Phys.*, **49**, 4601 (1978).
30. G. T. Davis, J. E. McKinney, M. G. Broadhurst and S. C. Roth, *J. Appl. Phys.*, **49**, 4998 (1978).
31. P. D. Southgate, *Appl. Phys. Lett.*, **28**, 250 (1976).
32. J. P. Luongo, *J. Polym. Sci.*, A-2, **10**, 1119 (1972).

33. D. K. Das-Gupta and K. Doughty, *J. Phys. D: Appl. Phys.*, **11**, 2415 (1978).
34. S. Weinhold, M. H. Litt and J. B. Lando, *Macromolecules*, **13**, 1178 (1980).
35. Y. Takahashi and H. Tadokoro, *Macromolecules*, **13**, 1316 (1980).
36. A. J. Lovinger, *Macromolecules*, **15**, 40 (1982).
37. G. T. Davis, *The Applications of Ferroelectric Polymers*, Eds T. T. Wang, I. M. Herbert and A. M. Glass, (Blackie, London, 1988), Chapter 4, p. 37.
38. T. Furukawa, M. Date and E. Fukada, *J. Appl. Phys.*, **51**, 1135 (1980).
39. R. G. Kepler and R. A. Anderson, *J. Appl. Phys.*, **49**, 1232 (1978).
40. T. Furukawa, M. Date, M. Ohuchi and A. Chiba, *J. Appl. Phys.*, **56**, 1481 (1984).
41. Y. Takase, A. Odagima and T. T. Wang, *J. Appl. Phys.*, **60**, 2920 (1988).
42. T. Yagi and X. Tatemoto, *Polym. J.*, **11**, 429 (1979).
43. T. Furukawa and T. T. Wang, *The Applications of Ferroelectric Polymers*, Eds T. T. Wang, I. M. Herbert and A. M. Glass, (Blackie, London, 1988), Chapter 5, p. 66.
44. F. I. Mopsik and M. G. Broadhurst, *J. Appl. Phys.*, **46**, 4204 (1975).
45. M. G. Broadhurst, G. T. Davis, G. E. McKinney and R. E. Collins, *J. Appl. Phys.*, **49**, 4992 (1978).
46. Y. Wada and R. Hayakawa, *Ferroelectrics*, **32**, 115 (1981).
47. R. Hayakawa and Y. Wada, *Report Prog. Polym. Phys. Jpn.*, **19**, 321 (1976).
48. C. K. Purvis and P. L. Taylor, *J. Appl. Phys.*, **54**, 1021 (1983).
49. K. Tashiro, M. Kobayashi, M. Tadokoro and E. Fukada, *Macromolecules*, **13**, 691 (1980).
50. R. Al-Jishi and P. L. Tayor, *J. Appl. Phys.*, **57**, 897 (1985).
51. R. Al-Jishi and P. L. Taylor, *J. Appl. Phys.*, **57**, 902 (1985).
52. J. B. Lando and W. W. Doll, *J. Macromol. Sci:—Phys.*, **132**, 205 (1968).
53. T. Furukawa, J. S. Wen, K. Suzuki, Y. Takashima and M. Date, *J. Appl. Phys.*, **56**, 829 (1984).
54. J. X. Wen, *Polym. J.*, **17**, 399 (1985).
55. S. Tasaka and S. Miyata, *J. Appl. Phys.*, **57**, 906 (1985).
56. K. Koga and H. Ohigashi, *J. Appl. Phys.*, **59**, 2142 (1986).
57. T. Yagi, M. Tatemoto and J. Sako, *Polym. J.*, **12**, 209 (1980).
58. M. Miyata, S. Yoshikawa, S. Tasaka and M. Ko, *Polym. J.*, **12**, 857 (1980).

4

Molecular Magnets

R J Bushby and J-L Paillaud

4.1   Introduction

The magnetic behaviour of solids is complex and many different types of magnetism can be distinguished: diamagnetism, paramagnetism, antiferromagnetism, ferrimagnetism, ferromagnetism, metamagnetism, speromagnetism, spin glass, asperomagnetism, etc.[1] In all cases, the ultimate carrier of the effect is the same, the magnetic moment of the electron. The many different types of magnetic phenomena arise because there are so many different ways in which these moments can be coupled together. In principle, the development of molecular magnets[2]—systems where the unpaired electrons are associated with discrete molecules, allied to recent advances in chemical synthesis and in methods for aligning and emplacing molecules—offers a hitherto unknown degree of control over this coupling and hence in the design and production of magnetic materials. In practice we are still learning how to do this and most experimental work has concentrated on the simplest target: the production of bulk ferro- and ferrimagnets. After a brief résumé of the essential characteristics of ferro- and ferrimagnets and of the most important mechanisms of spin-coupling, this chapter describes the principle approaches being made to the production of molecular magnets. As in any developing science, some progress is attributable to careful planning but much is purely empirical and serendipitous. Some of the systems discovered by the latter route are very interesting but not yet properly understood. They are described at the end of the chapter together with a brief discussion of possible applications.

4.2   Basic Concepts

When a sample is placed in a magnetic field, the field within the material will differ from the free-space value. If the substance is diamagnetic, containing only spin-paired electrons, the density of the magnetic lines of force is reduced within the sample. However, for a paramagnetic substance (one that contains unpaired electrons), the density of magnetic lines of force within the sample is intensified. Figure 4.1a shows the temperature dependence for the susceptibility of a paramagnetic material which can follow either the Curie law (4.1) or the Curie–Weiss law (4.2) in which C and θ are constants and T is the absolute temperature.

$$\chi = C/T \tag{4.1}$$

$$\chi = C/(T - \Theta) \tag{4.2}$$

For a 'pure' paramagnet (non-interacting spins) Θ is zero and the Curie law applies but if there is *local* ferromagnetic (spin-parallel) coupling Θ is positive whereas *local* antiferromagnetic (spin-antiparallel) coupling gives a negative value for Θ. For substances that show bulk ferromagnetism a transition occurs at a temperature known as the Curie temperature (T_c) leading to a phase in which there is *long-range* parallel ordering of the spins. Below this temperature the susceptibility rises to a very high value (Fig. 4.1b). The equivalent transition in a bulk antiferromagnet leads to *long-range* antiparallel ordering of adjacent spins and the transition temperature leading to the antiferromagnetic phase is known as the Néel temperature (T_N). In a bulk ferrimagnet, below the Curie temperature, there is also *long-range* antiparallel ordering of the spins but ferrimagnetic substances are composed of sublattices with two types of magnetic ions with different moments, in such a way that this antiparallel ordering results in a net moment (Fig. 4.2). Although, for either a ferro- or ferrimagnet below the Curie temperature, there is *long-range* ordering of the spins, a sample may still not behave 'like a magnet' since this ordering occurs within 'domains'. The domains themselves are randomly oriented and cancel each other out. However, application of a magnetic field will magnetize the sample. This occurs initially by movement of the walls between domains (Bloch walls) so that some domains grow at the expense of others. At the beginning (point O in Fig. 4.3) there is no net magnetization. As the applied field is increased there is at first a reversible movement of the Bloch walls (O to A) but beyond point A movement of the Bloch walls becomes irreversible and when the field is turned off the magnetization curve shows hysteresis and the sample retains some magnetization.

4.3   Magnets Based on Transition Metal Complexes

The mechanism of spin-coupling in molecular magnets differs from that in metals like iron. It is normal to distinguish two main mechanisms, direct

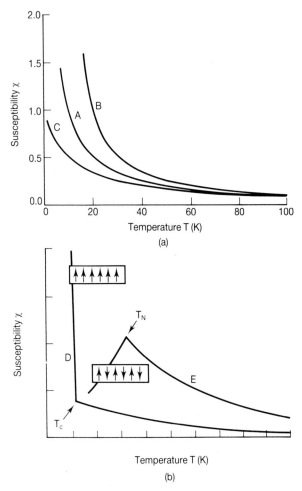

Fig. 4.1 (a) Graphical representation of the Curie and Curie–Weiss law (4.1) and (4.2). Curve A = Curie law (4.1), curve B = Curie–Weiss law (4.2) with $\theta = +10$, curve C = Curie–Weiss law (4.2) with $\theta = -10$ (in the three cases the Curie constant = 10). (b) Schematic representation of the susceptibilities of an antiferromagnetic and a ferromagnetic polycrystalline substance as function of temperature (arbitrary scale). Curve D = thermal variation of the susceptibility for a ferromagnet. Note that for a temperature above T_c the Curie–Weiss law ($\theta > 0$) applies and also that $T_c > \theta$. Curve E = thermal variation of the susceptibility for an antiferromagnet. Note that for a temperature above T_N the Curie–Weiss law ($\theta < 0$) applies and also that $T_N > \theta$.

exchange and indirect or 'superexchange'. To chemists the most familiar example of direct exchange is Hund's rule of maximum multiplicity. This 'rule' predicts that atomic carbon will have a triplet ground state. Four of its six electrons are spin-paired (in $1s$ and $2s$ orbitals) but the two remaining electrons, which occupy degenerate $2p$ orbitals, are predicted to have parallel spins (maximum multiplicity). Hund's rule can also apply to

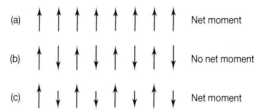

Fig. 4.2   Schematic representation of spin coupling in (a) a ferromagnet below the Curie temperature (b) and antiferromagnet below the Néel temperature (c) a ferrimagnet below the Curie temperature. The last example is intended to represent two types of metal ions associated with different moments. Note that, whereas these representations are of one dimensional materials, bulk ferromagnetism requires coupling of spins in three dimensions.

molecular and intermolecular situations but it needs to be applied with caution. Whatever the system (atomic, molecular, or intermolecular) the interaction which lies behind the rule is only significant if the half-filled orbitals are orthogonal but represent electron distribution which overlap significantly in space. This 'orthogonal but co-extensive' requirement is always met for co-centred atomic orbitals and the case for atomic carbon is illustrated in Fig. 4.4. It is not, however, always fulfilled for interactions between molecular orbitals and so the rule can break down. Whereas one of the main themes of research into high spin organic compounds has been the limitation of Hund's rule in situations where direct exchange dominates those involved in the study of polynuclear transition metal complexes have been faced by a seemingly very different problem. At an early stage it became clear that quite strong coupling of spins could occur over distances which were too great to be attributed to direct interactions between half-filled metal ion d-orbitals. In these cases the interaction is mediated

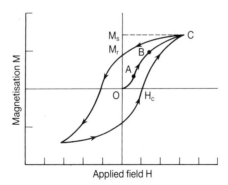

Fig. 4.3   Typical behaviour of a bulk ferro- or ferrimagnet sample in a magnetic field. M = magnetization, H = applied field, M_s = saturation magnetization (maximum value), M_r = remanent magnetization (remaining magnetization when the field is turned off), H_c = coercive field (field required to 'zero' the magnetization of the sample). The point B on the curve represents the formation of a single domain and the portion of the curve B to C corresponds to a rotation of the domain wall when the direction of the applied field and the 'easy' direction of magnetization of the sample differ.

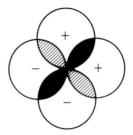

Fig. 4.4   Singly occupied 2*p* orbitals of atomic carbon. Note that the orbitals are co-extensive, they represent electron distributions which overlap, but they are orthogonal, i.e. the actual overlap integral is zero since the regions of positive overlap ■ and regions of negative overlap ▨ exactly cancel.

through the ligands and the mechanism of spin-interaction is known as superexchange. A classical example is that of copper acetate (Fig. 4.5). The crystal structure of this compound shows that it contains isolated pairs of copper ions which never-the-less interact strongly through exchange forces, each pair forming a low energy singlet and high energy triplet state

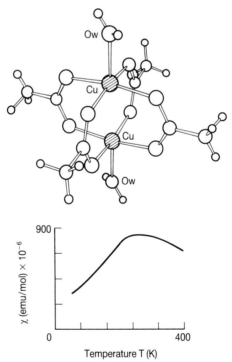

Fig. 4.5   Structure and susceptibility/temperature dependence behaviour of $Cu(CH_3CO_2)_2 \cdot H_2O$. Ow = water of crystallizaton.

(spin-spin coupling $J = -149\ cm^{-1}$). At first, it was argued that there was direct bonding between the metal atoms and that a δ bond was formed by a lateral overlap of two $d_{x^2-y^2}$ orbitals. The large separation (2.616 Å) makes this unlikely, however, and the fact that the analogous quinoline adduct of the trifluoroacetate has a significantly greater separation (2.886 Å), but almost the same interaction ($J = -155\ cm^{-1}$) renders the argument untenable. Clearly the interaction is mediated through the acetate ligands.

These ligand-mediated 'superexchange' interactions have been the subject of a detailed systematic study by the group of O. Kahn.[3a] The sign and the strength of the interaction varies widely from system to system but, just as in the case of direct exchange, it depends ultimately on the symmetry of the 'magnetic' orbitals and their overlap density. In this respect it is instructive to compare the complexes $CuVO(fsa)_2en \cdot CH_3OH$ [compound 1] and $CuCu(fsa)_2en \cdot CH_3OH$ [compound 2] [$(fsa)_2en^{4-}$ is the bichelating ligand derived from the Schiff base N,N-(2-hydroxy-3-carboxybenzylidene)-1,2 diaminoethane] (Fig. 4.6). In each of these compounds the magnetic sites [Cu(II), S = 1/2 and V(IV)O, S = 1/2] interact to give rise to a singlet and a triplet state. In compound [2] the 'magnetic'

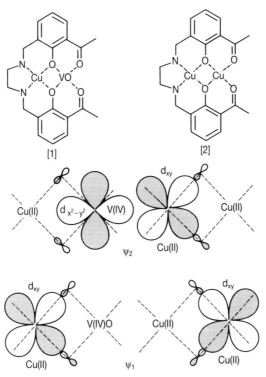

Fig. 4.6   Structures of the complexes $CuVO(fsa)_2en \cdot CH_3OH$ [**1**] and $CuCu(fsa)_2en \cdot CH_3OH$ [**2**], together with a description of the relevant 'magnetic' orbitals ψ_1 and ψ_2 expressed as linear combinations of d orbitals of the metal and p orbitals of the bridging oxygen atoms.

orbitals have the same symmetry and overlap. This kind of nonorthogonal relationship inevitably leads to a singlet ground state ($J = -650 \, \text{cm}^{-1}$). In compound [1], on the other hand the 'magnetic' orbitals are orthogonal and coextensive. This stabilizes the triplet which becomes the ground state ($J = +120 \, \text{cm}^{-1}$). Based on such systematic studies of bimetallic complexes, Kahn has devised a fruitful general strategy for the production of molecular magnets.[3] This involves the synthesis of bimetallic chains $[M_1, M_2]_n$ in which M_1 and M_2 are metal ions of differing spin, antiferromagnetically coupled, and held together by bridging ligands. This leads to a net moment for each chain and in some cases these 'chain moments' couple in a ferromagnetic manner. Among the first examples studied were $Mn(II)Cu(II)(pba),3H_2O \cdot 2H_2O$ [compound 3] and $Mn(II)Cu(II)$ $(pbaOH) \cdot 3H_2O$ [compound 4] [pba = 1,3-propylene bis (oxamato)] Fig. 4.7a). Within the chains of these compounds there is an antiferromagnetic interaction between Cu(II), $S = 1/2$ and Mn(II), $S = 5/2$. Compound [3] crystallizes in such a way that in neighbouring chains Cu is adjacent to Cu and Mn adjacent to Mn, and below 4.2 K these interact to give an antiferromagnetic coupling of the moments of the chains (Fig. 4.7b). By an ingenious structural modification, in which a hydroxy group is introduced into the pba ligand, it is possible to produce a structurally very similar material [compound 4] but in which the closest interchain interactions are Cu–Mn not Cu–Cu and Mn–Mn (Fig. 4.7c). Below $T_c = 4.6 \, \text{K}$ the moments of the chains now couple in a ferromagnetic manner. This is then a true 'molecular magnet' and shows, for example, the hysteresis behaviour typical of soft magnets (Fig. 4.3). The dihydrate corresponding to compound [4] shows $T_c = 30 \, \text{K}$, possibly because the removal of water facilitates a stronger inter-chain coupling.[3b] Closely related to Kahn's compounds are the edta complexes $M_1M_2(edta)(H_2O)_n$ [edta = ethylenediamine tetraacetate] studied by Coronado, Drillon et al. (1991) ($M_1, M_2 = Ni^{II}, Co^{II}, Mn^{II}, Zn^{II}$). Except for $MnMn(edta)(H_2O)$ which is a weak ferromagnet, $T_c = 1.489 \, \text{K}$, most are antiferromagnets, $T_N < 2 \, \text{K}$ but these compounds have proved important in developing our understanding of this type of molecular magnet.[4] The 3-d coordinated polymers $M^tM(M'edta) \cdot 4H_2O$ where $M^t = Co^{II}$, M, M' $= Co^{II}, Ni^{II}$, more simply written $[M^t, M, M']$ are also weak bulk ferromagnets, $[Co, Co, Co]$ $T_c = 0.1 \, \text{K}$, $[Co, Co, Ni]$ $T_c = 0.5 \, \text{K}$, $[Co, Ni, Ni]$ $T_c = 0.44 \, \text{K}$.[5]

When the ligand in an organometallic complex is itself a free radical there may be a ferromagnetic or antiferromagnetic coupling between the spin on the ligand and that on the metal ion. The same criteria, however, apply: orthogonality of co-extensive orbitals is required for a ferromagnetic interaction. In most cases, however, the orthogonality requirement is not met and, as in the $Mn(hfac)_2(proxyl)_2$ complex shown in Fig. 4.8, the interaction is antiferromagnetic. In this example there is direct overlap of the metal d_{XZ} and nitroxide π^* orbitals. Interactions of this type have been exploited by Gatteschi who has made a series of compounds using the bridging bidentate nitroxide ligands NITR [2-alkyl-4,4',5,5'-tetramethyl-4,5-dihydro-1H-imidazole-1-oxyl-3-oxide]. This can be used to build polymeric chains in which there are alternating spins on the metal and on

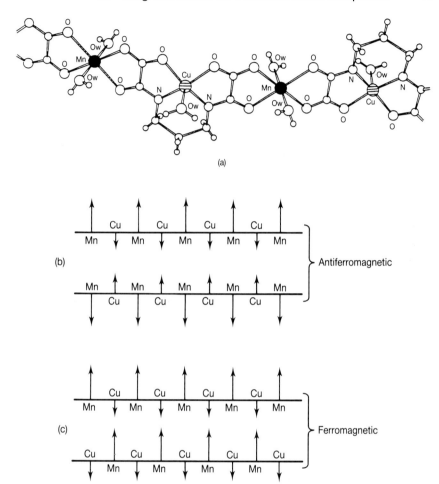

Fig. 4.7   (a) Segment of the repeating chain structure of compound [**3**] Mn(II)Cu(II)pba, 3H₂O·2H₂O; (b) schematic representation of spin coupling between adjacent chains in [**3**] below 4.2 K; and (c) for [**4**] below 4.6 K.

the ligand. Above 20 K the complexes with Mn(hfac)₂ [hfac = hexafluoroacetylacetonate] R = iPr, Et and nPr (Fig. 4.9) behave as typical ferromagnetic chains but at low temperatures [T_c = 7.6, 8.1, 8.6 K for R = iPr, Et, nPr] a three-dimensional ferromagnetic ordering occurs and the magnetic moment rapidly increases.[6] The interactions between the chains are apparently simply dipolar in nature and, since the energy of such dipolar interactions is proportional to the spin, the transition temperatures for the Mn complexes are higher than those for lower spin state metallic ions. Thus for Ni(hfac)₂NITMe, T_c = 5.3 K (for Ni[II] S = 1).

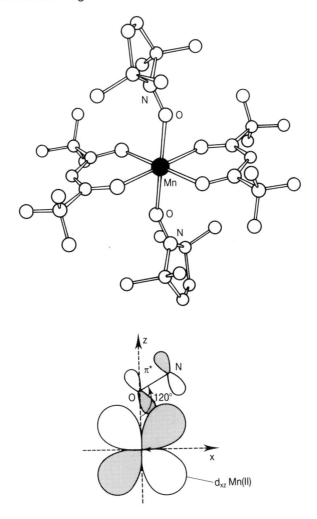

Fig. 4.8 The Mn(hfac)$_2$(proxyl)$_2$ complex. The interaction between the spin of the manganese $S = 5/2$ with two nitroxide ligands each $S = 1/2$ gives an $S = 3/2$ ground state, two $S = 5/2$ and one $S = 1/2$ states. The diagram illustrates the nonorthogonal nature of the π^*/d_{xz} interaction.

In 1967, at a Robert Welch Foundation conference, Mulliken delivered a lecture on charge-transfer complexes. In the discussions which followed McConnell proposed a detailed recipe for using such charge-transfer complexes to produce molecular ferromagnets, a recipe seemingly quite different to those we have described so far.[7] Although both lecture and discussions were published they are not widely available, and McConnell's contribution received little attention and since he did not publish his ideas elsewhere they remained largely unknown. However, McConnell's recipe

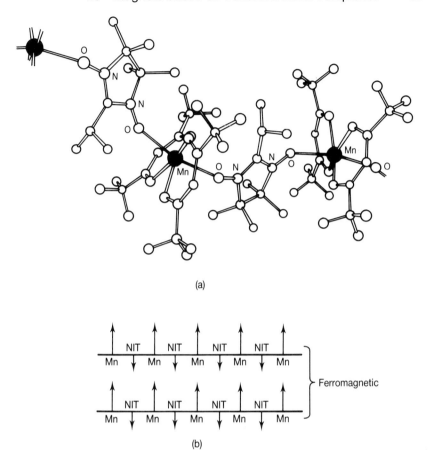

(a)

(b)

Fig. 4.9 Segment of the repeating chain structure of compound [**5**] Mn(hfac)₂NITiPr (hydrogens omitted); (b) schematic representation of the spin-coupling between chains.

was eventually developed by others. For a charge-transfer complex to be a molecular magnet four main criteria must be fulfilled: (1) Formation of a complex of the type $A^{\cdot-}, D^{\cdot+}$ where $A^{\cdot-}$ is the radical anion of the acceptor and $D^{\cdot+}$ the radical cation of the donor, (2) Crystallization as mixed $(A^{\cdot-}, D^{\cdot+})_n$ not discrete $(A^{\cdot-})_n$ and $(D^{\cdot+})_n$ stacks, (3) Significant admixture of the excited state A^{2-}, D^{2+} and (4) Either A^{2-} or D^{2+} to be a ground state triplet species. The basis of the ferromagnetic coupling of the spins on $A^{\cdot-}$ and $D^{\cdot+}$ under these conditions is easy to visualize using a simple resonance picture. If D^{2+} for example has a triplet ground state admixture of the double charge transfer state can only occur if the radical ion spins are also ferromagnetically coupled:

$$\underbrace{A^{-\uparrow}, D^{+\uparrow}}_{S\,=\,1} \quad \longleftrightarrow \quad \underbrace{A^{2-}, D^{2+\uparrow\uparrow}}_{S\,=\,1}$$

The seminal example of this type of molecular magnet is the decamethyl-ferrocene radical cation/tetracyanoethylene radical anion complex studied in detail by Miller, Epstein *et al.*[8] It is a bulk ferromagnet, albeit one with a low Curie temperature ($T_c = 4.8$ K). According to the McConnell mechanism the 'driving force' behind its ferromagnetism is the triplet ground state character of [DMeFc]$^{2+}$ (Fig. 4.10). This leads to ferromagnetic coupling

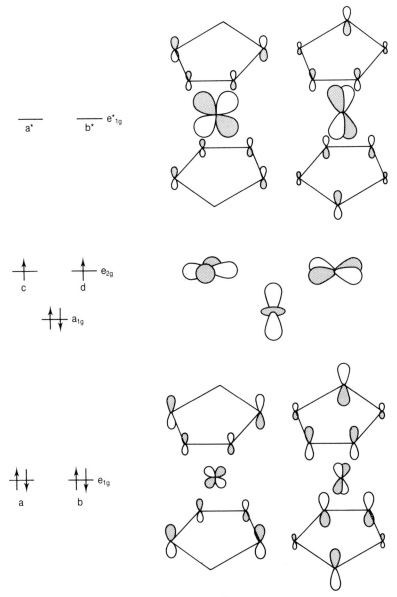

Fig. 4.10   Suggested orbital occupancy of [DMeFc]$^{2+}$ triplet ground state.

between [DMeFc]·+ and [TCNE]·− both within the stack and between out-of-registry adjacent stacks giving the full three-dimensional coupling of spins required for bulk ferromagnetism:

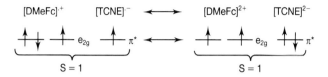

This is not, however, the only mixing of excited states that needs to be considered in such complexes and it should be noted that a similar mechanism provides ferromagnetic coupling between [DMeFc]·+ and [DMeFc]·+ ions in adjacent stacks:

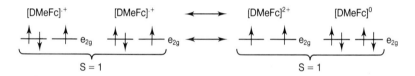

However, the coupling between [TCNE]·− ions must be antiferromagnetic:

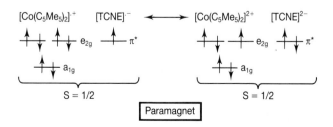

Fortunately, as can be seen from Fig. 4.11 the relative sizes and shapes of the ions maximize [DMeFc]·+/[TCNE]·− and [DMeFc]·+/[DMeFc]·+ contacts and minimize [TCNE]·−/[TCNE]·− contacts. Although the application of the McConnell mechanism to this system has been challenged,[9] it certainly provides a simple rationalization which correctly predicts the sign of the spin coupling in this and in related complexes across the periodic table.

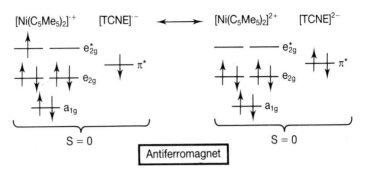

$[Ni(C_5Me_5)_2]^{\cdot+}$ $[TCNE]^{\cdot-}$ ⟷ $[Ni(C_5Me_5)_2]^{2+}$ $[TCNE]^{2-}$

$S = 0$ $S = 0$

Antiferromagnet

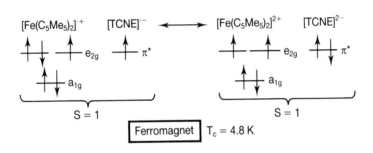

$[Fe(C_5Me_5)_2]^{\cdot+}$ $[TCNE]^{\cdot-}$ ⟷ $[Fe(C_5Me_5)_2]^{2+}$ $[TCNE]^{2-}$

$S = 1$ $S = 1$

Ferromagnet $T_c = 4.8\ K$

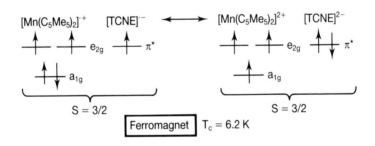

$[Mn(C_5Me_5)_2]^{\cdot+}$ $[TCNE]^{\cdot-}$ ⟷ $[Mn(C_5Me_5)_2]^{2+}$ $[TCNE]^{2-}$

$S = 3/2$ $S = 3/2$

Ferromagnet $T_c = 6.2\ K$

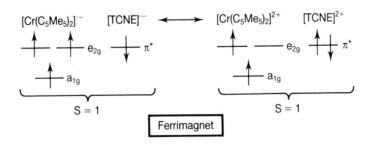

$[Cr(C_5Me_5)_2]^{\cdot-}$ $[TCNE]^{\cdot-}$ ⟷ $[Cr(C_5Me_5)_2]^{2+}$ $[TCNE]^{2+}$

$S = 1$ $S = 1$

Ferrimagnet

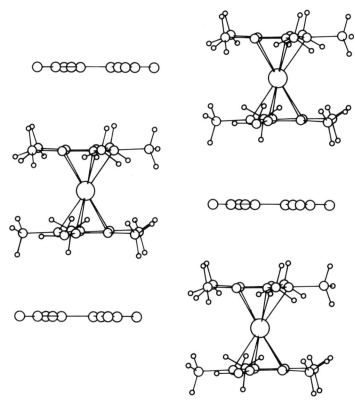

Fig. 4.11 Out-of-registry adjacent stacks in the decamethylferrocene/tetracyanoethylene charge transfer complex.

4.4 Organic Ferromagnets

The McConnell 'recipe' is quite general and, although the charge transfer complexes listed above are based on transition metals, it should be possible to design a similar bulk ferromagnet that is purely organic. This problem has been carefully and systematically investigated by the group of Breslow.[10] They have made many charge transfer complexes based on donors such as [compound 6] and [compound 7]:

[6]

[7]

By matching first and second redox potentials it is possible to design charge transfer complexes of these with quinone acceptors so that the basic structure is of the $A^{\cdot-},D^{\cdot+}$ type with some admixture of the double charge transfer state A^{2-},D^{2+}. So far, however, none of these has proved ferromagnetic. These complexes were designed with the expectation that the dications of [6] and [7] would have a triplet ground state, and that this would 'drive' the ferromagnetic coupling of the spins. However, whereas these dications do indeed give triplet ground state salts in dilute glassy media, it seems that in the lower symmetry environment of the charge transfer complex the singlet state falls below the triplet state, and so the basis for ferromagnetic coupling of the spins is lost.[11] Efforts to produce organic ferromagnets continue, however. Other than those based on McConnell-type charge transfer complexes most of these have aimed to make polymeric high spin π-multiradicals, systems sometimes referred to as topological magnets. Using the simplest level of approximation to describe π-molecular orbitals, Hückel Molecular Orbital (HMO) theory, it is seen that for most π systems with $2n$ π-electrons there will be n bonding orbitals which are doubly occupied in the ground state and n antibonding orbitals that are vacant. A simple example, shown in Fig. 4.12, is butadiene. For certain topologies of the π system (cyclobutadiene and trimethylenemethane in Fig. 4.12), however, HMO theory yields $(n-1)$ bonding orbitals, $(n-1)$ antibonding orbitals and a degenerate pair of non-bonding orbitals. Such systems are always unstable and they are potentially high-spin. Whether such systems are or are not high-spin depends on the magnitude of the exchange interaction, which in turn depends on how far these non-bonding molecular orbitals are co-extensive in space. In the case of trimethylene methane ψ_2 and ψ_3 are orthogonal but

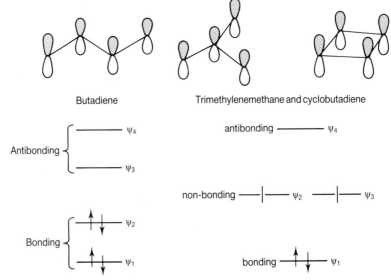

Fig. 4.12 Results of HMO calculations on simple π systems containing four elements.

	ψ_2	ψ_3		
 Trimethylenemethane	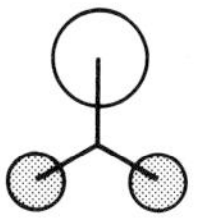	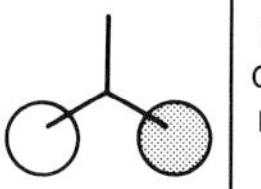	ψ_2 and ψ_3 orthogonal, coextensive, non-disjoint	Triplet ground state
 Cyclobutadiene	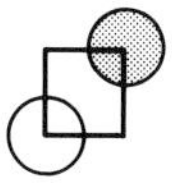	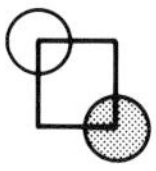	ψ_2 and ψ_3 orthogonal, non-coextensive, disjoint	Singlet ground state

	ψ_4	ψ_5		
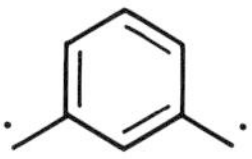 Metaquinodimethane	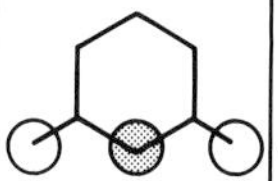	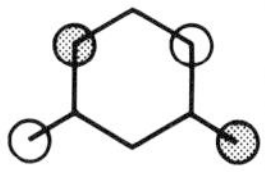	ψ_4 and ψ_5 orthogonal, coextensive, non-disjoint	Triplet ground state

coextensive (they share atoms in common); the exchange interaction is large and the triplet state lies well below the singlet state. Such systems are also called 'non-disjoint' or referred to as 'robust triplets' since the separation of states is such that—even when the natural three-fold symmetry of trimethylenemethane is lifted by unsymmetrical substitution, destroying the strict degeneracy of ψ_2 and ψ_3, the triplet ground state nature is retained. In the case of cyclobutadient ψ_2 and ψ_3 are spatially distinct, they are said to be 'disjoint' (they have no atoms in common) and there is a small exchange interaction. In the case of the square geometry shown above the singlet and triplet states lie close in energy. A small change in geometry lifts the degeneracy of ψ_2 and ψ_3 and brings the energy of the singlet well below that of the triplet state. Most attempts to make topological magnets have been based on derivatives of two 'robust triplets', trimethylenemethane and metaquinodimethane.

The original suggestion of Mataga[12] and Ovchinnikov[13] was that organic ferromagnets could be produced by extending such systems into a continuous conjugated π system. The band structure of such a polymer, compared to that of polyacetylene (see Chapter 7), is shown in Fig. 4.13. 'Localized' orbitals for the odd electrons in the superdegenerate band can be expressed as a set of Wannier functions[14] which are coextensive leading to ferromagnetic coupling of spins. Many conjugated polymers which are potentially high-spin, can be written on paper, but it is very much easier to do this than it is to actually make them in the laboratory! Nevertheless the theoretical speculations of Ovchinnikov, Mataga and others stimulated synthetic organic chemists to produce some interesting high-spin molecules

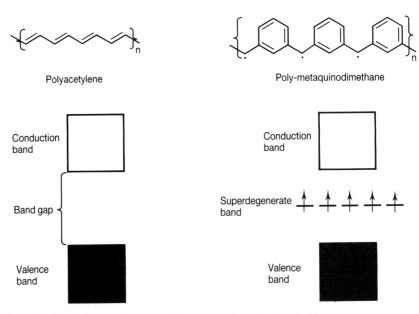

Fig. 4.13   Band structure for polyacetylene and poly-metaquinodimethane.

like the undecet polycarbene[15] [compound 8] and the pentet meta-quinodimethane derivative[16] [compound 9]:

Ar = Bu^{t}C$_6$H$_4$—

[8]

[9]

Unfortunately, attempts to build up this success and to make true polymers with bulk ferromagnetic properties have been frustratingly unproductive. For example the carbene [8] is generated by deazetization of the corresponding pentadiazo compound by photolysis in a frozen matrix at 77 K. Attempts to photolyze analogous polydiazo-polymers result in very incomplete deazetization since the substrate rapidly turns black! An alternative and experimentally more attractive route to ferromagnetic organic polymers was proposed by Fukutome.[17] It is based on established technologies for producing 'polarons' by oxidative doping of conjugated

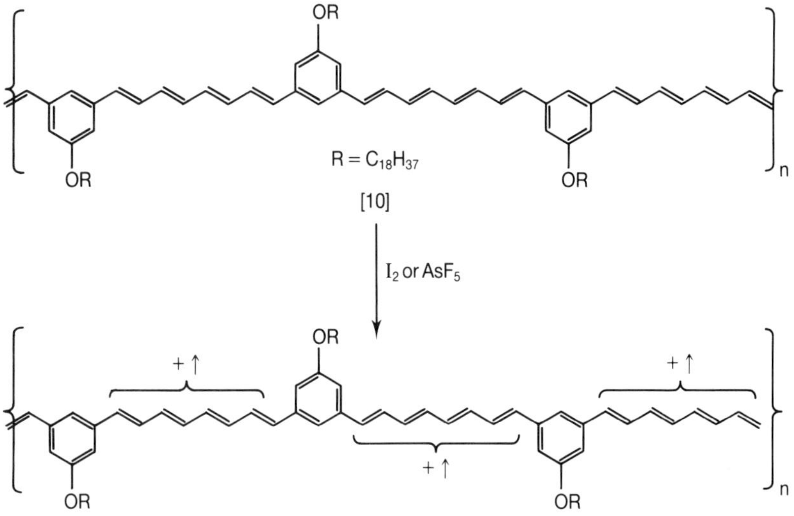

R = C₁₈H₃₇

[10]

I₂ or AsF₅

systems, and the observation that meta-phenylene and other 'coupling units' can induce a ferromagnetic alignment of adjacent polaron spins. Hence iodine or arsenic pentafluoride doping of the polymer [compound 10] gives a system which is not a bulk ferromagnet but in which there is local ferromagnetic coupling.[18]

In some ways the oxidatively doped polychloropolyphenylenes [compound 11] are similar except that in these systems it is orthogonal geometric relationship of adjacent phenylene units rather than the topology of the π system (the coupling unit) that ensures that the 'orthogonal but coextensive' arrangement of the magnetic orbitals is met.[19]

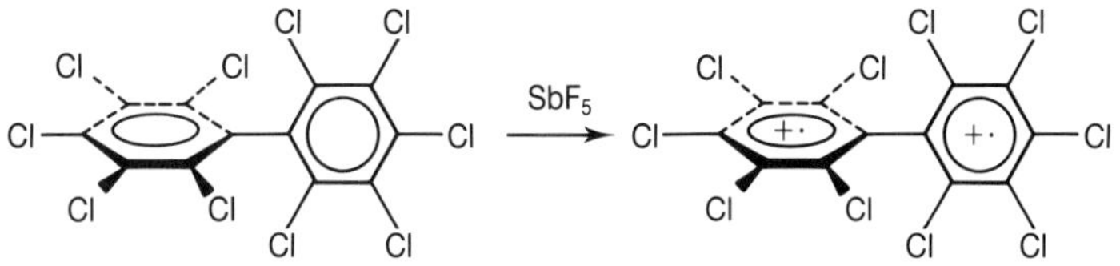

[11] S = 1

4.5 Conclusions

Some of the first claims to have made molecular magnets were for materials that were not well defined chemically. Some of these claims proved to be irreproducible[20] or even false.[21] This gave the area an unfortunate, somewhat unsound reputation and it has made workers very cautious in accepting new claims, particularly if the origin of the observed bulk phenomena, the purity or structure of the substance, or the quality of the physical measurements are in any doubt. However, science is often led by experiment rather than theory and all such claims must be seriously assessed. Present claims of bulk ferromagnetic properties for the indigo polymer [compound 12][22] and for certain polyaryl resins[23] need to be substantiated and so perhaps do those for the fullerene/TCNE complex,[24] which it is claimed is a bulk ferromagnet, $T_c = 16$ K. One system that does appear to be a genuine organic bulk ferromagnet [$T_c = 0.6$ K] is the

[12]   [13]

nitroxide [compound 13]. The mechanism of spin-coupling in the crystal-line solid is still not clear but it may be a three-dimensional version of that found in galvinoxyl.[25]

One system where neither the structure nor the mechanism of spin coupling are clear but which seems to be a genuine bulk ferromagnet and which has the highest Curie temperature yet recorded for a molecular magnet is $V(TCNE)_x \cdot (CH_2Cl_2)_y$ with $x \approx 2$ and $y \approx 1/2$ $[T_c > 350 \text{ K}]$. This is made by reacting bis(benzene)vanadium with tetracyanoethylene in dichloromethane.[26]

In view of the effort that has been expended in making molecular magnets it is pertinent to ask whether any are likely to prove of practical value. The properties of some of the molecular materials described in this chapter are summarized in Table 4.1 and compared to those of conven-tional magnets. As this table makes clear molecular magnets are not as good as conventional magnets on a 'magnet per unit weight or unit volume' basis. In almost all cases the spin-bearing unit is going to be larger and heavier. For more specialist applications, such as data storage, factors such as coercivity become important and in this respect molecular systems seem more promising. It may also be that molecular systems will find uses based on their solubility, ease of fabrication or optical transparency. It would, however, be wrong to see molecular magnets as replacements for conven-tional materials. The science and technology of these is firmly established. Molecular magnets offer new opportunities. As other chapters of this book make clear, molecular systems offer unique possibilities in the creation of novel low-dimensional, layered or otherwise microscopically structured material. It should therefore be possible to control the dimensionality and microscopic structure of molecular magnets in a unique way. The design,

Table 4.1

Magnetic material	Saturation magnetization		Coercivity H_c, G	T_c (K)
	emuG/mol	emuG/cm^3		
Fe	12195	1713	1	1043
CrO$_2$	8800	512	575	387
Mn^{II}Cu^{II}(pbaOH)(H$_2$O)$_2$	22300	–	60[a]	30
Mn^{II}(hfac)$_2$(NITEt)	19375	48	$\geqslant$460[b]	8.1
[Fe^{III}(C$_5$Me$_5$)$_2$]$^{.+}$[TCNE]$^{.-}$	16300	47	1000[c]	4.8
V^{II}(TCNE)$_x \cdot$(CH$_2$Cl$_2$)$_y$	6000	–	60[d]	>350

a Measured at 4.2 K.
b Measured at 4.2 K.
c Measured at 2 K.
d Measured at 4.2 K and room temperature.

synthesis and exploration of the properties of such systems presents a fascinating challenge.

References

1. C. M. Hurd, *Contemp. Phys.*, **23**, 469–93 (1982).
2. *Magnetic Molecular Materials*, Eds D. Gatteschi, O. Kahn, J. S. Miller and F. Palacio, NATO ASI Series, Series E: Applied Sciences, **198**, 1–411 (1991).
3a. O. Kahn, *Structure and Bonding*, **68**, 89–167 (1987).
 b. K. Nakatani, P. Bergerat, E. Codjovi, C. Mathonière, Y. Pei and O. Kahn, *Inorg. Chem.*, **30**, 3977–78 (1991).
4. J. J. Borrãs-Almenar, R. Burriel, E. Coronado, D. Gatteschi, C. J. Gómez-Garcia and C. Zanchini, *Inorg. Chem.*, **30**, 947–50 (1991) see also E. Coronado and M. Drillon in Reference 2 (references therein).
5. F. Sapinã, E. Coronado, D. Beltran and R. Burriel, *J. Am. Chem. Soc.*, **113**, 7940–44 (1991).
6. A. Caneschi, D. Gatteschi, R. Sessoli and P. Rey, *Acc. Chem. Res.*, **22**, 392–98 (1989).
7. R. W. Mulliken, *Proc. Robert A. Welch Found. Conf. Chem. Res.*, **11**, 109–50 (1967).
8. J. S. Miller, A. J. Epstein and W. M. Reiff, *Chem. Rev.*, **88**, 201–20 (1988).
9. C. Kollmar, M. Couty and O. Khan, *J. Am. Chem. Soc.*, **113**, 7994–8005 (1991).
10. R. Breslow, *Pure Appl. Chem*, **54**, 927–38 (1982).
11. J. S. Miller, D. A. Dixon, J. C. Calabrese *et al.*, *J. Am. Chem. Soc.*, **112**, 381–98 (1990).
12. N. Mataga, *Theor. Chim. Acta*, **10**, 372–76 (1968).
13. A. A. Ovchinnikov, *Theor. Chim. Acta*, **47**, 297–304 (1978).
14. T. Hughbanks and K. A. Yee in Reference 2.
15. I. Fujita, Y. Teki, T. Takin *et al.*, *J. Am. Chem. Soc.*, **112**, 4074–75 (1990).
16. A. Rajca, *J. Am. Chem. Soc.*, **112**, 5890–92 (1990).
17. H. Fukutome, A. Takahashi and M. Ozaki, *Chem. Phys. Lett.*, **133**, 34–38 (1987).
18. D. A. Kaisaki, W. Chang and D. A. Dougherty, *J. Am. Chem. Soc.*, **113**, 2764–66 (1991).
19. J. Veciana, J. Videl and N. Jullian, *Mol. Cryst. Liq. Cryst.*, **176**, 443–50 (1989).
20. I. Johannsen, A. I. Nazzal, S. S. P. Parkin and P. Batail, *J. Appl. Phys.*, **63**, 2962–65 (1988).
21. J. H. Zhang, A. J. Epstein, J. S. Miller and C. J. O'Connor, *Mol. Cryst. Liq. Cryst.*, **176**, 271–76 (1989).
22. H. Tanaka, K. Tokuyama, T. Sato and T. Ota, *Chem. Lett.*, 1813–16 (1990).
23. M. Ota, S. Otani, K. Kobayashi and M. Igarashi, *Chem. Lett.*, 1175–78 (1989).
24. P. M. Allemand, K. C. Khemani and A. Koch *et al.*, *Science*, **253**, 301–03 (1991).
25. M. Tamura, Y. Nakazawa and D. Shioni *et al.*, *Chem. Phys. Lett.*, **186**, 401–04 (1991).
26. J. M. Manriquez, G. T. Yee, R. S. McLean, A. J. Epstein and J. S. Miller, *Science*, **252**, 1415–17 (1991).

5

Organics for Nonlinear Optics

G H Cross

5.1 Introduction

The operation of the active elements in a photonics system most commonly utilize some form of nonlinear optical (nlo) process. To modulate the phase or amplitude of light, for example, requires modulations to the index of refraction of the material from which the device is fabricated. At the heart of the operation of most modulators in use or under development, the 'electro-optic' response provides the necessary nonlinear response. Other operations may require a change in the frequency of the transmitted light, to produce second and third harmonic beams, for example, from an infra-red laser. Future optical data storage systems will require shorter wavelength sources than are at present available in order to increase the information density, and harmonic generation is one possible route to short wavelength lasers. A great demand is therefore envisaged for devices which can perform these operations using only low power lasers and small supply voltages. Fibre telecommunications systems will offer the largest area of use for nonlinear optical devices. It will be here where low cost, mass produced, devices will be needed. The eventual choice of materials and fabrication methods will reflect these needs. The inroads into the many areas of technology that synthetic polymers have made in the twentieth century amply demonstrates that organic materials could well be materials of choice.

This is not the sole reason, however, that organics are being considered for implementation in devices for photonics. It has become clear from research over the last 15 to 20 years that large optical nonlinearities exist in organic molecules and that this, coupled with the powerful versatility of synthetic chemistry in some instances, gives organics a lead over their inorganic counterparts. The source of the large nonlinearity lies in the polarizability of the π charge cloud in conjugated systems. Large displacements of charge for small applied fields results in easy modulation of the optical properties of the material, such as the refractive index. In addition, these induced polarizations, being purely electronic in nature (rather than thermal for example), have response times much shorter than 1 picosecond in many cases. It is thus possible, at least in principle, to fabricate devices whose bandwidth is limited only by device design and the speed of the drive electronics. These advantages are the sole preserve of organic materials at present. They may, or may not, alone be sufficient to impel the wide-scale use of organics in future photonics systems, but the study of nonlinear optical processes in organic materials will surely continue. This chapter sets out to introduce these processes and to provide a correlation between chemical structure and nonlinear optical responses at the molecular level.

5.2 Basic Concepts

A rigorous description of nlo phenomena would require attention to the details of the tensor properties of the dielectric susceptibilities and the vector form of the applied fields (see Chapters 2 and 3). Since these aspects do not add significantly to the understanding of the sources of the nonlinearity, and for reasons of brevity, all equations will be presented in their scalar forms. Some mention of tensors is made only in Section 5.6. For a more detailed, yet clear, presentation of the full use of notation in nonlinear optics, the reader is referred to Boyd, 1992 and references therein.[1]

The polarization produced in a material by an applied field, E, is usually regarded as a linear interaction. That is, charge displacements are supposed to be a linear function of the applied field. The polarization per unit volume, P, is then given by

$$P = \varepsilon_0 \chi E \quad \mathrm{Cm}^{-2} \tag{5.1}$$

in which χ, the linear susceptibility, describes the ease of polarization and is related to the relative permittivity, ε_r, of the medium thus

$$\chi = \varepsilon_r - 1 \tag{5.2}$$

χ is, accordingly, a dimensionless quantity. For the macroscopic phenomena, we adopt the MKSA system of units here.[2] If we recall that $\varepsilon_r = n^2$ at optical frequencies, an important inter-relationship is indicated between the susceptibility and refractive index.

There are two conditions under which the approximation implicit in equation (5.1) breaks down. The first is under the conditions of high field strengths such as those available from high power lasers. The second is in materials where the ease of polarization of the charges is large enough for higher order susceptibility terms to be important at perhaps even modest field strengths. The polarization must, in these circumstances, be considered in terms of a perturbation expansion as a Taylor series in powers of the applied field thus

$$P_{TOT} = P_0 + \varepsilon_0 \sum_{i}^{\infty} \frac{1}{i!} \chi^{(i)} E^i \tag{5.3}$$

where P_{TOT} is the total polarization, P_0 is any permanent polarization in the medium and i is an integer. The summation represents the induced polarization in the medium due to the applied electric field(s), E^i. For most practical purposes, only the first three terms in the sum are of interest and then the induced polarization, P_{ind}, becomes

$$P_{ind} = \varepsilon_0 (\chi^{(1)} E + \chi^{(2)} E^2 + \chi^{(3)} E^3) \tag{5.4}$$

In equation (5.4), the factorial terms arising from the expansion in (5.3) have been subsumed into the susceptibilities in line with common practice.

It is worth noting also that in writing the fields in terms of such a simple power series masks the fact that fields of differing frequencies and amplitudes can be combined to give a nonlinear response. Here we are to overlook this point, and regard the nonlinearities as arising from degenerate wave mixing in the material. The form of equation (5.4) is shown schematically in Fig. 5.1.

Noting that charge displacement is directly linked with applied field, an alternative way of viewing the nonlinearity is to show the potential energy of the system as a function of charge displacement (Fig. 5.2). This description is in fact that of a one electron anharmonic oscillator where the potential energy of the electron is related to the restoring force due to the Coulomb attraction with the nucleus. The potential energy deviates from its quadratic increase with displacement to become a slower function of displacement as the field strength increases.

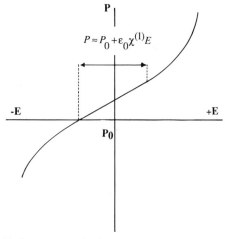

Fig. 5.1   The relationship between polarization and applied field in a nonlinear medium.

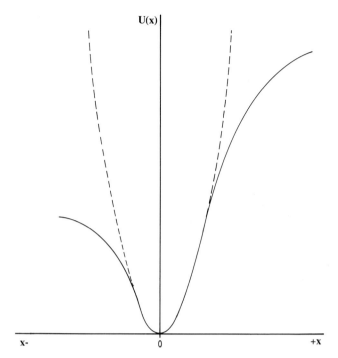

Fig. 5.2 Potential energy function versus charge displacement along a coordinate, x, for a material exhibiting second and third order nonlinearity. The parabolic curve (dash line) represents the energy of the system in the absence of nonlinearity.

Figure 5.2 shows that the potential energy is asymmetrically related to the direction of charge displacement and therefore to the applied field direction. This is due to the inclusion of the $\chi^{(2)}$ term in equation (5.4) which is finite only for noncentrosymmetric materials. This point will be discussed in due course but for now we can examine the consequences of this asymmetry on the polarization response in the material.

Consider a time varying optical field of frequency ω thus

$$E(t) = E_0 \cos(\omega t) \tag{5.5}$$

The component of polarization in the material arising through the $\chi^{(2)}$ term is then (from equation (5.4))

$$P^{(2)} = \varepsilon_0(\chi^{(2)} E_0^2 \cos^2(\omega t)) \tag{5.6}$$

which upon expansion yields the following components:

$$P^{(2)} = \tfrac{1}{2}\varepsilon_0(\chi^{(2)} E_0^2 + \chi^{(2)} E_0^2 \cos(2\omega t)) \tag{5.7}$$

The polarization has a DC component and one at 2ω. The result is that an optical field is radiated from the material having a frequency twice that of the fundamental field. Harmonic generation is thus a natural consequence of nonlinear optical interactions (third harmonics arise through the $\chi^{(3)}$ term).

Of relevance to applications in telecommunications, the electro-optic response is also a consequence of noncentrosymmetry in a material and again arises out of the second order susceptibility, $\chi^{(2)}$. Suppose the fields applied now are from a low power laser beam (amplitude, $E_0(\omega)$) and a slow (nearly DC) electric field, $E(0)$. The total field applied is now

$$E_{TOT} = E_0 \cos(\omega t) + E(0) \qquad (5.8)$$

in which $E(0) \gg E_0(\omega)$ and equation (5.6) becomes:

$$P^{(2)} = \varepsilon_0 \chi^{(2)} [E_0 \cos(\omega t) + E(0)]^2 \qquad (5.9)$$

which upon expansion and neglect of the resulting quadratic terms gives a term

$$P^{(2)} = 2\varepsilon_0 \chi^{(2)} E_0(\omega) E(0) \qquad (5.10)$$

showing a linear relationship between polarization and the applied DC field.

The susceptiblity describing this polarization term has a direct relationship with the more familiar electro-optic coefficient r. Before presenting this, reference is made to an important assumption which is very often valid for the nonlinearities in organic materials. This holds that the susceptibilities at optical frequencies (i.e. those which give rise to harmonic generation) are of the same magnitude as those at DC. This assumption can generally be made since the polarization response to field, in organic materials, is dominated by displacement of bound charge. Further, a good approximation is made if the low frequency relative permittivity is equated with n^2; i.e. there is an assumption of zero dielectric dispersion. In such circumstances it is possible to relate $\chi^{(2)}$ to r through[3]

$$\chi^{(2)} = -\frac{n^4 r}{2} \qquad (5.11)$$

Second harmonic generation and the electro-optic response are members of a family of effects collectively referred to as second order nonlinear optical responses. The first is one example of the general process of three-wave mixing. In second harmonic generation, two waves of identical (i.e. degenerate) frequency at ω combine to produce a new re-radiated wave at 2ω. The material acts as a kind of catalyst for the process in that it acts only to facilitate the mixing. Energy is neither given up to, nor extracted from, the medium. To represent these and other nlo processes, a notation has been adopted which identifies the fields involved. For SHG and the electro-optic response the coefficients may be distinguished by writing $\chi^{(2)}(-2\omega; \omega, \omega)$ and $\chi^{(2)}(-\omega; \omega, 0)$, for SHG and electro-optic susceptibilities, respectively. The frequency arguments in parentheses and to the right of the semicolon depict the applied fields (but give no information about their relative magnitude or phase), whereas the field resulting from the interaction (which may be regarded as being emitted,

hence the negative sign) is given to the left of the semicolon. All 'catalytic' processes are termed 'parametric' processes.

Third order nonlinear optical processes are typified by four-field mixing. Simplest of all is third harmonic generation, mediated through $\chi^{(3)}(-3\omega; \omega, \omega, \omega)$ but of more general use are the rather exotic $\chi^{(3)}$ effects known collectively as 'self-action' responses. Among these, self-focusing and self defocusing are the most important. These arise where the refractive index of the material is sensitive to the intensity of incident light. The intensity dependence may be accounted for in a modified description of the refractive index thus

$$n = n_0 \pm n_2 I \tag{5.12}$$

in which n_0 is the conventional (linear) refractive index and n_2 is a coefficient known as the nonlinear refractive index which is given in units of $m^2 W^{-1}$. I is the intensity of the incident light ($W m^{-2}$). There is a simple relationship between $\chi^{(3)}$ and n_2, valid for all cases where the index changes produced are only small, when we find[4]

$$n_2 = \frac{\chi^{(3)}}{n_0^2 c \varepsilon_0} \tag{5.13}$$

where $\chi^{(3)}$ is in units of $m^2 V^{-2}$.

The nonlinear refractive index finds applications in directional coupler devices.[5] Here two optical waveguides are brought into close proximity such that optical power oscillates from one guide to the other along the length of interaction. At the end of the path, the waveguides diverge and power is transmitted into only one of the arms. By changing the intensity of the incoming beam, the period of the power oscillations between the guides is modified through the refractive index changes produced, and thus the light can be switched alternately between the two diverging output waveguides.

There is an important distinction between the occurrence of second and third order effects. As discussed earlier second order effects can only occur in noncentrosymmetric media whereas third order responses are ubiquitous. As a general rule, third order effects are simply the nonlinear counterparts of the linear optical properties of the material. We have already discussed nonlinear refraction for instance, but nonlinear absorption, i.e. intensity dependent absorption, is also a third order response, arising out of the imaginary part of $\chi^{(3)}$. Since energy is lost to the material, this is termed a 'dissipative' rather than a 'parametric' process. The $\chi^{(2)}$ responses are all parametric in nature although that does not mean that absorption is unimportant. As will be discussed, the proximity of electronic excited states in the material (seen as absorption bands) to the energy of the incident and resultant waves is important from the point of view of enhancing the nlo coefficients, but a careful balance between nonlinear enhancement and absorptive loss must be found.

We will now turn to the microscopic origins of the nonlinear response and discuss the molecular contributions to the macroscopic nlo coefficients.

5.3   Molecular Nonlinear Optics

The microscopic origins of the observed macroscopic nlo responses are simple to identify in organic materials. They are essentially isolated molecules in the case of non-ionic crystals or molecularly doped host materials, but may also comprise segments of polymeric chain. Each molecule or segment is considered to act as a source for the nonlinear polarization and, as long as the separation between adjacent sources is much less than the wavelength of the applied fields (the assumption of spatial invariance), the bulk coefficients may be formulated as a sum over the microscopic nlo sources, with an appropriate correction made for molecular orientational order.

There is an expression describing the microscopic nonlinearity which is analogous to that of equation (5.4) wherein each individual source unit has a polarization represented (to third order) by

$$p = p_0 + \alpha E_l + \beta E_l^2 + \gamma E_l^3 \tag{5.14}$$

The term p_0 accounts for any permanent polarization (i.e. dipole moment) and the terms α, β and γ are the microscopic counterparts of $\chi^{(1)}$, $\chi^{(2)}$ and $\chi^{(3)}$ of equation (5.4). Since we are now concerned with the polarization of species embedded in the body of the material, some account needs to be made of the local environment. Thus E_l denotes the *local* field acting on the molecule or subunit. To a reasonable approximation in many cases, the inter-relationship between the magnitude of the fields applied and those 'felt' at the molecule are expressed through the Lorentz and Onsager field correction factors. These relate applied fields of optical frequency, $E(\omega)$ or of low frequency (including DC), $E(0)$ to the local field through

$$E_l(\omega) = f(\omega) E(\omega) \tag{5.15}$$

and

$$E_l(0) = f(0) E(0) \tag{5.16}$$

in which the Lorentz–Lorenz factor is given by

$$f(\omega) = \frac{n^2 + 2}{3} \tag{5.17}$$

and the Onsager factor by

$$f(0) = \varepsilon_r (n^2 + 2)/(2\varepsilon_r + n^2) \tag{5.18}$$

Interestingly, when these are evaluated, they show that the local field is actually *greater* at the microscopic site than that applied from the external source. In principle then, the nonlinearity could be enhanced (although to a rather modest degree!) through having materials of higher refractive index or relative permittivity.

5.4   Linear and First Nonlinear Polarizabilities, α and β

In the presence of a molecular dipole moment the nonlinear moiety is by definition noncentrosymmetric and thus β is finite. This then is the fundamental requirement in molecular structure design for second order nlo responses. The charge displacements produced upon the application of a field oscillate around their lowest energy equilibrium positions in an anharmonic manner. The anharmonic oscillator model accounts for this in the extra term, max^2, in the following equation of motion[1]

$$m\frac{d^2x}{dt^2} + m\tau\frac{dx}{dt} + m\omega_0^2 x + max^2 = -eE_l(t) \tag{5.19}$$

where charges displaced along a direction, x, are subject to some damping, τ (absorption perhaps) and possess a single mode of natural oscillation, ω_0. The terms a and m are constants of the system and specifically, ma is related to the second order nonlinear coefficient, β in equation (5.14). During forced oscillation due to the applied local field, the charges are subject to a restoring force, F_r, where

$$F_r = -m\omega_0^2 x - max^2 \tag{5.20}$$

Integrating equation (5.20) with respect to x gives the potential energy of the system as a function of displacement thus

$$U(x) = \int F_r \,.\, dx = \tfrac{1}{2}m\omega_0^2 x^2 + \tfrac{1}{3}max^3 \tag{5.21}$$

The direction of charge displacement, x, follows (inversely for electrons) the direction of the applied field. Taking the first term on the RHS of equation (5.21) it is clear that regardless of the direction of the applied local field (and thus regardless of the *sign* of x) the energy change is the same. This is thus a consequence of the symmetrical component of the polarization response which in equation (5.14) is embodied in the α term (and indeed the γ term to be discussed later). The situation with the second term on the RHS of equation (5.21) is that the sign of the applied field *does* matter. Depending on the sign, the contribution to the total energy of the system is either positive or negative leading to the asymmetrical relationship between U and x evident in Fig. 5.2.

In using the form of equation (5.19) to describe the polarization response of a molecule we are making the assumption that a single normal mode of the system (at a frequency ω_0) governs the characteristics of the α and β coefficients as a function of frequency. This is a good assumption where measurements of the coefficients are made to the long wavelength side of the major electronic absorption band for the molecule or subunit. Where only a single dominant mode is considered, a 'two-level' model can be invoked with the excitation depicted in Fig. 5.3. The excitation process is described both in terms of its frequency, ω_0, and in terms of its strength. This latter property relates to the probability of transitions of electrons to the excited state (n) from the ground state (g) and may be grouped

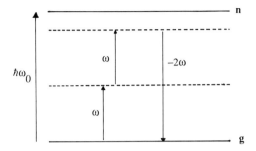

Fig. 5.3 Schematic energy level diagram of the two-level model approximation. The process of second harmonic generation is shown by arrows between real (solid lines) and virtual (dash lines) states, $\hbar\omega_0$ is the energy of the parity allowed 2-level transition.

together into the product of the oscillator strength, f, and the change in dipole moment occurring upon excitation, $\Delta\mu_{gn}$. With the inclusion of these terms and those connected with frequency dispersion (i.e. how close the fields are in frequency to ω_0) an expression for $\beta(-2\omega; \omega, \omega)$ may be formulated as follows[6]

$$\beta(-2\omega; \omega, \omega) = \frac{3e^2\hbar^3}{2m} \frac{\omega_0 f \Delta\mu_{gn}}{[(\hbar\omega_0)^2 - (\hbar\omega)^2][(\hbar\omega_0)^2 - (2\hbar\omega)^2]} \; Cm^3V^{-2} \quad (5.22)$$

The expression exhibits two poles. The first occurs where the fundamental field coincides with the absorption maximum (i.e. at ω_0) and the second where $2\omega = \omega_0$. The magnitude of β clearly becomes very large where these frequencies coincide but, in both cases, there is significant absorption of either the fundamental fields, or second harmonic field, respectively.

All of the foregoing may be exemplified in the characteristics of one of the most studied of molecules for second order non-linear optics, *para*-nitroaniline, *p*-NA.

_p_-NA, a Model NLO Molecule   The structure of *p*-NA (Figure 5.4(a)) shows all the characteristics which have come to typify the now large class

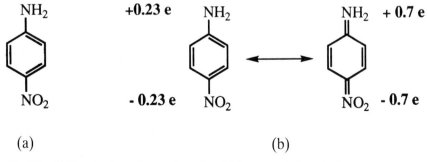

(a)   (b)

Fig. 5.4 (a) The structure of *para*-nitroaniline; (b) the aromatic-to-quinoid resonance representative of the single normal mode, ω_0, in the two-level scheme.

of organic molecules designed for $\chi^{(2)}$ effects. The general scheme for these systems requires that an electron donating group (examples of which are $-NH_2$, $-N(CH_3)_2$) and an electron accepting group ($-NO_2$, CN, etc.) are connected via a conjugated (i.e. alternating single and double bond) hydrocarbon system. The difference in the electron affinities of the groups and the polarizable nature of the π bonds in the conjugated system causes a permanent dipole moment to establish over the molecule in a general direction aligned between the attached groups. An applied field may then further polarize the molecule, but due to the electronic biasing, the polarization of electronic charge is favoured in the direction of the accepting group and hindered in the opposite direction. An AC field (such as an optical field) thus induces a partially rectified polarization response which contains Fourier components at DC, ω and 2ω. Notice that these responses are those predicted using oscillating fields in equation (5.4) as discussed earlier. The principal excitation which results in an absorption band in the near UV region for p-NA may be represented by the aromatic-to-quinoid resonance depicted in Fig. 5.4(b). The change in charge asymmetry upon excitation is also shown.[7]

Design for Enhanced β   Equation (5.22) makes the important point that it is a large *change* in dipole moment and not the *absolute* dipole moment that is important for a large β coefficient. In spite of this, a trend towards larger microscopic second-order nonlinearity is seen in a series of molecules where the difference in electron affinities of donor and acceptor groups increases.

Table 5.1 shows the values for β measured in solution of a homologous series of *para*-disubstituted benzenes where the electron donating strength of the donor substituent increases from left to right, and where the accepting group $-NO_2$ is placed below $-CN$ to indicate its enhanced electron affinity over the latter.[8] The enhancement in the β value is evident as one views the table from top left to bottom right. Notice also that as the polarity of the molecule increases so also does the wavelength of the principal absorption band (in parentheses).

In addition to increasing the charge asymmetry, an increase in the conjugated path length between the substituents also increases the nonlinearity. Table 5.2 shows the results of measurements of two series of

Table 5.1 *Values measured for β_{zzz} (the component along the molecular axis) for a series of para-disubstituted benzenes in solution. The values are given in units of esu $\times 10^{30}$ (Note: 1 esu $= 3.711 \times 10^{-21}$ Cm^3V^{-2}). The values in parentheses represent the wavelengths of maximum absorption for the molecules (i.e. $\lambda_{max} = 2\pi c/\omega_0$).*

	Donor (D)			
Acceptor (A)	$-CH_3$	$-OCH_3$	$-NH_2$	$-N(CH_3)_2$
$-CN$	2.92	4.83	13.34	14.24
	(232)	(247)	(269)	(297)
$-NO_2$	9.12	17.35	47.67	52.75
	(280)	(314)	(378)	(418)

Table 5.2 *β values relative to that of p-NA of longer conjugated homologues*

Molecule	β(normalized)
 H₂N— —NO₂	1
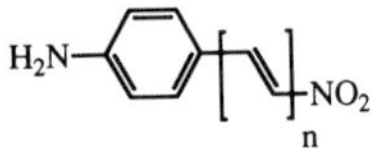	$n=1$ 5.2 $n=2$ 11 $n=3$ 17
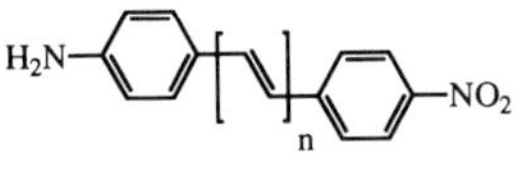	$n=1$ 5.5 $n=2$ 9.2 $n=3$ 15 $n=4$ 20

molecules where the substituent groups are kept the same but where there is a monotonic increase in their separation.[9] The largest effects seem to be due to the addition of ethylenic bonds between the substituents whereupon a superlinear relationship between molecular length on β obtains. This observation serves to illustrate the general empirical rule that aromatic contributions to the conjugation scheme lead to a greater localization of π electrons. This lowers the polarizability of the molecule. Notice, for example in Table 5.2, that although the conjugation length in the stilbene type chromophores is longer for a corresponding value of *n*, the β value is insensitive to the aromatic ring contribution.

A combination of synthetic methods (i.e. those which increase the polarity and increase the length) are often used to optimize the nonlinearity of new molecules. Both have the possible drawback of moving the absorption band of the molecule further into the visible and near infra-red region. If this can be tolerated, however, devices such as electro-optic modulators, operating in the near infra-red region where there is little concern over the colour of a material, may benefit from the collective response of an assembly of molecules of high β. In this spirit it is pertinent to mention a molecule whose β is very large, by virtue of a minimization of aromaticity, but whose absorption characteristics rule out operation in the visible. The material is a dye, N,N-dimethyl indoaniline whose resonance forms are depicted in Fig. 5.5. Upon excitation there is no net change in either quinoid or aromatic character. This makes the molecule highly polarizable and produces a large β value of 190×10^{-30} esu $(796 \times 10^{-40} \, m^4 V^{-1})$.[9]

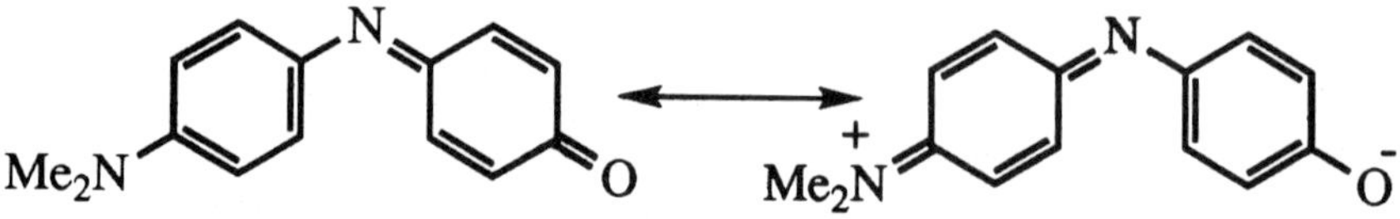

Fig. 5.5 The resonance forms of N,N-dimethyl indoaniline. Notice that the aromaticity remains unchanged between the two canonical forms.

5.5 Second Nonlinear Polarizability, γ

The nonlinearities which arise through the γ term in equation (5.14) do not require asymmetry of structure. The polarizations induced, nevertheless, result from anharmonic charge oscillations, and lead to deviations from the parabolic dependence of potential energy on displacement. In this case, however, the deviation is symmetric with regard to displacement direction.

In the absence of any particular structural requirements there remains only one rather universal rule for achieving high molecular third-order response and that is that the charge cloud must be easily polarizable. This general statement can be focused into comparisons between materials of certain classes. The most common are those containing π conjugated electronic systems. The relationship between conjugation length and the magnitude of β was discussed in the preceding section, and some evidence has been put forward which shows that similar trends arise for γ.

Free Electron Models   A conjugated chain, i.e. a system of alternating single and double bonds, almost by definition denies its description in terms of a free electron model, (FEM). This is because there is an assumption in the FEM that the potential energy which electrons experience along a chain of lattice sites is constant. Nevertheless the free electron model can be used as a useful intuitive starting point for discussion.[10]

It will be convenient first to introduce the concept of bond alternation. This will describe the extent to which the π electrons are localized on alternate bonds in the conjugation path.

The $2p_z$ atomic orbitals on adjacent carbon atoms in a chain overlap to a greater or lesser extent, and the overlap is characterized by the resonance integrals β_1 and β_2. β_1 is taken to characterize the orbital overlap in energy terms between adjacent π orbitals of the 'single' bond. β_2 is that for the 'double' bond. Taking the single electron wavefunctions on atoms m and $m-1$ in a chain as ϕ_m and ϕ_{m-1}, respectively, we have

$$\beta_{1,2} = \int \phi_{m-1}^* H \varphi_m . d\tau \tag{5.23}$$

where H is the Hamiltonian operator and τ represents the volume over which overlap occurs.

We might then define the alternation parameter[11]

$$\nu = \beta_1 / \beta_2 \tag{5.24}$$

to describe the degree of localization. When the parameter $\nu = 1$ it is clear that a free electron model may be applicable since there is no periodic modulation of charge density in this case.

Chains with complete electron delocalization are extremely rare in organic materials since a periodic charge modulation (a Peierls distortion) is energetically more favourable. It is known, however, that in a type of cyanine dye which exhibits symmetry between its two resonance forms, such a situation may exist.[12] Figure 5.6 depicts the dye in its two resonance states. There is clearly no energy difference between the two forms, i.e.

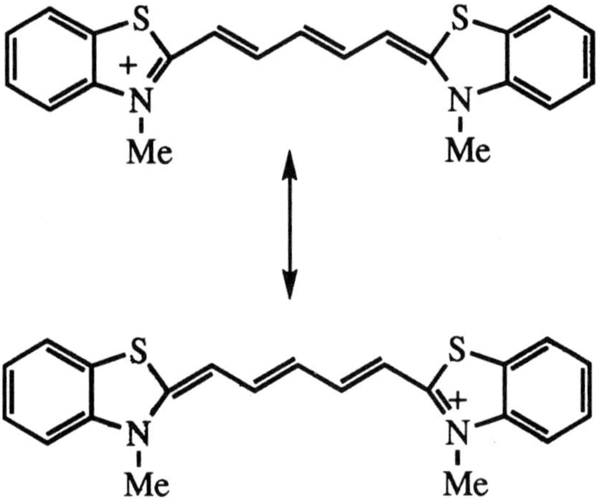

Fig. 5.6 Resonance forms of the symmetric cyanine dye whose degeneracy ensures a close approximation to a free electron model description of its chain structure.

they are degenerate. In such cases the free electron model predicts the following relationship between γ and molecular length

$$\gamma = L^5 e^4/a (\hbar\omega_0)^3 \quad Cm^4V^{-3} \tag{5.25}$$

in which L is the molecular extension, a is the (equidistant) C–C bond length and ω_0 is the resonance energy. Such a strong dependence of γ on chain length might appear at first sight to offer almost limitless enhancements in polymers where small bond alternation can be maintained. Because bond alternation is always present, however, such enhancements are not observed. Polymers having the highest nonlinearity such as polyacetylene appear to consist of chains where the γ per chain has reached a saturated value at 50 carbon sites, i.e. much shorter than the actual chain lengths.

Semiconductor Models   Both experimental work on oligomers[13] and the results of more recent and detailed calculations[14] show that the precise dependence of γ on chain length may be better expressed through

$$\gamma \propto N^{a(N)} \tag{5.26}$$

Equation (5.26) suggests that the exponent, a which relates γ to the number of repeat groups, N, in a chain is itself a function of chain length (i.e. a function of N). Beyond a certain length, a tends to unity and the real enhancement in the nonlinearity, i.e. an increase in the γ coefficient *per repeat unit*, saturates.

   The reasons for this are again related to those factors which the FEM notably excludes. These are: bond alternation which leads to a periodic potential, heteroatomicity, which also adds to the shaping of the lattice potential and, significantly, electron correlations. These factors combine to give the material a characteristic 'delocalization length'. Early theoretical studies, whilst neglecting Coulomb correlations, introduced the concept of delocalization length and used a description of conjugated chains in terms of one-dimensional band theory in which Bloch functions characterize the electron wavefunction. These represent the electronic wavefunctions as plane waves which are modulated according to the lattice potential along the chain.

Using these arguments, Cojan *et al.* introduced the delocalization parameter, N_d which is defined in terms of the alternation parameter, v (equation (5.24)) as follows[11];

$$N_d = (1+v)/(1-v) \tag{5.27}$$

Not to be confused with the number of repeat units in a chain as in (5.26), N_d, is a dimensionless quantity which describes the degree of delocalization of π electrons, and only becomes representative of a delocalization length when combined with the lattice parameter (or repeat length). A typical example for N_d assumes that $v = 0.75$ and thus $N_d = 7$ repeat units. The general result obtained from band theory arrives at the following 'scaling laws'

$$\gamma \propto N_d^6 \propto (2E_F/E_g)^6 \tag{5.28}$$

where E_F and E_g are the Fermi energy and optical energy gap of the one-dimensional semiconductor, repectively. It should be emphasized that such an inverse sixth power dependence of the nonlinearity on band gap would be apparent only in oligomeric materials where simple, (i.e. single/double) bond alternation pertains, and in the absence of Coulomb correlations.

Quantum Chemical Models More exact accounts of the atomic and molecular wavefunctions are required to cope with heteroaromatic materials and even to refine the predictions of γ for linear chain polyenes. For small molecules, *ab initio* quantum chemical calculations may be used to predict the nonlinearity. For larger systems, however, and thus to investigate the scaling of γ with chain length in oligomers, the otherwise impractical computation may be reduced by using empirical data for some of the parameters, or by parameterizing some of the wavefunctions. This may be done by combining the eigenvalues obtained from *ab initio* studies of representative atoms or larger units.[14] Whichever method is used, however, the problem is to calculate the energies and transition dipole moments ('strengths') of the system and then to combine these parameters as a sum-over-states, SOS. A condensed form of the equation then used to calculate $\gamma(-3\omega; \omega, \omega, \omega)$ is as follows

$$\gamma(-3\omega; \omega, \omega, \omega) = \frac{e^4}{4\hbar^3} \frac{r_{gn_3} r_{n_3 n_2} r_{n_2 n_1} r_{n_1 g}}{(\omega_{n_3 g} - 3\omega)(\omega_{n_2 g} - 2\omega)(\omega_{n_1 g} - \omega)} \; Cm^4 V^{-3} \tag{5.29}$$

r_{gn_3} etc. refer to the transition moments between the ground state, g, and the virtual intermediate states, n_1, n_2 and n_3. The terms in the denominator illustrate the resonance phenomena associated with third harmonic generation. There are three poles in the expression. Those where 3ω and ω equal $\omega_{n_3 g}$ and $\omega_{n_1 g}$, respectively, correspond to resonance with symmetry allowed transitions, i.e. between those states of opposite parity. These are usually easily seen in the UV/Visible absorption spectrum of the molecule under study. The third pole, at $2\omega = \omega_{n_2 g}$ corresponds to a 2-photon absorption process and represents resonance with a parity forbidden level. Whilst all three may give rise to an enhancement of γ, there is the penalty of absorption associated with any such enhancement. It is thus better to compare the so-called 'static' values of γ, i.e. those where $\omega \to 0$. The differences between materials is then traceable to the sum of the transition dipole moments whose values depend on the chemical and electronic configurations in the molecule.

Figure 5.7 shows the evolution of the theoretical length-normalized values of γ for polyenes, polypara-phenylene vinylenes (PPV) and

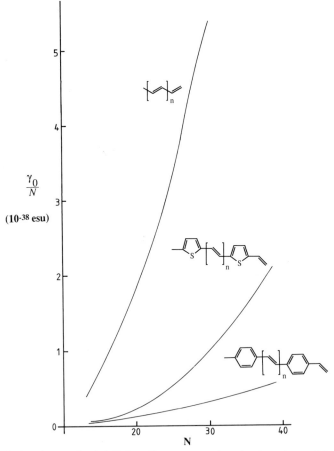

Fig. 5.7 The length normalized static $(\omega = 0)\gamma$ versus number of carbon atoms, N, in the chain for three conjugated material types.

polythienylevinylenes (PTV) as the number of carbon atoms in the chain increases.[14] Notice that this is equivalent to showing the dependence of the bulk nonlinearity, $\chi^{(3)}$, (see Section 5.6) on the number of π electrons in the chain. By dividing out the number of carbon atoms, Fig. 5.7 shows the real enhancement to the bulk nonlinearity; by showing the *real* enhancement to γ with chain length.

All three material types show an enhancement in γ/N up to at least 40 carbon units (although not shown for the polyene). In the region of length calculated and shown, γ/N does not reach saturation. Evidence from further theoretical studies,[15–17] has shown that γ/N should saturate at around 60 carbon sites in the polyenes (80 in the work of Yu[17]) and also at around 60 carbon sites for PPV.[15]

The differences in the absolute values of γ for the three types reflects their relative degrees of aromaticity. Accordingly one can find from the data the following ratio. For chains containing 30 π electrons

$$\gamma(\text{Polyene}) = 5\gamma(\text{PTV}) = 16\gamma(\text{PPV})$$

The presence of the aromatic rings in PPV serves to reduce the nonlinearity by localizing the π charge cloud. Polyacetylene has the largest third order nonlinearity among the conjugated polymers $(2 \times 10^{-10}\,\text{esu}^{18})$ *and* it is the most reactive; i.e. the least stable. This connection between reactivity and the nonlinearity is perhaps not unexpected since both rely on displacement of charge and many reactions rely on substitution at π bonded carbon sites.

Figure 5.8 shows the evolution of the optical gap, E_g, as the chain length increases. The bathochromism (red shift) towards longer chains is expected but the relative sizes of the gap among the three material types is of

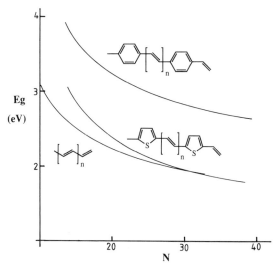

Fig. 5.8   The evolution of the band gap versus number of carbon atoms, N, in a chain for the three material types shown.

interest. For 30 π electrons, for instance, the gap for the polyene and for PTV is about the same and yet their nonlinearities differ by a factor of 5. Under the semiconductor theory given earlier (in Section 5.5), one might have predicted identical values for the nonlinearity. The differences in aromaticity and consideration of the electron–electron interactions, taken into account when using quantum chemical methods, shows their enhanced predictive power.

Non Polymeric Materials In spite of all the successful theoretical studies, it still seems certain that experimental studies will lead the search for materials of acceptably high nonlinearity. The reasons for this lie in the ubiquity of third order nonlinear responses and thus in the sheer diversity in material types considered.

Interest in γ (and thus $\chi^{(3)}$) is in no way limited to conjugated polymers. While these may offer important advantages, such as ease of processing, there is nothing to insist that the largest nonlinearities will occur in a polymeric material. In fact the choice of a material for use in any nonlinear optical application rests on a complex set of inter-related physical properties, not least of which is the combination of $\chi^{(3)}$ and α, the linear absorption coefficient. It remains an irrefutable fact that large $\chi^{(3)}$s are often associated with large values for α (and indeed with large values of *nonlinear* absorption). It has become necessary therefore to look for materials where a suitable trade-off may be obtained between these conflicting properties. As a recent example, there has been a study on a

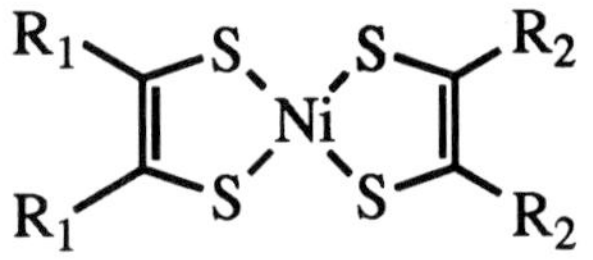

Fig. 5.9 The general structure of the nickel dithiolene complexes. The substituents R_1 and R_2 are given in Table 5.3.

series of nickel dithiolene complexes (Fig. 5.9) both in solution and in thin film form, dispersed into a host polymer.[19] The principal feature of these materials is that by suitable choice of substituent groups it is possible to tune the absorption band in the near infra-red region. It may thus be possible to reach a suitable trade-off between absorption and nonlinearity.

Table 5.3 contains data on some of the substituted dithiolenes where the

Table 5.3 Measured $\chi^{(3)}(-\omega; \omega, -\omega, \omega)$ values for solutions of nickel dithiolenes where the principle lowest electronic excitation energy is tunable

R_1	R_2	λ_{max} (nm)	α (m^{-1})	$\chi^{(3)}$ (m^{2}V^{-2}) × 10^{20}	$\chi^{(3)}/\alpha$ (m^{3}V^{-2}) × 10^{22}
$-CH_2CH_3$	$-CH_2CH_3$	770	6	0.68	11
$-C_6H_5$	$-(CH_2)_7CH_3$	800	15	1.7	11
MeC_6H_4	MeC_6H_4	900	283	33	12
$MeOC_6H_4$	$MeOC_6H_4$	935	1816	370	20

$\chi^{(3)}$ for solutions of 10^{18} molecules cm^{-3} have been determined from measurements of the nonlinear refractive index (see equations (5.12) and (5.13)). Also given is the absorption coefficient of the solution at 1.064 μm. The trends which became apparent are

(i) that the measured $\chi^{(3)}$ increases as the value of λ_{max} approaches the wavelength of the laser used (1.064 μm).
(ii) that even as α increases, it is still possible to enhance the practical value of the materials.

The first point demonstrates the influence of resonance on the enhancement of the third order response. Although higher order absorption processes (such as two-photon absorption) may also have a role to play in $\chi^{(3)}$ phenomena, it is clear in this case that the one-photon allowed transition dominates the enhancement in this 'near resonance' region. The second point hints at the use of a figure-of-merit system by which materials can be assessed. There are a number of ways in which figures-of-merit (FOMs) may be formulated. In general, it is only wise to use any one of them if comparing materials for use in a specific device. Here we use $\chi^{(3)}/\alpha$ simply to demonstrate an important point regarding the utility of materials in $\chi^{(3)}$ applications. That is, that by minimizing α, even at the expense of a reduction in $\chi^{(3)}$ one might approach a usable trade-off. This argument extends to the interesting observation in silica optical fibres. Silica itself has only a rather low $\chi^{(3)}$ coefficient $(2.8 \times 10^{-22} m^2 V^{-2})$ but because of its excellent transparency $(\alpha \simeq 10^{-5} m^{-1})$ has the best FOM of any material. The penalty is, of course, that rather long interaction lengths (fibre devices) are required to perform active functions.

5.6 Macroscopic Assemblies

The macroscopic nonlinear susceptibilities are made up of the volume normalized vector sum of the microscopic susceptibilities. Thus the simplest way to represent the relationship between β and γ and their macroscopic counterparts, $\chi^{(2)}$ and $\chi^{(3)}$ is by putting the following

$$\chi^{(2)} = N\langle\beta\rangle F \tag{5.30}$$

and

$$\chi^{(3)} = N\langle\gamma\rangle F \tag{5.31}$$

in which N is the number density of nonlinear moieties, F is the combination of local field factors required for the process in use or under study, and the angular brackets imply an orientational distribution of microscopic entities from which the N-bodies may be projected onto the macroscopic reference frame. This distribution is dependent upon the form of the macroscopic assembly whether, for instance, as a crystal, LB film, or polymer. The latter example is one where the principles of the orientational distribution may be easily seen and indeed introduces an important branch of organic nonlinear optics, that of polar ordered polymers.

Poled Polymers   The simplest material system of this class consists of polar molecules, doped into an amorphous polymer host. Concentrations of up to 20–30% by weight have been achieved in certain doped polymers which gives number densities in the region of 10^{26}–10^{27} m^{-3}.[20] To form the required noncentrosymmetric system, the polymer is heated to its glass transition temperature, and an electric field is applied to bring the dipoles into partial alignment. Thermodynamic equilibrium is established very quickly under these conditions and the orientational probability function, f derives from a Boltzmann-like term thus

$$f = \exp(-\mu_z E_Z \cos\theta / kT) \tag{5.32}$$

in which E_z is the poling field, μ is the dipole moment and kT, the thermal energy of poling. The term θ represents the angle between the dipole and the applied field. The system adopts a macroscopic polar moment and the β coefficient, aligned collinearly with the dipole thus contributes to a macroscopic susceptibility in the direction of the poling field. Carrying out an evaluation of the statistical distribution of the system [for example,[3]] yields the following results

$$\langle \beta_{zzz} \rangle = \beta_{zzz} \langle \cos^3\theta \rangle = \beta_{zzz} \frac{\mu_z E_Z}{5kT} \tag{5.33}$$

where it has been necessary to make reference to the tensorial nature of β. The tensor component of interest here is that where the fields acting and the polarization produced are all aligned along the molecular z axis, by virtue of the assumption that the molecules are one-dimensional. This, for the *para*-substituted aromatics of Table 5.1 is along a line between the donor and acceptor groups. The result, equation (5.33), gives a value for the average microscopic susceptibility contributing to $\chi^{(2)}_{ZZZ}$ in which the upper case subscripts now refer to the laboratory reference frame. We can now express the bulk susceptibility in terms of the averaged microscopic contribution.

$$\chi^{(2)}_{ZZZ} = \frac{NF\beta_{zzz}\mu_z E_Z}{5kT} \tag{5.34}$$

Here, Z is the direction along which the poling field was applied.

5.7   Conclusions

The foregoing shows that the details of the macroscopic susceptibilities, measured and utilized in devices, depend ultimately on the nature of the microscopic groups from which the material is composed, and then on their orientational distribution. Research in all areas of organic nonlinear optics is concerned with optimizing these susceptibilities at both the microscopic level, through novel synthetic chemistry, and where enhancements may be obtained by engineering, on improvement in the macroscopic properties.

References

1. R. W. Boyd, *Non-linear Optics*, (Academic Press, 1992).
2. I. M. Skinner and S. J. Garth, *Am. J. Phys.*, **58**, 177 (1990).
3. K. D. Singer, M. G. Kuzyk and J. E. Sohn, *J. Opt. Soc. Am. B*, **4**, 968 (1987).
4. R. W. Munn, in *Principles and Applications of Non-linear Optical Materials*, Eds R. W. Munn and C. W. Ironside, (Chapman and Hall, 1993).
5. G. I. Stegeman, C. T. Seaton and R. Zanoni, *Thin Solid Films*, **152**, 231 (1987).
6. J-L. Oudar and D. S. Chemla, *J. Chem. Phys.*, **66**, 2664 (1977).
7. D. S. Chemla, J-L. Oudar and J. Zyss, *L'Echo des Recherches (Eng.)*, 47 (1981).
8. P. N. Prasad and D. J. Williams, *Introduction to Nonlinear Optical Effects in Molecules and Polymers*, (John Wiley, 1991).
9. L-T. Cheng, in *Organic Molecules for Non-linear Optics and Photonics*, Eds J. Messier, F. Kajzar and P. Prasad, (Kluwer Academic, 1991).
10. K. C. Rustagi and J. Ducuing, *Opt. Commun.*, **10**, 258 (1974).
11. C. Cojan, G. P. Agrawal and C. Flytzanis, *Phys. Rev. B*, **15**, 909 (1977).
12. H. Kuhn, *Forsch. Chem. Org. Mat.*, **16**, 169 (1958).
13. M-T. Zhau, B. P. Singh and P. Prasad, *J. Chem. Phys.*, **89**, 5535 (1988).
14. Z. Shuai and J. L. Bredas, *Phys. Rev. B*, **46**, 4395 (1992).
15. Z. Shuai, and J. L. Bredas, *Phys. Rev. B*, **44**, 5962 (1991).
16. A. F. Garito, J. R. Heflin, K. Y. Wong and O. Zamani-Khamiri, in *Non-linear Optical Properties of Polymers*, Eds A. J. Heeger, J. Orenstein and D. R. Ulrich, MRS Symp. Proc. No. 109, (MRS, Pittsburg, 1988).
17. J. Yu and W. P. Su, *Phys. Rev. B*, **44**, 13315 (1991).
18. W-S. Fann, S. Benson, J. M. J. Madey, S. Etemad, G. L. Baker and F. Kajzar, *Phys. Rev. Lett.*, **62**, 1492 (1989).
19. S. N. Oliver, C. S. Winter, J. D. Rush, A. E. Underhill and C. Hill, *Proc. SPIE*, **1337**, 81 (1990).
20. Y. Karakus, D. Bloor and G. H. Cross, *J. Phys. D.: Appl. Phys.*, **25**, 1014 (1992).

6

Photochromism

P J Martin

6.1   Introduction

Photochromic materials reversibly change their absorption spectra, that is, their colour, and related optical properties in response to incident optical illumination. The first reports of photochromic behaviour date back to the nineteenth century[1] but it is only recently that the phenomenon has been investigated systematically. These investigations have resulted in the emergence of a few high quality systems with applications in, for example, photochromic sunglasses, actinometry, optical information processing and data storage. This review will focus upon the organic photochromic systems, particularly those of relevance to molecular electronics. Such systems have been referred to as the optical equivalents of electronic flip-flops.

6.2   Basic Concepts

A general photochromic reaction can be represented by the equation

$$A \underset{h\nu,\Delta}{\overset{h\nu}{\rightleftarrows}} B \tag{6.1}$$

where A represents the initial, 'unswitched' photochromic material and B the 'switched' material; in most systems, A and B represent single molecules. Typically, A is switched to B by illuminating it with light of the appropriate wavelength, often in the near uv for organic materials, whilst B reverts back to A by means of a thermal process. More rapid reverse switching can be achieved by illumination at a second wavelength, normally near an absorbance peak of B. The initial unswitched state, A, is usually less coloured than the product B in which case the system is said to be positively photochromic.

Equation (6.1) represents ideal photochromic behaviour but in practice photochromic systems all exhibit 'fatigue' to a greater or lesser extent. The time required to switch between colourless and coloured species gradually increases and the maximum coloration is reduced after repeated cycling between A and B. Fatigue is caused by the unwanted side reactions A and B undergo under the effect of light or heat, the situation often being aggravated by interference of the fatigue products in the primary photochromic process. Fatigue effects can become noticeable after any number of switching cycles but some organic systems can survive more than 10^4 cycles without damage.

There are a number of parameters in addition to fatigue which are important in photochromic systems, including reaction rates and quantum efficiency. Many photochromic systems exhibit a simple first-order switching response although timescales on which the reactions operate can vary from microseconds to months. The rates of the forward and reverse reactions need to be taken into account, and for many applications thermal stability in the dark is desirable. The switching efficiencies of photochromic reactions are frequently characterized by their quantum yield, Φ, defined as the ratio of the number of molecules reacting to the number of photons absorbed. In photochromic materials the efficiency of coloration ('colorability') is of most practical interest and hence a pseudoquantum yield is sometimes used as a figure of merit, defined as the product of the actual quantum yield and the molar absorption coefficient, ε. When a photochromic material is exposed to its actinic light, that is the light causing the photoreaction, it colours until equilibrium with the various decay processes is reached. The equilibrium absorbance thus depends in part upon the same factors that influence the thermal fading rate, a lower fading rate giving a higher absorbance value. In solution the maximum absorbance is frequently proportional to the concentration of the photochromic material, until at high concentrations aggregation phenomena and excited-state annihilations begin to dominate.

The positions of the absorbance peaks and the contrast between switched and unswitched forms are also important. For example, to prevent competition between the forward and reverse reactions the switched material should not absorb strongly at the actinic wavelength of the unswitched material. Moreover, ideally it should be possible to interrogate the state (switched or unswitched) of the material without altering it in the process.

Some photochromic materials will also colour thermally and electrochemically, although these properties are not the subject of this review. Thermochromism is not uncommon since the colourless and coloured forms of a material will often exist together in thermal equilibrium, the proportion of the coloured form increasing with temperature. Electrochromism tends to be displayed by materials whose photochromism is

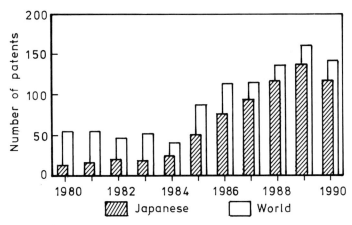

Fig. 6.1 Patents relating to photochromism granted per year between 1980 and 1990. Reproduced by permission from Ashwell *et al.*[2]

based on charge transfer, such as the viologens and the copper and silver salts of TCNQ (7,7,8,8-tetracyano-*p*-quinodimethane).

6.3 Data Storage and Other Applications

In the ten years between 1980 and 1990 the number of patents granted per year relating to photochromism increased almost threefold and considerably more in Japan (see Fig. 6.1).[2] The upsurge of interest has been sparked off in part by the realization that materials can be produced which will satisfy the requirements of some commercial applications, in particular data storage and retrieval for which a large market exists.

The demand for greater capacity storage is growing concomitantly with the increasing sophistication of computers. At present the bulk of this storage capacity is still supplied by purely magnetic media such as magnetic tape and disks, but by utilizing optically based technologies the area required to hold one bit of information can be reduced to approximately λ^2, where λ is the wavelength of the illuminating light. There are a number of possible optical recording media and some existent systems include magneto-optical disks (read, write and erasable) and CD-ROMs (compact disk, read only memories), both holding approximately 10^8 bits/cm^2. Photochromic materials can offer (although not necessarily simultaneously) increased capacity (to $>10^{12}$ bits/cm^2), a simple readout mechanism, the possibility of optical replication and in some cases, selective erase and rewrite facilities. The potential of molecular and submolecular level information storage is shown by DNA (deoxyribonucleic acid) which has an estimated capacity of 10^{21} bits/cm^3.

Information can be recorded by switching a photochromic material between colourless and coloured states, the advances in semiconductor lasers easing the problem of finding a compact light source of suitable wavelength. A retrieval mechanism can detect changes in absorbance or refractive index, the latter being more suited to nondestructive readout. Information storage does however put stringent constraints on the thermal stability and fatigue resistance of the photochromic material.

Data storage by photochromic systems usually requires the chromophores to be embedded in a thin film. Inorganic systems typically rely on light doping of crystals or glass, and hence suffer from the disadvantage that relatively thick samples (of the order of mm) are required to give appreciable contrast between switched and unswitched forms; this in turn limits the spatial resolution obtainable. Organic thin films may be constructed by a number of methods including evaporation, spin coating, which results in dilution of the chromophores in a polymer matrix, and the Langmuir–Blodgett (LB) method (see Chapter 10). The last has the advantage that no dilution of the chromophores is necessary other than that required to render them amphiphilic for LB deposition. The LB film thickness, the molecular alignment and the local microenvironment can also be controlled to adjust the system's switching, fatigue and thermal fading characteristics.

Some systems offer the possibility of increased capacity through the storage of multiple bits per pixel. Photochromic materials with narrow and close but nonoverlapping absorption bands would be selected and sequentially deposited, to give an LB film with a number of absorption peaks each of which would be switched independently. Information in this multi-frequency optical memory (MFOM) is thus wavelength multiplexed and can be written and read using a series of lasers and detectors. The LB technique allows the stacking of a number of uniform, thin recording layers which is difficult by conventional coating methods, and the LB film's thinness circumvents problems associated with the lasers' depth of focus.

Another technique for the application of photochromic materials in the field of data storage involves spectral hole burning which promises $>10^{12}$ bits/cm^2.[3,4] Dye molecules are doped into a host matrix, often an amorphous polymer, which provides a wide range of different microenvironments for the chromophores. Each individual dye molecule in the host matrix exhibits a narrow, homogeneously-broadened absorbance whose linewidth approaches the natural linewidth as the temperature falls towards absolute zero, freezing out phonon-related relaxation processes. However, at low temperatures the spectrum of the solid solution as a whole remains inhomogeneously broadened, due to the distribution of local chromophore environments. The broad inhomogeneous absorbance is effectively composed of a number of very narrow zero-phonon lines, each corresponding to a subset of microenvironments giving rise to a particular transition energy. By irradiating the material with light from an intense, highly monochromatic laser, such as a tunable dye laser, a 'hole' can be burnt in part of the spectrum. The dye molecules corresponding to a particular subset of local microenvironments are 'site selectively' photochromically switched so that they absorb outside the original inhomogeneously-broadened absorption band.

The commonly used dyes such as phthalocyanine, porphyrin derivatives and cyanine are not photochromic *per se* but show photochromic properties when in their amorphous matrix. For example, octaethylporphyrin and related compounds undergo hydrogen tautomerism on irradiation. This effectively rotates the molecules 90° with respect to their surroundings thus shifting their absorption maxima, a process known as photochemical hole burning. These photochemical holes persist provided that a low temperature is maintained. There are also systems relying on photophysical hole burning in which the spectral hole results solely from small changes induced in the local microenvironment, although this usually results in a 'photoproduct' which absorbs within the original, inhomogeneous-broadened band. Since the laser linewidth is usually less than the homogeneous linewidth, the increase in storage capacity that can be achieved by hole burning is equal to the ratio of the inhomogeneous linewidth to the homogeneous linewidth. This factor is increased by employing highly amorphous systems and can vary from 10–100 at 77 K to around 10^6 at 1 K.

The laser used to write the spectral hole can be used at much reduced intensity to detect it, but a superior method makes use of holographic techniques. The narrowness of the hole linewidths makes Stark and other effects clearly visible, and the storage capacity may be increased still further by applying an electric field across the dye-doped polymer to allow the information to be addressed by the voltage applied as well as by the laser frequency. Spectral hole burning has been reported in LB films in connection with optical data storage and investigations of the films' microstructures and fluorescence and vibrational properties.[5]

An alternative technique for achieving storage densities in excess of 10^{12} bits/cm^2 has been suggested by Liu *et al.*[6] In essence, the photochromically switched state of an azobenzene-based LB film is converted to a permanent and thermally stable state by electrochemical reduction. If this reduction and subsequent read processes can be carried out selectively using a glass needle electrode, positioned to the molecular resolution (≈ 1 Å) possible with scanning tunnelling microscope (STM) technology (see Chapter 12), then very high storage densities become possible. More details of this system are given later in the chapter; the principle may be adaptable to other electron-transfer-active photochromic systems. The optical analogue of the STM, the scanning near-field optical microscope (SNOM) may provide another way of increasing lateral resolution. A 50 nm size light source has been demonstrated using exciton transmitted radiation, and an improvement by a factor of 100 on a diffraction limited spot size is forecast.[7] Close to the source Förster exciton donation also provides increased efficiency of energy transfer. Photon tunnelling microscopy of photochromic systems could dispense with the need to scan the sample at the expense of reduced lateral resolution.

Aside from their potential as data storage media, photochromic materials also have a number of more imaginative applications. These include fashion, novelty items, security printing, light filtering of various sorts, actinometry, novel reprographic processes and holography. One of the best known applications of photochromics is in sunglasses, where silver halide doped glass darkens over a broad range of wavelengths in response to solar radiation. This inorganic system outperforms organics in that it can be switched more than 10^6 times before fatigue is noticeable, but the increasing use of plastic lenses has spurred the search for alternative materials. Spirooxazines have received some attention in this context although for commercial applications their limited fatigue lifetime still presents problems. Other related uses of photochromics include goggles to protect against intense flashes of light from high power lasers or nuclear explosions, and 'smart' windows which automatically adjust their optical density in accordance with ambient light levels. Applications have also been proposed in the fields of spatial light modulation, optical switching, phase conjugation and solar energy conversion.

6.4   Classes of Photochromic Materials

There are a great number of photochromic systems so that whilst compounds do not necessarily fall neatly into separate categories it is necessary to impose occasionally somewhat artificial divisions in order to simplify the taxonomy. According to the basic mechanism of the photochromic reaction in the forward direction, six separate classes can be identified:

- Hydrogen tautomerism
- Dissociation
- Dimerization
- *Cis-trans* isomerization
- Cyclization
- Charge-transfer

Since there are numerous examples in each category this review will concentrate only on those most relevant to molecular electronics. Good reviews which cover many other materials can be found in the references.[2,8,9,10]

Hydrogen Tautomerism   The most important group of materials in this category comprises the salicylidine anilines; Fig. 6.2 shows an amphiphilic derivative. These materials change colour typically from pale yellow to red under the influence of uv light and can be switched back by visible light.[11] The forward photochromic reaction consists of an enol-keto tautomerization followed by a *cis-trans* isomerization, and in general members of this class are *ortho* substituted aromatics which form orthoquinoid structures on uv irradiation. The properties of the N-salicylidene aniline derivatives are strongly influenced by steric effects, those having a planar structure, such as the N-salicylidene-α-aminopyridines, exhibiting thermochromism. Thermochromism, which yields the *cis* isomer and photochromism, which yields the *trans* isomer, are mutually exclusive in the solid state. A nitrogen atom at the α-position of the aniline ring results in a planar, thermochromic material whilst a CH group at the same position forces the molecule into a twisted conformation promoting photochromic behaviour.[12]

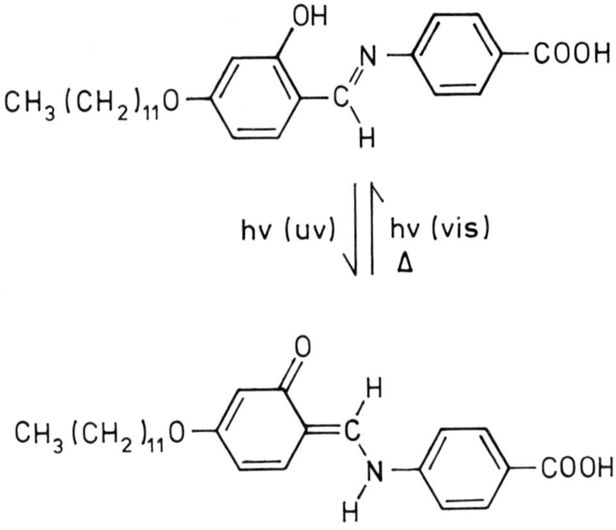

Fig. 6.2 Photochromic hydrogen tautomerism in N-(4-dodecoxysalicylidene)-4-carboxyaniline.[13]

In solution the thermal back reaction is so fast (rate constant, $k > 10^3 \, s^{-1}$) that photochromic behaviour is only observable at low temperatures. In the crystalline state the reaction is slower ($k \approx 10^{-3} \, s^{-1}$) and photochromism is observable at room temperature. Amphiphilic salicylidene anilines derivatives have been deposited as LB films, these demonstrating reversible photochromic behaviour.[13] The overall rate of thermal decoloration was similar to that of the crystalline state, although it was possible to modify this slightly by inserting spacer groups between the body of the molecule and the hydrophilic head; some evidence of H-aggregation was also noted. The rapid thermal decoloration currently limits the practical usefulness of these materials but a redeeming property is their excellent fatigue resistance—one highly pure sample of α-salicylidene aniline has been photochromically cycled around 50 000 times.[14]

Dissociation Materials in this category have photochromic properties based on bond cleavage, frequently resulting in the formation of coloured radicals. However they often suffer from fatigue and are sensitive to the presence of water and oxygen as well as lacking thermal stability. As photochromics they are therefore of limited usefulness although some have found other applications in photosensitive materials.

The various compounds can be grouped according to the dissociating bond, examples include C–C, C–N, C–Cl, C–S and N–N. One system which has received much attention is 2,3,4,4-tetrachloro-1-oxo-1,4-dihydronaphthalene which was among the first photochromic materials to be reported. Under uv light this compound undergoes C–Cl bond cleavage to turn purple thermally fading, in the solid state, after a few hours. A number of triarylmethane compounds also undergo heterolytic cleavage, some showing good colorization and a high quantum yield.

A few photochromic compounds have a homolytic cleavage mechanism. These include triarylimidazole dimers which if joined via a C–N bond change colour from pale yellow to purple when exposed to uv irradiation (C–C bonded dimers are not photochromic). Other examples are nitroso dimers (N–N bond cleavage) and 1,2,4,5-(tetrathiobenzyl)benzene which is photochromic in the solid state only, uv-induced cleavage of the C–S bonds turning the material pink.[10]

Dimerization   In general photochromic compounds relying on dimerization are highly conjugated systems, often comprising multiple aromatic rings with $\pi\pi^*$ lowest excited states. The 'dimerization' may be either intermolecular or intramolecular, the latter if the participating molecules are linked by a chain or more closely, as a cyclophane. The photochromic effects arise from a reduction in the conjugation on dimerization with a consequent blue-shift in the absorption band. Changes in colour may not be readily apparent as the materials frequently have little absorbance in the visible part of the spectrum. However the changes in refractive index, which occur away from the absorbance bands, may still be utilized, for example, for information storage or in holographic applications.

Among the materials which undergo photoreversible intermolecular photodimerization are the derivatives of anthracene, naphthalene and phenanthrene, cinnamates and heterocycles such as α-pyridone and acridizinium salts. Intermolecular reactions, whilst efficient in single crystals, are often slow and inefficient in liquid solutions and in guest–host polymer matrices. LB films however offer an ideal environment in which to position molecules with the required physical proximity and alignment, although as yet there are few examples existent. The first report of this technique was the reversible intermolecular photodimerization of [p-(octyloxy)cinnamylidene]acetic acid in Y-type LB films, as shown in Fig. 6.3.[15] The *cis-trans* isomerization observed in solution was suppressed and the reaction proceeded by means of a [2 + 2] photocycloaddition. The narrow, H-aggregated peak (full width at half maximum, 35 nm) at 270 nm rapidly diminished under visible irradiation, its return being promoted by 254 nm uv light. Unfortunately the spectra of the switched and unswitched forms overlapped reducing reversibility and the system also suffered from fatigue.

Constraining the chromophores to be close to one another in bichromophore and cyclophane compounds increases the efficiency of the photoreactions. Many intermolecular systems have been linked in this way to form intramolecularly dimerizing structures. Cyclophanes, in which the molecules are kept in strict registry with one another, are the most elegant examples.[16] Although these systems are frequently susceptible to fatigue, particularly through oxidation, many exhibit high thermal stability of the switched form and some have potential for photochromic data storage.

***Cis-trans* Isomerization**   This section will deal with those compounds

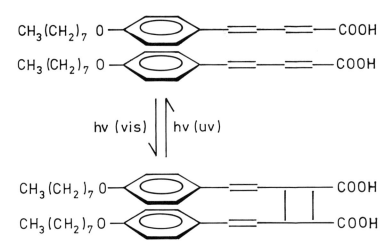

Fig. 6.3   Photochromic dimerization of [*p*-(octyloxy)cinnamylidene]acetic acid (*trans, trans*).[15]

for which *cis-trans* (Z-E) isomerization is the only photochromic response. Many materials however undergo *cis-trans* isomerization as part of a photochromic mechanism which is classified under a different heading. For example, salicylidene aniline, which rotates about the C–N bond after proton transfer, and the spiropyrans (see below).

Cis-trans isomerization is responsible for photomorphogenetic responses that is, light-influenced developmental responses, in the higher plants and for vision in animals. A number of photomorphogenetic responses, all stimulated by 660 nm light and inhibited by 730 nm light, have been linked to the reversible photoisomerism of the chromophore of the phytochrome protein. In the visual process 11-*cis*-retinal photoisomerizes to all-*trans* retinal (see Fig. 6.4) which in turn induces a conformational change in the rhodopsin protein, otherwise known as visual purple, to which it is attached. This allows the permeation of Ca^{2+} across the photoreceptor cell membrane, eventually triggering a nerve impulse to the brain. The photoisomerism of amphiphilic azobenzene derivatives has been used to construct similar artificial photoresponsive membrane systems, and the *cis-trans* isomerization of Langmuir films of azobenzene-containing lipids has been used to investigate photoinduced changes in molecular packing.

The photochromic protein bacteriorhodopsin (BR), found in the purple membrane of *Halobacterium halobium*, has been actively investigated with a view to a number of applications including holography, optical RAM (random access memory) and optical information processing (see Chapter 15).[17] These proteins can be genetically as well as chemically engineered to modify their properties and have a high quantum yield and low fatigue (upwards of 10^5 switching cycles), although their thermal relaxation time is short (at most, seconds). In common with the majority of photochromics in this class the *trans* isomer of bacteriorhodopsin is the more stable of the two, so that photoinduced *trans-cis* isomerization with

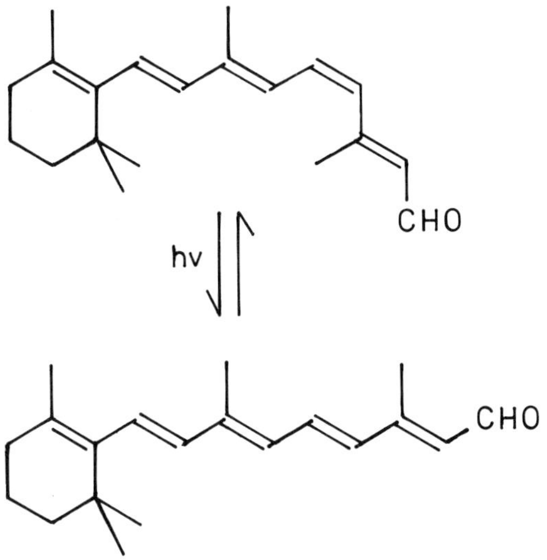

Fig. 6.4   *Cis-trans* photoisomerization of 11-*cis*-retinal.

thermal decay to the initial state is observed. Multilayer purple membrane LB films and films of BR in a phospholipid matrix have provided favourable environments to investigate the mechanism of the BR photocycle. In LB films the BR orientation and the film thickness and homogeneity can be controlled, and the photocycle intermediates have a relatively longer life than in the purple membrane in aqueous suspension.[18] The reaction centre of the photosynthetic bacteria *Rhodopseudomonas viridis* is a membrane protein-pigment complex able to generate electrical energy from light. Although not strictly photochromic it is noteworthy that these reaction centres have been deposited as Z-type LB films whilst retaining their functionality. The films provided a steady photovoltage and photocurrent under near-IR illumination. Transient photocurrents have been observed from purple membrane BR LB films under green light irradiation.

Photochromic *cis-trans* isomerizations which occur in solution are often suppressed in the semi-rigid environment of LB films. The *trans-cis* photochromism of stilbenes, which in any case is prone to fatigue reactions, in LB films exhibits only these fatigue reactions. In the case of some thioindigo derivatives, in both spread Langmuir films and in LB assemblies only *cis-trans* photoisomerism is observed despite the occurrence of the reverse photoisomerization in solution. In LB films of spiropyran monolayers the observation has been made that, unless the spirapyran layer is the top layer, the film's photochromic changes are lessened and slowed.

A versatile group of materials is the azobenzenes, LB films of which have attracted much attention. In these uv illumination of the *trans* isomer at the wavelength of the $\pi\pi^*$ band near 330 nm causes this band to diminish and a less intense $n\pi^*$ band of the *cis* isomer near 440 nm to grow.

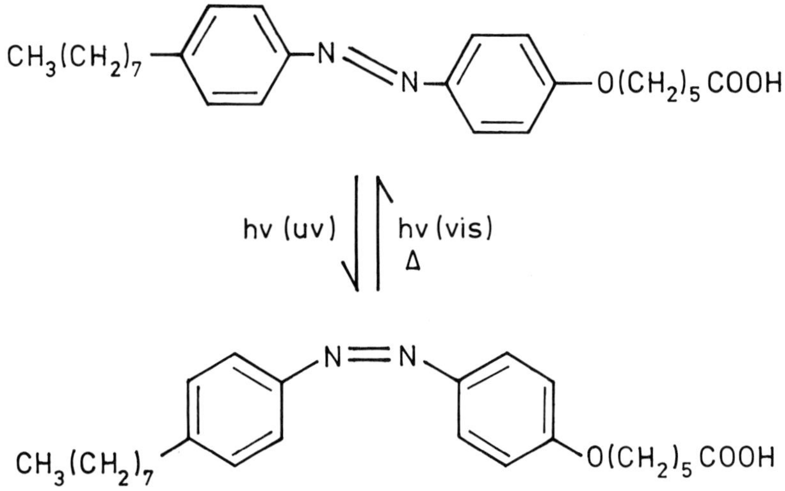

$CH_3(CH_2)_7$ —〈benzene ring〉— $N = N$ —〈benzene ring〉— $O(CH_2)_5 COOH$

hv (uv)   hv (vis)
           Δ

$N = N$

$CH_3(CH_2)_7$ —〈benzene ring〉   〈benzene ring〉— $O(CH_2)_5 COOH$

Fig. 6.5   Photoisomerization of 4-octyl-4′-(5-carboxy-pentamethyleneoxy)azobenzene.

Heat, uv irradiation at 250 nm and irradiation at >400 nm, that is, irradiation of the nπ* band, recovers the *trans* isomer.[10,19] An amphiphilic azobenzene derivative is shown in Fig. 6.5. As with other materials, in LB films the isomerization of amphiphilic derivatives of these compounds is heavily influenced by the film structure and whilst *cis-trans* photoisomerization is normally observed *trans-cis* isomerization is often inhibited. This can be explained by the bulky nature of the intermediate states (between the two isomers) whose molecular areas are greater than that of the *trans* isomer.

Some novel solutions have been proposed to circumvent the problem of reversibility. One group has increased the area-per-molecule available for switching by complexing the amphiphilic azobenzene derivatives with ionic polymers introduced into the subphase.[20] The area available was controlled by the pH of the subphase, this influencing the degree of dissociation of the terminal carboxyl group of the substituted azobenzenes, and hence the degree to which the ionic polymer was incorporated into the floating monolayer. It proved possible to achieve good LB deposition with sufficient free volume in the films to allow reversible photoisomerization.

A separate approach incorporated azobenzene derivatives (without long alkyl chains) into the cylindrical cavity provided by amphiphilic β-cyclodextrin molecules.[19,21] An example of one of the resulting non-covalently bonded guest–host complexes is shown in Fig. 6.6. The additional free volume in the cavity permitted *trans-cis* isomerization although still to a limited degree compared with solution experiments. It was postulated that the available free volume was still not sufficient. The driving force for the complexation, which takes place in chloroform solution, was identified as a salt formation between the amino group of the

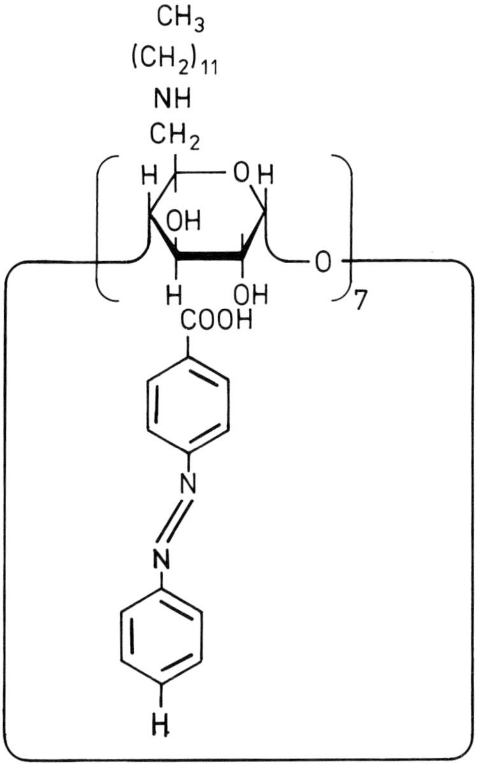

Fig. 6.6   A β-cyclodextrin molecule incorporating *p*-phenylazobenzoic acid. Such photochromic molecular host–guest complexes have deposited as LB films.[19]

host and the acidic group of the guest. The process has been described in terms of molecular recognition, the size of the cyclodextrin cavity and the substitution on the azobenzene determining the quantity of azobenzene incorporated into the cavity and hence into the LB films.

A photoinduced optical anisotropy (POA) effect in LB films of azobenzenes has also received some attention. It appears that under irradiation with linearly polarized visible light an optically induced recrystallization process tends to align the molecules perpendicular to the light's electric vector. The anisotropic molecular absorption of light is considered to have a small but not negligible effect on the molecular thermal equilibrium. This is a form of POA distinct from that induced by photochromic switching in spiropyran LB films (see below).

The azobenzene systems so far described are not ideal for photochromic data storage, not least because the *cis* isomer lacks thermal stability and will revert to the *trans* form, often in a matter of hours or less. One solution to this problem has been to incorporate the molecules into liquid crystals (LCs) and such materials have been used to construct phase holograms relying on photoinduced refractive index variations. Sawodny *et al.* have constructed multilayer LB films of homo- and copolymers with azobenzene side groups, the homopolymer films exhibiting LC-like X-ray reflectivity patterns.[22] Irradiation of the homopolymer LB LC films with 360 nm light caused permanent alteration to their structure, contrasting with the fully reversible behaviour of the copolymer films in which the chromophores were more dilute. The startling mechanical and thermal stability of such LC systems is apparent from investigations of LB films of amphiphilic polymers bearing mesogenic azobenzene chromophores.[23] The copolymer studied in most detail exhibited a highly ordered, aggregated structure at room temperature, with a phase change to an LC-like condition at 60°C. This LC phase was stable up to the highest temperature investigated, 150°C (in contrast with the bulk material which became isotropic at 84°C) and the ordered, aggregated structure was recovered on cooling. In an alternative approach to an LC LB assembly, polymers consisting of a stiff poly-L-glutamate backbone with flexible azobenzene-bearing side chains—so called 'hairy rods'—have been deposited.[24] The azobenzene moieties could be switched from an initial *trans* state to *cis* and back to *trans*, but in the process the layered structure of the films was lost although the preferred orientation of the main chains, induced by dipping, was retained. Other work demonstrates the potential of polyglutamate LB structures for data storage utilizing low-power, high-sensitivity read-out techniques based on surface plasmon microscopy.

Azobenzene derivatives have also been used for dramatic, reversible photocontrol of liquid crystal alignment. LC cells were constructed from low molecular weight nematic liquid crystals sandwiched between substrates coated with azobenzene derivatives, using the LB technique or other methods.[25,26] To obtain reversible photochromic behaviour poly-(vinyl alcohol) azobenzene derivatives were prepared for LB deposition. The substrates are known as command surfaces since each azobenzene molecule on a substrate is capable of influencing the alignment of $\approx 10^4$ LC molecules. The *trans* azobenzene isomer induces a homeotropic LC alignment (see Chapter 9), with the director perpendicular to the command surface, whilst the *cis* isomer induces a planar LC alignment with the director parallel to the command surface. It was found that the in-plane LC orientation could also be controlled since with sufficiently mobile chromophores, that is, with a minimum length of spacer between the dye and the polymer backbone, the director aligned perpendicular to the electric vector of the polarized uv light switching the azobenzenes. The longest spacer allowed orientation control using a single monolayer. The substituent at the *para* position opposite the spacer linkage, also plays an important role, only hydrophobic groups enabling the *trans* azobenzene isomer to induce a homeotropic LC alignment.

The electrochemical method of Liu *et al.* for stabilizing the *cis* isomer of azobenzene is particularly interesting.[6] It relies on the twin facts that (i) the *cis* azobenzene isomer can be reduced to a hydrazobenzene at greater anodic potentials than can the *trans* form, and that (ii) hydrazobenzene and its derivatives are oxidized exclusively to the thermodynamically stable *trans* isomer. Liu's group deposited LB films of 4-octyl-4′-(5-carboxy-pentamethylene-oxy)-azobenzene onto tin oxide-coated substrates. Provided that these LB films were first pretreated with 45 minutes of uv irradiation and 5 minutes of visible irradiation reversible, low-fatigue switching could be observed in under 60 s. The pretreatment was presumed to cause a gradual expansion of the films increasing the free volume available for chromophore isomerization.[27]

For optical writing of information a local *trans-cis* isomerization is induced then the *cis* isomer is electrochemically reduced to the stable hydrazobenzene derivative, leaving the *trans* isomer unaffected. Electrochemical writing would be by initial uniform irradiation of the entire film to convert it all to the *cis* isomer, followed by controlled electrochemical reduction using an ultra-fine counter electrode positionally controlled by scanning tunnelling microscope technology. The entire film could be erased by anodizing the film, reoxidizing the hydrazobenzene to the original *trans* isomer. The hydrazobenzene was found to be prone to oxidation by air, though this could be controlled by applying a small negative bias to the film. Some dissolution of the film in the aqueous electrolyte was also observed, although it was suggested that this might be overcome by using a solid electrolyte. Several hundred switching cycles were performed and the technique holds much potential and may be adaptable to other systems.

The Maxwell displacement current due to azobenzene *trans-cis* photoisomerization in an LB film has been measured, although H-aggregation retricted its magnitude. A more promising approach to molecular electronic switching is indicated by the construction of a compound incorporating a photochromic 4-octylazobenzene 'antenna'. This is linked by an alkyl chain to a pyridinium$^+$(TCNQ)$_2^-$ charge transfer complex. The structure of this compound is shown in Fig. 6.7. On deposition it yielded an LB film whose

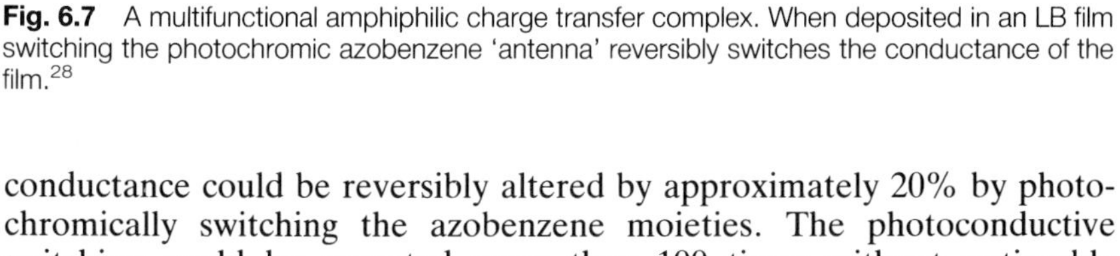

Fig. 6.7 A multifunctional amphiphilic charge transfer complex. When deposited in an LB film switching the photochromic azobenzene 'antenna' reversibly switches the conductance of the film.[28]

conductance could be reversibly altered by approximately 20% by photochromically switching the azobenzene moieties. The photoconductive switching could be repeated more than 100 times without noticeable fatigue.[28] The conductance change may be due to a change in the molecular overlap in the TCNQ stacks driven by the azobenzene isomerization. Covalently linking an azobenzene to a bipyridinium component has allowed the construction of a 'photo-triggered switchable molecular machine'.[29] The N-[α-(p-methyl-*trans*-azobenzene)]-N'-methyl-4,4'-bipyridinium diad forms a complex with the dye eosin, Eo^{2-}, with the association constant of the *cis* isomer approximately 12-fold greater than that of the *trans* isomer. Thus in solution, with appropriate concentrations of the diad and eosin, light induced *trans-cis* isomerization reversibly 'gears' the eosin onto the diad to form an intermolecular complex.

Cyclization The photochromism of this group of materials results from the presence of a conjugated ring which is photochemically opened and closed. Such electrocyclization reactions can be categorized according to the size of the ring opening or closing: 1,3-electrocyclization reactions include oxiranes and aziridines and 1,5-electrocyclizations include cyclic systems which photoreact, forming betaines and 1,3,5-hexatriene/cyclohexa-1,3-diene whose higher derivatives encompass, among other things, the fulgides and spiropyrans.

Fulgides, that is derivatives of bismethylenesuccinic anhydride, are commercially available in a number of congeners and are particularly noteworthy because of their fatigue resistance and readily definable properties including, for example, a variety of colours.[30] With no reactive intermediates only two significant fatigue reactions were identified, a thermal 1,5-hydrogen shift and a photochemical 1,7-hydrogen shift and fatigue was virtually eliminated by replacing this labile hydrogen with a methyl group. Improvements in coloration and quantum yield have been made by *inter alia* replacing the fulgide benzene ring by a furan ring, and by replacing the isopropylidene group (Me–C–Me) by a bulky adamantylidene group. The quantum yield of photobleaching in these materials shows an unusual, linear dependence on the wavelength of the actinic light. An 'improved', fatigue-resistant fulgide is shown in Fig. 6.8.

Electrocyclization reactions occur through conrotatory and disrotatory modes of motion whereby the terminal sections of the ring-open molecule, or those same atoms in the cyclic compound, rotate together or in opposing senses respectively. The mode of motion involved in ring opening and closure for photochemical and thermal reaction pathways can be predicted from considerations of orbital symmetry using the Woodward–Hoffman selection rules, whose predictions depend upon the number of π-electrons in the cyclic structure.[31] This formalism has been used to predict that in fulgides the photochemical ring closure operates by means of a conrotatory mode, whilst photochemical and thermal ring opening occur by means of con- and disrotatory modes respectively.

Fulgides have applications in actinometry, data storage and other areas. A fulgide derivative has been incorporated into acrylate and methacrylate liquid crystal copolymers to produce materials with thermally irreversible photochromic properties which have been shown to be capable of holographic pattern storage.[32]

Spiropyrans
In this review the behaviour of spiropyrans will be described in some

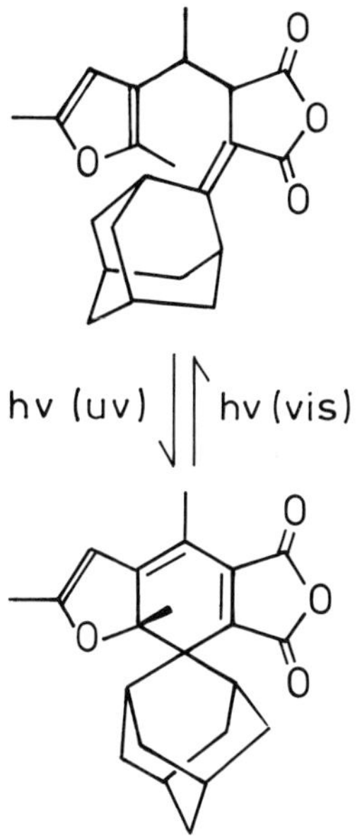

Fig. 6.8 Photocyclization in the fatigue-resistant, high quantum yield, commercial fulgide Aberchrome 670.[30]

depth. They are a particularly important group of materials and one of the most thoroughly studied and in this latter regard can be viewed as a case study. In LB films spiropyrans are of interest because of their narrow H- and J-aggregate absorption bands which cover a range of wavelengths and offer some potential for the realization of practical MFOMs. The majority of the compounds are positively photochromic, under uv irradiation changing from the pale yellow spiropyran (SP) form to a coloured, often blue photomerocyanine (PMC) form. Reversion to the SP form can occur thermally or photochemically, under visible light irradiation. Thermo-chromism and negative photochromism have also been observed.

A much-studied spiropyran is BIPS (1′,3′,3′-trimethyl-spiro-[2H-1-benzopyran-2,2′-indoline]), a generalized derivative of which is depicted in Fig. 6.9. The molecule in its spiropyran form consists of an azaheterocycle and a pyran ring held at approximately 90° to one another by the central sp^3 spiro carbon. The photomerocyanine dye has two important resonance forms, a predominant zwitterionic structure and a non-polar quinoid-type structure. When switching takes place the C–O bond of the pyran ring breaks, allowing conjugation through the molecule and giving rise to a $\pi\pi^*$ absorption band in the visible. Immediately after the bond breaks the molecule is in a twisted *cisoid* configuration but it usually relaxes to a planar *transoid* configuration because of steric effects.[10]

Spirooxazines are chemically similar to spiropyrans but can show greater resistance to fatigue and an improved quantum yield. Figure 6.10 shows a

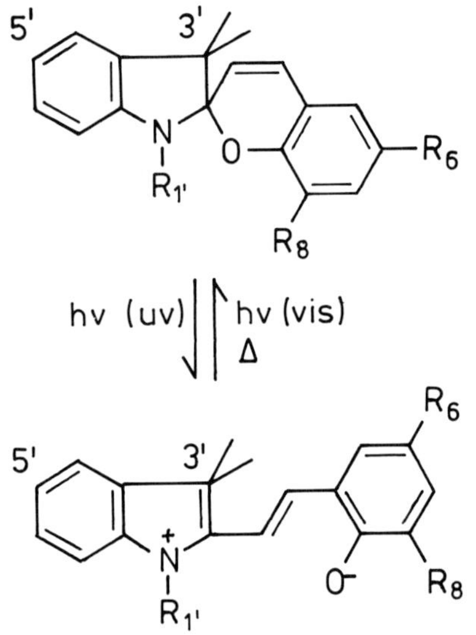

Fig. 6.9 Photochromic switching in a generalized amphiphilic spiropyran. Refer to Table 6.1 and to the text for a list and discussion of the various derivatives that have been synthesized.

typical example—the pyran ring is replaced by a 3H-oxazine ring, often fused to a larger aromatic system than is common in spiropyrans. Little work has been published on spirooxazines, possibly for reasons of commercial secrecy and because of synthetic difficulties.

In solution spiropyran photocoloration and photobleaching are usually first-order processes. Studies of decoloration kinetics over short timescales have identified more complex behaviour, however, attributed to the decay of various closed and open-ring intermediates such as the PMC *cis* isomer. Photoinitiated ring-closure generally proceeds more slowly than ring-opening, although typically more than an order of magnitude faster than the thermal dark reaction. Thermal fading rates are strongly dependent upon the substitution and rate constants can vary between $\approx 10^3\,\mathrm{s}^{-1}$ and $\approx 10^{-6}\,\mathrm{s}^{-1}$, the latter for 1',4',4',5',5'-pentamethyl-spiro[(2H-1)-3-methyl-6-nitro-8-methoxy-benzopyran-2,2'-oxazolidine] in ethanol.[33] In order for the PMC to SP transition to take place the PMC must first adopt the *cis* conformation, after which the phenolate anion can attack the positive

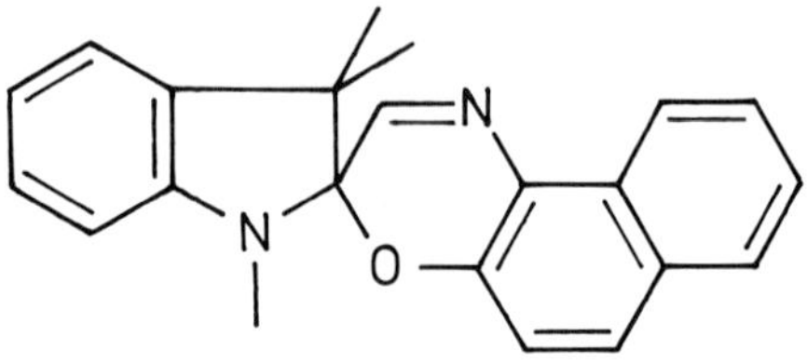

Fig. 6.10 A typical spirooxazine molecule.

charge on the ring nitrogen. The former is the rate-determining step and its ease is dictated by the steric and electronic effects of the various substituents. Spiropyrans have been incorporated into polymer matrices but in this environment unequivocal interpretation of their behaviour becomes difficult.[10]

A typical BIPS derivative has a half-life of only around 10^2 coloration-bleaching cycles, the most important contribution to fatigue coming from photoinitiated homolytic, as opposed to heterolytic, cleavage of the C–O bond resulting in a diradical. Homolytic cleavage can however be made likely by electron donating substituents which tend to increase the C–O bond polarization, and $>10^4$ switching cycles are possible with some spiropyrans. In spirooxazines it is thought that photodegradation may occur differently, indirectly via relatively inaccessible molecular states.

To achieve a coherent overview of the properties of amphiphilic spiropyran derivatives first Langmuir and then LB films will be treated. A number of amphiphilic derivatives of BIPS have been synthesized for LB deposition, Table 6.1 listing the majority of these. Henceforth, for convenience, they will be referred to by their acronyms. A basic BIPS structure should be assumed unless the full name of the compound is given. Thus, for example, SP16 denotes 1'-hexadecyl-3,3'-dimethyl-6-nitro-spiro[2H-1-benzopyran-2,2'-indoline].

Whilst neither the PMC nor the SP isomers of the shorter chained congeners SP1 and SP7 will form stable Langmuir films, SP12, SP16 and SP18 will form films as the PMC isomer, that is, under uv irradiation.[34] Since most of the amphiphilic spiropyrans are positively photochromic, the SP isomer is obtained by spreading from solution in the dark or under green light. In the case of SP16, on compression the colourless SP isomer only begins to exert appreciable surface pressure at a very small area-per-molecule, this only because of the formation of microcrystals on the

Table 6.1 *Amphiphilic derivatives of the BIPS spiropyran*

Spiropyran	R^6	R^8	$R^{1'}$
SP1	NO_2	H	CH_3
SP7	NO_2	H	C_7H_{15}
SP12	NO_2	H	$C_{12}H_{25}$
SP16	NO_2	H	$C_{16}H_{33}$
SP18	OH	H	$C_{18}H_{37}$
SP144	NO_2	H	$C_{18}H_{37}$
SP145	NO_2	CH_2OH	$C_{18}H_{37}$
SP147	NO_2	OH	$C_{18}H_{37}$
SP1801	NO_2	OCH_3	$C_{18}H_{37}$
SP1802	NO_2	$CH_2OC_2H_5$	$C_{18}H_{37}$
SP1822	NO_2	$CH_2OCOC_{21}H_{43}$	$C_{18}H_{37}$
MSP1822	5'-methoxy-SP1822		
SP0512	NO_2	$CH_2OCOC_{11}H_{23}$	C_5H_{11}
SP16NAP	1,3-dimethyl-3-hexadecyl-8'-nitrospiro-(indoline-2,3'-(3H)-naphtho(2,1-b)-pyran)		
SP10THIO	3'-methyl-3-decyl-6-nitro-spiro[2H-1-benzopyran-2,2'-benzothiazoline]		

subphase. However, reversible spreading can be observed—if the uncompressed SP isomer is irradiated by uv light a deep blue oil is formed, exerting a surface pressure which grows with a typical first-order photochromic response and thermally decays in the dark.[35] Similarly, a reversible 400% change in area-per-molecule can be observed at constant surface pressure.

The isotherms of a number of amphiphilic spiropyrans can be improved by mixing the spiropyrans with other long-chain surface active molecules such as tripalmitine (TP) and octadecane (HC18), and a large number of such heterosystems have been investigated.[36] Polymeropoulos and Möbius found that a 1:6 molar ratio of SP18:TP stabilized monolayers of the SP isomer.[37] The 20 Å^2 area-per-spiropyran corresponding to the cross-sectional area of a single alkyl chain suggested that the TP molecules arranged themselves to fit exactly over the chromophore. Similarly isotherms of the PMC isomer of SP1801 showed a steeper rise to an increased collapse pressure when the spiropyran was mixed in a 1:3 molar ratio with HC18.[36] With 'two-legged' spiropyrans such as SP1822 and SP0512 stable monolayers are obtained from both SP and PMC isomers at low surface pressures, but the best stability is obtained by mixing, for example SP1822 with HC18 in a 1:2 mole ratio.[35,38,39] In all these heterostructure Langmuir films the molecular areas suggest a flat chromophore orientation over which four alkyl chains tessellate.

The PMC isomers of a number of surface active spiropyrans form aggregates in both Langmuir and LB films. These are usually pictured as linear stacks of molecules and give rise to an intense absorption peak in the uv-visible spectrum superimposed upon the broad absorption due to unaggregated molecules. They are of great interest since the narrowness of the peaks makes the LB films suitable candidates for MFOMs. Figure 6.11

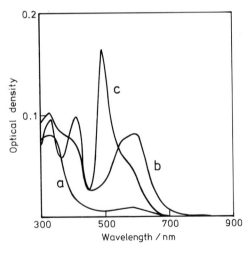

Fig. 6.11 Absorption spectra for a heterostructure LB film of 1:2 MSP1822:Octadecane. (a) Spiropyran form; (b) photomerocyanine form; (c) H-aggregated photomerocyanine form. Reproduced by permission from Hibino *et al.*[39]

shows the absorption spectra for the SP, PMC and H-aggregated PMC (H-PMC) forms of a heterostructure LB film of $1:2$ MSP1822:HC18. In the case of J-aggregation the molecular dipoles align antiparallel, and the absorption peak shows a bathochromic (red) shift relative to the unaggregated species; in the case of H-aggregation the molecular dipoles align parallel and the shift is hypsochromic (blue). Converse changes are observed in the fluorescence spectra.[36] These aggregates can be highly stable and it is thought from observations on spiropyran homopolymers that photochromic switching in the aggregate stacks occurs from the ends of the stacks inwards.

Langmuir films of SP16 show progressive replacement of PMC by J-aggregated PMC (J-PMC) as the surface pressure increases above 10 mN m^{-1}. A reversible photochromic reaction still occurs in the J-PMC although the half-life for thermal decay is very much increased, to $\approx 10^4$ s.[34] Polarized absorption spectroscopy indicates that at low surface pressures the long axes of the chromophores are randomly orientated, though tilted with respect to the air–water interface. As the aggregates are formed the chromophores take up a nearly flat orientation parallel to the direction of compression in all parts of the Langmuir film.[34,40] For SP16 this macroscopic ordering and thermal stability disappears on LB deposition. In LB films it was found that the J-PMC decoloration half-life was $\approx 10^3$ times shorter than in the Langmuir films, and that once destroyed aggregates could not afterwards be reformed.

The behaviour of the 'two-legged' spiropyrans SP1802 and SP1822 is similar in that these materials also form J-aggregates on the subphase at higher surface pressures. The PMC chromophores however remain flat but randomly orientated at all surface pressures although there is a reversible photochromic reaction.[40,41] The aggregation and other properties of a number of amphiphilic spiropyrans in Langmuir and LB films are summarized in Table 6.2 at the end of this section. It is interesting to note that stability of aggregates in the Langmuir film does not necessarily imply stability in the LB film and vice-versa.

Early work on the properties of spiropyran LB films found that the rate of dark decoloration in a $1:6$ SP18:TP heterostructure was approximately 2.7×10^{-4} s^{-1} at room temperature; LB films of pure SP16 have similar rate constants.[37,38,42] An ingenious method relying on the fluorescence quenching of a thiacyanine-based dye monolayer was employed to measure the photoinitiated reaction rates in this heterostructure which were determined as 1.2×10^{-1} s^{-1} and 6.7×10^{-2} s^{-1} for ring opening and ring closure respectively.[37] Photoreaction quantum yields, Φ, were then calculated from

$$\Phi = (kN)/(QA) \qquad (6.2)$$

where k is the reaction rate constant, N the number of reacting molecules per unit surface area, Q the number of light quanta per unit surface area per second and A the absorption of the irradiated layer. Quantum yields of $\Phi_{op} = 0.45$ for ring opening and $\Phi_{cl} = 0.03$ for ring closure were found, the lower quantum yield for ring closure reflecting the condition that for this to take place the two parts of the spiropyran molecule must geometrically align.

Dark decoloration and quantum yields have been measured by Minami *et al.* for a 1:6 SP18:TP Y-type LB film with similar results: $k = 1.44 \times 10^{-4} \, \text{s}^{-1}$, $\Phi_{op} = 0.44$, $\Phi_{cl} = 0.02$.[43] This group also deposited LB films using other matrix molecules to vary the chromophore microenvironment, and used time-resolved fluorescence spectroscopy to probe the reaction mechanism. A nonplanar *cisoid* precursor, X, of the PMC isomer was identified and it was found that the larger the chromophore reaction cage, the faster the PMC to SP photoisomerization.

The influence of the chromophore microenvironment has been investigated in more detail, in naphthoxazine-based LB films, by Anisimov *et al.*[44] In these films the photocoloration and thermal fading behaviour could only be successfully described by assuming a statistical distribution of rate constants for the forward and reverse reactions, attributed to a variety of local molecular environments. Chromophores with a low reactivity in the forward direction were also shown to have a low reactivity in the reverse direction. This was explained by assuming that after a change in molecular conformation the local microenvironment relaxes, faster than the reverse ring closure, allowing the reaction to proceed. A 'frozen' microenvironment would link a high energy barrier for a reaction in one direction with a low energy barrier for a reaction in the converse direction, and hence a low forward reactivity with a high reverse reactivity. The actual energy barrier for the photochromic reactions was thought to depend to a first approximation on the density of the local microenvironment.

IR spectroscopy has been used to study photochromic switching in a 1:1 SP16:Behenic acid LB heterostructure. Although the results were inconclusive due to the complicating effects of associated molecular reorganization, it is of interest that to study the PMC structure a photoreversible, negatively photochromic state was temporarily induced by protonating the film with hydrochloric acid vapour, thus forming the yellow merocyanine hydrochloride salt.

As has already been mentioned, J-aggregates in SP16-based LB films rapidly decay but in LB films of other spiropyrans aggregation confers exceptional stability. At room temperature Y-type LB heterostructures of 1:2 SP1822:HC18 show the usual photochromic behaviour with a half-life of $7.2 \times 10^4 \, \text{s}$ for the PMC thermal decay.[38] However, when subjected to uv irradiation at temperatures above 35°C a single, sharp J-aggregate absorption appears in the uv-visible spectrum dominating the monomer band. The J-PMC absorption exhibits a half-life $\approx 10^4$ times longer than that of the monomeric spiropyrans, although disaggregation to the SP form can be brought about by heating the film above 50°C or by irradiation with light at the wavelength of the aggregate absorption peak.[45] This latter effect has been used to align the J-aggregate stacks in an SP1822-based LB film by photochromically switching the film with polarized light from a HeNe laser. The stacks with a transition dipole moment aligned parallel to the laser polarization are preferentially disaggregated, preferentially reforming perpendicular to the polarization direction, nucleated by the remaining stacks.[45]

LB heterostructures of the related derivative MSP1822 can form either H- or J-aggregates.[39] In a 1:2 mole ratio with HC18 both Y and Z-type LB structures form H-aggregates above 40°C, the former stable for several months in the dark (see Fig. 6.11). In a 1:1 mole ratio with methyl stearate (MS) however the Y-type films form H-aggregates whilst the Z-type films form J-aggregates. The capricious behaviour of MSP1822 is not unusual— Miyata *et al.* have reported that SP1802 forms J-aggregates on a water subphase but H-aggregates or I-aggregates ('intermediate' aggregates with almost no shift in the absorption peak) when deposited as an LB film, depending upon the surface preparation of the substrate.[41]

The effect of the film structure and codeposited heteromolecule has also been investigated in SP1801-based LB films.[36] It was found that some heteromolecules, such as stearic acid (SA), promoted the spontaneous formation of J-aggregates in the dark, in effect negative photochromism. The polar environment in general and the carboxylic acid group of the stearic acid in particular were deduced to play an important role in stabilizing the PMC isomer and hence the J-aggregates. Half-lives of unaggregated mixed films varied between 25 and 100 minutes whilst those of the aggregated films were 10^4 times longer.

LB films of SP144, SP145 and SP147, derivatives incorporating OH groups, have been used to investigate the role of hydrogen bonding in aggregation.[46] IR studies of LB films of SP144, as well as demonstrating hydrogen bonding of the OH groups, showed that the PMC form in this material is *cisoid* rather than *transoid*. In SP145-based LB films intermolecular hydrogen bonding was important in the PMC and J-PMC structures but not in the SP structure, explaining this material's increased stability in comparison with SP18 and SP1801. A head-to-tail molecular configuration within the J-aggregate stacks was also deduced. A growth in the OH and NO_2 stretching bands of LB films of SP147 on H-aggregation suggested that hydrogen bonding was also important in this congener.

Stability in spiropyran-based LB films is not the exclusive preserve of aggregated systems. Improved stability has also been achieved in supermolecular PMC-phenylalanine assemblies through deposition from a phenylalanine-containing subphase. However, the potential of spiropyrans for data storage has focused attention on identifying thermally stable structures with narrow, non-overlapping aggregate absorption bands. Table 6.2 summarizes information on the various stable Langmuir (L) and LB forms of a number of published systems. In the case of heteromolecular systems the superscript indicates the molecular mixing ratio so that, for example, a '2' indicates a spiropyran:heteromolecule ratio of 1:2. The absorption maxima of the PMC LB films are listed followed by, where applicable, those of the aggregated films and the full width at half maximum (FWHM) of the aggregate absorption peak(s). Where data are unavailable or unclear a space is left or a question mark inserted. Several systems show aggregation with LB film absorption peaks ranging from 485 nm to 638 nm, some as narrow as 40 nm FWHM. Thermal decay half

Table 6.2 *Summary of the properties of some spiropyran-based Langmuir and LB films*

System	L film forms	LB film forms	Maxima (nm)	FWHM (nm)	Ref.
SP16	Not_SP/PMC/J	SP/PMC/(J)	563/620	95	34
SP18	Not_SP/PMC/J				34
SP18:TP[6]	SP/PMC	SP/PMC	576		37
SP1801	PMC	SP/PMC	590		36, 40
SP1801:HC18[3]	PMC	SP/PMC	610		36
SP1801:SA[1]	PMC	PMC/J	600/600	67	36, 40
SP1801:OD[1]		PMC/J	590/618	?	36
SP1801:TP[1]		SP/PMC	595		36
SP1802	PMC/J	SP/PMC/H/I	580/495/575	53/73	41
SP0512	SP/PMC				35
SP1822	SP/PMC/J	SP/PMC/J	545/600*	40*	38, 40
SP1822:HC18[2]		SP/PMC/J	583/618	50	38
SP1822:AA[2]		SP/PMC/J	?/615	40	45
MSP1822:HC18[2]		SP/PMC/H	?/485	51	39
MSP1822:MS[1]		SP/PMC/J/H	550/595/?	99/?	39
SP144		SP/PMC	614		46
SP145	J	SP/PMC/J	?/638	113	46
SP147	PMC	PMC/H	?/560	?	46
SP10THIO	Not_SP/PMC				35
SP16NAP	SP/PMC	SP only			42

*Langmuir film.

lives are not given but are typically around 30 minutes for a PMC LB film and up to 10^4 times longer for H/J-PMC LB films.

Finally, a flavour of other possible spiropyran applications is given by a multilayer LB 'optical gate' in which fluorescence emission was switched between dye bilayers by altering the wavelength of light irradiating the structure.[47] Spectral overlaps between successive thiacyanine, PMC and indocarbocyanine bilayers allowed Förster excitation energy transfer between the bilayers resulting in 725 nm fluorescence provided that 363 nm uv light illuminated the film. However under green 545 nm light the PMC bilayer was switched to the SP form and the sequence of energy transport was closed so that only fluorescence at 480 nm from the upper thiacyanine bilayer was observed.

Charge-transfer   There are only a few organic photochromic materials with a mechanism relying solely on charge-transfer (CT) (see Chapter 8). Copper and silver salts of TCNQ (7,7,8,8-tetracyano-*p*-quinodimethane) colour under irradiation by means of partial back charge-transfer, thermally reverting to the fully ionic state. However, of greater interest is the electro and photochromic behaviour of the 4,4'-bipyridinium derivatives (viologens) and a series of D-π-A (electron donor, π-bridge, electron acceptor) zwitterions.

The charge-transfer (CT) salts of viologen dications and (tetrakis[3,5-bis(trifluoromethyl)phenyl]borate), (TFPB)⁻ show photochromic behaviour in solution, in the solid state and in polymer films.[48] Amphiphilic derivatives of these viologen salts mixed in a 1:4 ratio with arachidic acid have been deposited as LB films. When illuminated in the region of the ion pair CT band ($\lambda_{ex} > 365$ nm) these films change colour from pale yellow to blue, thermally fading back to yellow with a half-life at room temperature of about 3 hours. Electron spin resonance (ESR) measurements indicated the formation of monocation radicals. Oxidative decomposition of the anion was ruled out and a CT mechanism was postulated:

$$(\text{Viologen})^{2+} 2(TFPB)^- \underset{\Delta}{\overset{h\nu}{\rightleftharpoons}} (\text{Viologen})^{+\cdot}(TFPB)^-(TFPB)^{\cdot}$$

(6.3)

Similar reactions have been observed in other viologen salts but the reverse electron transfer reactions are usually too rapid for photochromic behaviour to be visible. The relative stability of the TFPB-based system after initial, fast reverse processes, is ascribed to a potential energy barrier between the initial CT state and the light-induced redox state. This is probably linked to a change in the relative orientation and/or separation of the participants in the reaction and the bulky nature of TFPB. Photochromism can also result from uv irradiation of viologen salts, although this is prone to undesirable fatigue reactions. In general, systems based on charge-transfer should be able to respond faster to the actinic radiation and should be less susceptible to fatigue than those relying on a rearrangement of chemical bonds.

The isotherms of the salts show a number of phase transitions and suggest that in the solid-condensed phase (>15 mN m^{-1}) some of the viologen and TFPB ions are squeezed out of the film. Small-angle X-ray analysis of Y-type LB films indicated that they probably consist of vertical stacks of TFPB-viologen-TFPB 'sandwiches', with the arachidic acid molecules standing upright between the 3,5-bis(trifluoromethyl)phenyl groups of the TFPB molecules. The orientation of the cation radicals with respect to the substrate could be controlled by modifying the amphiphilic substituents. Currently these materials have yet to demonstrate the cyclical and thermal stability required for applications such as data storage although in a polymer film, below 0°C thermal fading was apparently greatly inhibited.

The TCNQ-derivative donor-(π-bridge)-acceptor zwitterions constitute a novel group of materials with charge-transfer related photochromic behaviour. These multifunctional molecules have a large second order optical nonlinearity and behave as molecular rectifiers, raising the possibility of photochromic control of these properties.[49–51] A number of congeners have been synthesized, dividing roughly into two main series depending upon whether the electron donor is a pyridinium or a quinolinium derivative. Typical examples from both these series, Z-β-(N-alkyl-4-pyridinium)-α-cyano-4-styryldicyanomethanide (R-P3CNQ) and Z-β-(N-alkyl-4-quinolinium)-α-cyano-4-styryldicyanomethanide (R-Q3CNQ) are shown in Fig. 6.12.

The zwitterionic nature of the compounds was determined by ir spectroscopy, solvatochromism and examination of the bond lengths and molecular

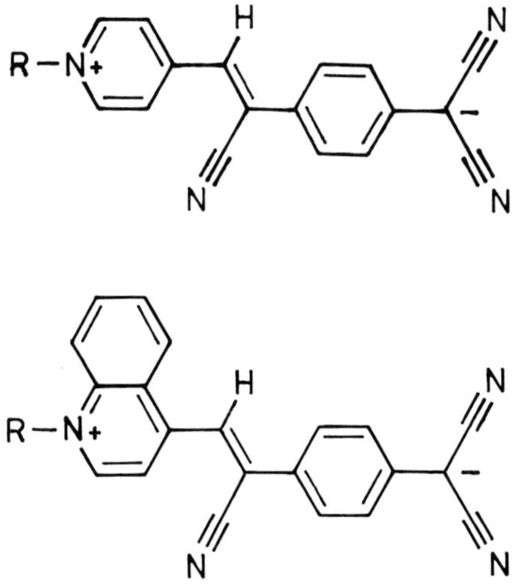

Fig. 6.12 Photochromic zwitterions: (top) Z-β-(N-alkyl-4-pyridinium)-α-cyano-4-styryldi-cyanomethanide (R-P3CNQ); (bottom) Z-β-(N-alkyl-4-quinolinium)-α-cyano-4-styryldicyano-methanide (R-Q3CNQ).

geometry determined from X-ray crystallography. In the crystalline state both intermolecular and intramolecular CT transitions are observed. Table 6.3 shows the unusually narrow width and relatively large range of wavelengths (480 nm to 670 nm) exhibited by the CT bands in LB films which makes the materials potential candidates for MFOMs. Figure 6.13 demonstrates this principle by superimposing the spectra of two LB films, one of $C_{16}H_{33}$-P3CNQ and one of $C_{16}H_{33}$-Q3CNQ.[52,53]

In solution the zwitterions are photochromic and electrochromic. They are bleached by irradiation at the CT band although the thermal colour return is erratic and concentration dependent. The LB films also photo-bleach but colour return is never observed—the highly coloured un-switched films and the bleached switched films have both been found to be stable for more than three years. LB films constructed from molecules in

Table 6.3 *Summary of the positions and widths of the CT bands of LB films of some of a series of donor-π-acceptor zwitterions*[52,54]

Zwitterion	λ_{max} (nm)	FWHM (nm)
$C_{16}H_{33}$-P3CNQ	495	54
C_6H_{13}-Q3CNQ to $C_{14}H_{29}$-Q3CNQ	614	74
$C_{15}H_{31}$Q3CNQ to $C_{20}H_{41}$-Q3CNQ	565	44
Mixed $C_{16}H_{33}$-P3CNQ/$C_{16}H_{33}$-Q3CNQ	Tunable: 495 to 565	In the range 44 to 68
$C_{16}H_{33}$-Q3CNQ(benz)	623	
$C_{16}H_{33}$-Q3CNQ(Cl$_2$)	545	
$C_{16}H_{33}$-Q3CNQ(Br$_2$)	545	
$C_{16}H_{33}$-Q3CNQ(F$_4$)	480	
Ph-CH$_2$-Q3CNQ	670	140

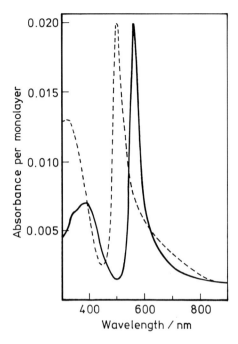

Fig. 6.13 Superimposition of the spectra of two photochromic zwitterionic LB films showing the potential for a multi-frequency optical memory. The solid line corresponds to $C_{16}H_{33}$-Q3CNQ ($\lambda_{max} = 565$ nm), the broken line to $C_{16}H_{33}$-P3CNQ ($\lambda_{max} = 495$ nm). Reproduced by permission from Ashwell.[52]

the R-P3CNQ series switch much faster than those constructed from the R-Q3CNQ series, although a benzyl (Ph) analogue in the latter series, Ph-CH$_2$-Q3CNQ, has also been found to switch quickly. Although the switching is permanent the stability this confers makes the zwitterions suitable for data storage applications in WORM (write once, read many) devices. The charge distribution on the molecules tends to stabilize Z-type LB structures, and for the quinolinium series the tilt angle of the chromophore has been found to be bivalued, depending upon the alkyl chain length. The mechanism of the photochromic behaviour has not been unambiguously determined but is thought to rely on a photoinduced distortion of the intermolecular charge transfer interaction.

   A novel feature of the zwitterions is the ease with which the absorption peaks of the LB films can be tuned. This may be accomplished by substitution, by altering the length of the hydrophobic alkyl chain, in the case of the quinolinium-based series and most interestingly by forming heteromolecular films.[54,55] LB films deposited from pure $C_{16}H_{33}$-P3CNQ have their absorption maximum at $\lambda_{max} = 495$ nm whilst those deposited from pure $C_{16}H_{33}$-Q3CNQ have $\lambda_{max} = 565$ nm. Those deposited from mixtures of $C_{16}H_{33}$-P3CNQ and $C_{16}H_{33}$-Q3CNQ do not show the expected double absorption peak, but instead a single peak whose position varies

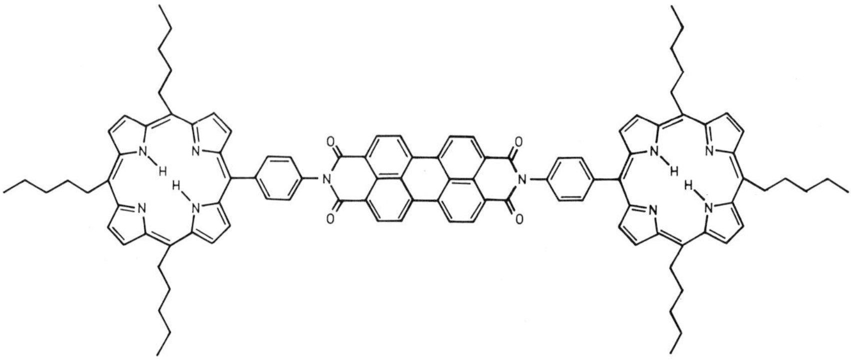

Fig. 6.14 The molecular structure of HP-PBDCI-HP, an intensity-dependent optical switch.   HP = *meso*-tripentyl-monophenylporphyrin,   PBDCI = N,N'-diphenyl-3,4,9,10-perylenebis(dicarboximide).[56]

linearly with the composition ratio between the two λ_{max} extrema. The single absorption peak is sufficiently sharp to be able to exclude the possibility that one peak is simply masking another. This elegant method for tuning an LB film's absorption may be applicable in other systems and could prove useful, for example, in matching a system to a particular laser frequency.

Also worthy of mention is a donor-acceptor-donor compound, HP-PBDCI-HP which exhibits light intensity dependent photochromic switching.[56] The molecule, whose structure is shown in Fig. 6.14, consists of two free-base *meso*-tripentyl-monophenylporphyrins, HP rigidly attached to N,N'-diphenyl-3,4,9,10-perylenebis(dicarboximide), PBDCI. The porphyrin donor in the molecule has a lowest excited singlet state, HP[1]* accessible through laser light irradiation at 585 nm. This irradiation induces rapid (≈ 10 ps) photoreduction of the PBDCI to give HP$^+$-PBDCI$^-$-HP which has a strong absorbance at 713 nm due to the PBDCI$^-$ ion and a lifetime of approximately 110 ps. Excitation of the system by intense laser light at 585 nm leads to two-photon absorption generating two excited singlets HP[1]*. The resulting HP$^+$-PBDCI$^-$-HP[1]* rapidly (≈ 60 ps) decays to HP$^+$-PBDCI^{2-}-HP$^+$ (lifetime ≈ 5 ns), the PBDCI^{2-} ion absorbing strongly at 546 nm. Such a system could be used, for example, to construct an intensity dependent molecular switch. The first electron transfer can also be initiated by using irradiation at 526 nm to stimulate the lowest excited singlet state of PBDCI. Retaining the excitation at 585 nm to accomplish the second electron transfer should enable the system to execute logical AND and OR functions.

6.5 Conclusions

The major groups of photochromic materials have been reviewed, particularly with regard to novelty and potential for applications within molecular electronics. In the field of information storage LB films of the spiropyrans and their derivatives and of the zwitterions show considerable promise, whilst fulgide and azobenzene liquid crystal derivatives have properties which make them interesting for high-density information storage using amplitude or phase holograms. Photochemical hole-burning is another emerging technique which, apart from providing a 'site-selective' spectroscopic probe, offers the possibility of storage densities greater than 10^{12} bits/cm^2, a factor of 10^4 improvement on the best currently available devices. The photoelectrochemical reduction of azobenzene derivatives combined with STM-based addressing techniques offers similarly high information densities. LB films of biological materials is also a growing area of research, as these can provide a controlled environment for studying simplified photobiological model systems and for building biomimetic visual and photosynthetic structures. The application of photochromic materials to molecular electronics is still in its infancy and the future promises to be exciting.

References

1. M. Fritzsche, *Comptes Rendus Acad. Sci.*, **64**, 1035 (1867).
2. G. J. Ashwell, I. Sage and C. Trundle, in *Molecular Electronics*, Ed. G. J. Ashwell, (Research Studies Press, Taunton, England, 1992), p. 15.
3. U. P. Wild and A. Renn, in *Photochromism*, Eds H. Dürr and H. Bouas-Laurent, (Elsevier, 1990), p. 930.
4. H. Suzuki, *Adv. Mater.*, **5**, 216 (1993).
5. J. Hala, F. Adamec, M. Ambroz *et al.*, *Thin Solid Films*, **223**, 178 (1993).
6. Z. F. Liu, K. Hashimoto and A. Fujishima, *Nature*, **347**, 658 (1990).
7. R. Kopelman, K. Lieberman and A. Lewis, *Mol. Cryst. Liq. Cryst.*, **183**, 333 (1990).
8. G. H. Brown, Ed., *Photochromism*, Techniques of Chemistry, Vol. III, (Wiley Interscience, New York, 1971).
9. H. Dürr, *Angew. Chem. Int. Ed. Engl.*, **28**, 413 (1989).
10. H. Dürr and H. Bouas-Laurent, Eds, *Photochromism. Molecules and Systems*, Studies in Organic Chemistry 40, (Elsevier, Amsterdam, 1990).
11. E. Hadjoudis, in *Photochromism*, Eds H. Dürr and H. Bouas-Laurent, (Elsevier, 1990), p. 685.
12. E. Hadjoudis, M. Vitorakis and I. Moustakali-Mavridis, *Mol. Cryst. Liq. Cryst.*, **137**, 1 (1986).
13. S. Kawamura, T. Tsutsui, S. Saito, Y. Murao and K. Kina, *J. Am. Chem. Soc.*, **110**, 509 (1988).
14. R. V. Andes and D. M. Manikovski, *Appl. Opt.*, **7**, 1179 (1968).
15. A. Yabe, Y. Kawabata, A. Ouchi and M. Tanaka, *Langmuir*, **3**, 405 (1987).
16. M. Usui, T. Nishiwaki, K. Anda and M. Hida, *Chem. Lett.*, 1561 (1984).
17. N. Hampp, C. Bräuchle and D. Oesterhelt, *Biophys. J.*, **58**, 83 (1990).

18. M. Ikonen, J. Peltonen, E. Vuorimaa and H. Lemmetyinen, *Thin Solid Films*, **213**, 277 (1992).
19. A. Yabe, Y. Kawabata, H. Niino *et al.*, *Thin Solid Films*, **160**, 33 (1988).
20. K. Nishiyama, M. Kurihara and M. Fujihira, *Thin Solid Films*, **179**, 477 (1989).
21. M. Matsumoto, M. Tanaka, R. Azumi *et al.*, *Thin Solid Films*, **210/211**, 803 (1992) and references therein.
22. M. Sawodny, A. Schmidt, W. Knoll, C. Urban and H. Ringsdorf, *Thin Solid Films*, **210/211**, 500 (1992).
23. T. L. Penner, J. S. Schildkraut, H. Ringsdorf and A. Schuster, *Macromolecules*, **24**, 1041 (1991).
24. H. Menzel, B. Weichart and M. L. Hallensleben, *Thin Solid Films*, **223**, 181 (1993).
25. K. Aoki, Y. Kawanishi, T. Seki, T. Tamaki, M. Sakuragi and K. Ichimura, *Thin Solid Films*, **219**, 226 (1992).
26. T. Seki, M. Sakuragi, Y. Kawanishi *et al.*, *Langmuir*, **9**, 211 (1993) and references therein.
27. Z. F. Liu, B. H. Loo, R. Baba and A. Fujishima, *Chem. Lett.*, 1023 (1990).
28. H. Tachibana, A. Goto, T. Nakamura *et al.*, *Thin Solid Films*, **179**, 207 (1989).
29. I. Willner, Y. Eichen, A. Doron and S. Marx, *Isr. J. Chem.*, **32**, 53 (1992).
30. H. G. Heller, *IEE Proc.*, **130**, 209 (1983) and references therein.
31. R. B. Woodward and R. Hoffmann, *Angew. Chem. Int. Ed. Engl.*, **8**, 781 (1969).
32. I. Cabrera, A. Dittrich and H. Ringsdorf, *Angew. Chem. Int. Ed. Engl.*, **30**, 76 (1991).
33. M. Maguet, F. Garnier and R. Guglielmetti, *J. Chem. Res. (M)*, 1519 (1982); M. Maguet, F. Garnier and R. Guglielmetti, *J. Chem. Res. (S)*, 145 (1982).
34. E. Ando, K. Moriyama, K. Arita and K. Morimoto, *Langmuir*, **6**, 1451 (1990).
35. D. A. Holden, H. Ringsdorf, V. Deblauwe and G. Smets, *J. Phys. Chem.*, **88**, 716 (1984).
36. E. Ando, J. Hibino, T. Hashida and K. Morimoto, *Thin Solid Films*, **160**, 279 (1988).
37. E. E. Polymeropoulos and D. Möbius, *Ber. Bunsenges. Phys. Chem.*, **83**, 1215 (1979).
38. E. Ando, J. Miyazaki, K. Morimoto, H. Nakahara and K. Fukuda, *Thin Solid Films*, **133**, 21 (1985).
39. J. Hibino, K. Moriyama, M. Suzuki and Y. Kishimoto, *Thin Solid Films*, **210/211**, 562 (1992).
40. E. Ando, M. Suzuki, K. Moriyama and K. Morimoto. *Thin Solid Films*, **178**, 103 (1989).
41. A. Miyata, D. Heard, Y. Unuma and Y. Higashigaki, *Thin Solid Films*, **210/211**, 175 (1992).
42. M. Morin, R. M. Leblanc and I. Gruda, *Can. J. Chem.*, **58**, 2038 (1980).
43. T. Minami, N. Tamai, T. Yamazaki and I. Yamazaki, *J. Phys. Chem.*, **95**, 3988 (1991).
44. V. M. Anisimov, A. M. Vinogradov, E. I. Kuznetsov, I. R. Mardaleishvili, S. A. Senakhov and M. V. Alfimov, *Izv. Akad. Nauk SSSR Ser. Khim.*, **9**, 1783 (1991).
45. Y. Unuma and A. Miyata, *Thin Solid Films*, **179**, 497 (1989).
46. A. Miyata, Y. Unuma and Y. Higashigaki, *Bull. Chem. Soc. Jpn.*, **64**, 1719 (1991).
47. I. Yamazaki, N. Tamai and T. Yamazaki, *Proc. Meml. Conf. Prof. S. Tazuke, 1990, Sapporo, Jpn.*, in *Photochem. Processes in Organized Mol. Syst.*, (North Holland, Amsterdam, 1991).

48. T. Nagamura and Y. Isoda, *J. Chem. Soc. Chem. Commun.*, 72 (1991) and references therein.
49. G. J. Ashwell, E. J. C. Dawnay, A. P. Kuczynski and P. J. Martin, *Proc. SPIE Int. Soc. Opt. Eng.*, **1361**, 589 (1991).
50. A. S. Martin, J. R. Sambles and G. J. Ashwell, *Phys. Rev. Lett.*, **70**, 218 (1993).
51. G. J. Ashwell, J. R. Sambles, A. S. Martin, W. G. Parker and M. Szablewski, *J. Chem. Soc. Commun.*, 1374 (1990).
52. G. J. Ashwell, *Thin Solid Films*, **186**, 155 (1990).
53. G. J. Ashwell, E. J. C. Dawnay, A. P. Kuczynski and M. Szablewski, in *Advanced Organic Solid State Materials*, Eds L. Y. Chiang, P. Chaikin and D. O. Cowan, *Mater. Res. Soc. Symp. Proc.*, **173**, 507 (1990).
54. G. J. Ashwell, E. J. C. Dawnay, A. P. Kuczynski *et al.*, *J. Chem. Soc. Faraday Trans.*, **86**, 1117 (1990).
55. G. J. Ashwell, E. J. C. Dawnay and A. P. Kuczynski, *J. Chem. Soc. Chem. Commun.*, 1355 (1990).
56. M. P. O'Neil, M. P. Niemczyk, W. A. Svec, D. Gosztola, G. L. Gaines III and M. R. Wasielewski, *Science*, **257**, 63 (1992).

7

Physics of Conductive Polymers

A P Monkman

7.1 Introduction

The concept of an organic 1-dimensional 'metal' has great attraction for both the theoretical and experimental physicist, not to mention the industrialist. Think of the humble yoghurt pot. How quickly can you vacuum mould a plastic yoghurt pot? How long would it take a skilled copper beater to make the same from copper? Thus what if you could combine the processibility of a plastic with the electrical properties of a metal? Rational enough. Combine this with the vast amount of physics that the conductive polymers exhibit as well and you start to understand why this area of molecular electronics is such a rapidly expanding field.

7.2 Basic Concepts

One can trace the routes of physicists' interests in molecules containing carbon π bonds, Fig. 7.1, back to the early twentieth century. These early scientists were interested in naturally coloured compounds such as β-carotene, the orange pigment found in carrots. The chemists of the day could also synthesize a number of small analogue molecules containing π bonds. In organic molecules two basic types of carbon bond hybridization sp^2 and sp^3 can be found. In ethane, the four valence electrons of each of the carbon atoms are paired to the valence electrons of the four other atoms it forms bonds to, e.g. three hydrogen atoms and one carbon atom. Thus in this molecule all the valence (bonding) electrons are tightly bound in covalent bonds termed σ bonds. Since these bonds are strong, it requires a large amount of energy to disrupt them or excite an electron into a higher

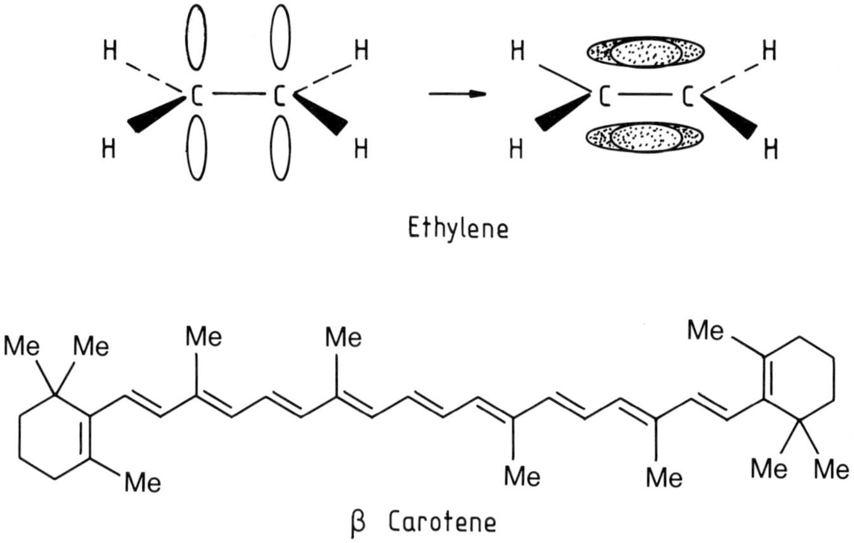

Ethylene

β Carotene

Fig. 7.1 Carbon π bond formation from out of plain carbon p_z orbitals in the simple molecule ethylene. β-carotene, the orange pigment in carrots contains many such π bonds.

energy, non-bonding orbital, thus we consider these bonds to be localized (as well as the electrons forming those bonds). Now consider the molecule ethylene. Here each carbon atom in the molecule forms three σ bonds, two C–H bonds and one C–C bond. This leaves each carbon atom with one unbonded valence electron. To achieve full outer shells, the two carbon atoms form a second covalent bond between them, pairing up each remaining valence electron. To accommodate this second C–C bond, the carbon atoms are forced to alter the orbital structure of their p orbitals. These final two electrons exist in p_z orbitals (sp^2 hybridization) and form a more tenuous carbon–carbon bond where the electrons in the bond are much less tightly bound to the carbon nuclei. This is termed a π bond which requires less energy to perturb an electron from it and can be considered to be a more delocalized entity. Hence in the case of the ethylene molecule there exists a double carbon–carbon bond, e.g. one σ bond, one π bond.

From these basic 'building blocks' large molecules such as β-carotene can be made. Such molecules as β-carotene are said to be conjugated molecules, that is it has a bonding structure of alternating double then single carbon–carbon bonds. In β-carotene we consider the linear chain between the capping benzene rings as consisting of nine repeated alternating double bond single bond units. As the electrons in these π bonds are less tightly bonded to the carbon nuclei the electrons require less energy to be perturbed into excited states, or in solid state physics parlance to excite across the energy gap between the valence band and conduction band. Hence molecules containing a number of conjugated repeat units absorb light in the visible region of the electromagnetic spectrum, i.e. β-carotene

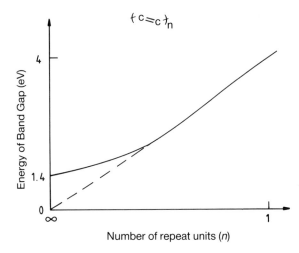

Fig. 7.2 Variation of electronic band gap (energy of absorption maxima) with the number of conjugated repeat units, n, in the molecule. The dash line represents the predicted extrapolation to zero band gap at infinite chain length, before the synthesis of polyacetylene showed that, even as $n \to \infty$, a band gap of 1.4 eV still persists.

absorbs in the blue and thus looks orange, giving orange carrots. It is the interaction with light that attracted the early chemists and physicists, mainly because there was a very large demand for new dyes.

In the early part of this century[1] workers showed that by measuring the energy, E, of the absorption maximum of a range of such conjugated molecules, and plotting this energy against the reciprocal of the number, n, of conjugated repeat units they found a straight line graph, Fig. 7.2. By extrapolating this line it can be seen that $E = 0$ at $n = \infty$. If you interpret this energy E as the band gap energy E_g, between the valance and conduction bands in semiconductors, $E_g = 0$ would imply a metal. Therefore it was surmised that by synthesizing a material with very large n, you would end up with an organic metal. Organic materials with a very large number of repeat units are called polymers hence, conjugated polymers. Thus, the holy grail was to be the synthesis of polyacetylene (PA), the most simple organic polymer known, which is just a polymer of repeat unit CH, see Fig. 7.3.

Many chemists attempted such a synthesis and in 1956 Natta et al.[2] succeeded in producing a black semicrystalline intractable powder, 'polyacetylene' (PA). Limited physics was undertaken on this material, but did show that intrinsically Natta's PA was semiconductive, not metallic. This was ascribed to the poor quality of the polymer. The real break-through in the field came in 1974 when Ito and Shirakawa[3] synthesized films of PA following a modified Natta route. These free-standing films enable physicists to use a vast array of characterization techniques to probe the polymer. However, even this material was found to be intrinsically semiconductive, with an energy (band) gap of 1.4 eV.[4] This situation was

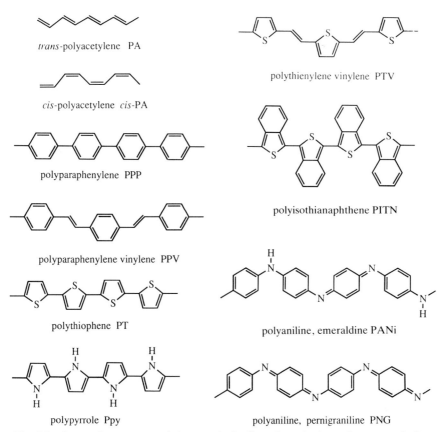

Fig. 7.3 Geometric structures of the generic families of conductive polymers, neglecting carbon–hydrogen bonds.

remedied in 1977 when Chiang *et al.*[5] following up earlier work by Berets and Smith,[6] used both arsenic pentafluoride and iodine to chemically oxidize 'Shirakawa' PA. This resulted in the conductivity of their samples increasing by as much as 11 orders of magnitude, to values comparable with mercury. Leading on from this breakthrough, many other small conjugated molecules were found to polymerize, yielding conjugated polymers which again were intrinsically insulating/semiconductive but become conductive upon chemical oxidation and reduction. Figure 7.3 shows a range of generic conductive polymers, but very many derivatives of these basic parent materials have also been synthesized.

The physics of these materials, since they are not perfect single crystals, but flexible polymer chains which may or may not be ordered, is very different to conventional semiconductors like Si and GaAs. This stems from their organic routes, the nature of the carbon–carbon bonds in the polymer backbone, and the more one-dimensional nature of the individual chains. From this very basic understanding we shall see how the physics of conductive polymers evolves, note its major new concepts and look at ways in which these new phenomena can be harnessed to give usable devices.

7.3   Polyacetylene

We are faced with a most intriguing situation. A polymer which chemically is as simple as it is possible to get, was predicted to be metallic, but was shown to be intrinsically semiconductive. Thus how can we describe its physics? The simplest way to treat PA is quantum chemically using a simple technique to derive molecular orbitals (MO) as a linear combination of atomic orbitals (LCAO), i.e.

$$\Phi_k = \frac{1}{\sqrt{N}} \sum_n e^{ikan} \phi_n \tag{7.1}$$

Here: Φ_k is the molecular wavefunction, a is the carbon atom spacing, k is the wavevector, ϕ_n is the atomic wavefunction, and N is the total number of atomic wavefunctions the summation is made over; n denotes a particular carbon site.

Such a wavefunction Φ_k can then be used in Schrödinger's equation to calculate the energy states of the system. Figure 7.4 shows pictorially what this summation is doing. This is analogous to the tight binding model, and so with increasing wavefunction overlap of the individual π wavefunctions with their neighbours the resulting energy bands will become broader, and in the limit of total overlap at infinite repeat unit number (i.e. the infinite polymer chain) the bands will merge producing a metal, e.g. total delocalization of the π electron cloud over the complete molecule. The solution of the Schrödinger equation with such a wavefunction yields energy levels of the form

$$E(k)^v = -2t \cos ka = -\varepsilon_k$$
$$E(k)^c = 2t \cos ka = \varepsilon_k \tag{7.2}$$

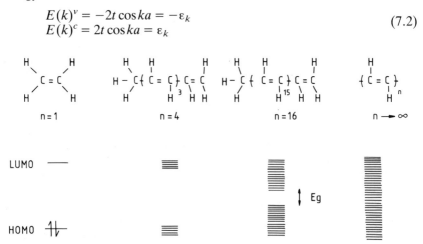

Fig. 7.4   Effect of increasing the number of conjugated repeat units in a molecule on its valence and conduction band structure.

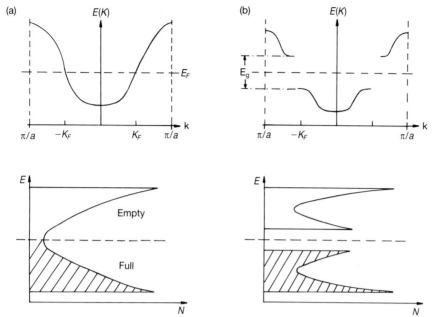

Fig. 7.5 Band structure and band filling for polyacetylene, (a) band structure for equal bond lengths; (b) dimerized case.

Here t is the transfer integral which arises from the nearest neighbour terms in calculating the energy of the system, and is a direct measure of the π wavefunction overlap between neighbouring carbon sites along the backbone, i.e. the degree of delocalization. The valence band is denoted by those states of negative energy and the conduction band by the positive energy states. This band structure along with theoretical band filling are shown in Fig. 7.5a. As each carbon atom donates one π electron per site (carbon atom) but Pauli's exclusion principal will allow two π electrons per site, e.g. one spin up, the other spin down, this energy band will be half filled, e.g. all negative states filled, but as the band structure is continuous the system as a whole is metallic.

This simple picture is wrong, as intrinsically PA is a semiconductor. Therefore one of the basic superstitions made in the calculation is incorrect. It is, in fact, the assumption that single and double bonds along the backbone are of the same length, in reality double bonds are slightly shorter than single bonds. The physical difference in length has been measured by NMR[7] and X-ray diffraction[8] at 0.03–0.04 Å with the average bond length being 1.4 Å. Hence why should such a small bond length difference produce such a marked effect on the electronic properties of the PA, and how is it caused? To answer these questions it is easiest to expand our formalism on PA a little but not straying very far from a tight binding model. This bond alternation was first suggested theoretically by Pople and Walmsley[9] and Salem.[10] Although their work was undertaken using data from short polyenes (a short conjugated chain) it is applicable to PA. However, the theory that best describes the theory of PA is ascribed to three workers, Su, Schrieffer and Heeger[11] and the theory is known as SSH theory. Rice also put forward the same view.[12] In the SSH picture the Hamiltonian of the system is made up of two components

$$H = H_{electronic} + H_{elastic} \tag{7.3}$$

$$H_{electronic} = - \sum_{n,s} t_{n+1,n} (C_{n+1,s}^{\dagger} C_{ns} + C_{ns}^{\dagger} C_{n+1,s}) \tag{7.4}$$

This term describes the π electrons of the PA with again $t_{n+1,n}$ being the transfer integral of π wavefunction overlap between adjacent carbon sites $n+1$ and n, and the Cs are creation and annihilation operators describing either the appearance or disappearance of π electrons of spin s $(\pm\frac{1}{2})$ at site n. This is a way of describing a π electron hopping from site $n+1$ to n, etc.

$$H_{elastic} = \frac{1}{2} \sum k (U_{n+1} - U_n)^2 \tag{7.5}$$

This describes the σ electrons between adjacent carbon sites (in the static case). In effect all this says is that the σ electrons act as springs, of spring constant k between neighbouring carbon sites, they play no part in the electronic properties of the PA (directly). The Us describe the displacement of carbon sites from their uniform position, see Fig. 7.6. Note, however, that this term does impinge on the electronic properties of the system indirectly. If two adjacent carbon sites are brought closer together the π wavefunction overlap between them increases, similarly if the sites are pulled apart t must decrease. This behaviour is shown schematically in Fig. 7.6, and a general linear form for t can be derived, as shown on the graph, since the displacements U are small,

$$t_{n+1,n} = t_0 - \alpha (U_{n+1} - U_n) \tag{7.6}$$

where t_0 is the transfer integral between two uniformly spaced carbon sites, and α the electron phonon coupling constant.

If we first take this SSH Hamiltonian and set all bond lengths equal and assume that the eigenstates of the system $|n\rangle$ can be expanded into atomic-like orbitals $|\phi\rangle$, i.e. a LCAO approach, such that

$$|n\rangle = \sum_n a_k(n) |\phi\rangle \tag{7.7}$$

We can solve Schrödinger's equation again to yield the eigenstates and dispersion relationship of the PA chain with U constant, we again find that

$$E(k) = E_0 - 2t \cos ka \tag{7.8}$$

thus the SSH is consistent with the previous model. Now we set the condition of bond alternation, or more explicitly we investigate chain dimerization, that is the chain repeat unit increases from a to $2a$, with the bond length criterion

$$U_n = (-1)^n U = U e^{\pm 2ink_F a} \tag{7.9}$$

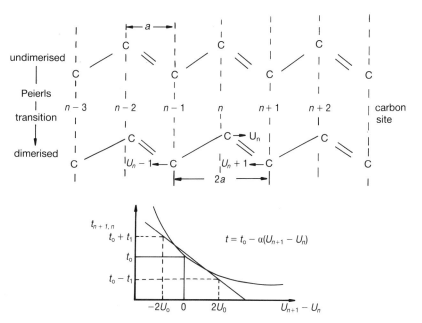

Fig. 7.6 Peierls' transition applied to polyacetylene and the effect that this dimerization has on π wavefunction overlap. The generalized form of t, the transfer integral, can be derived for the dimerized case as a function of carbon atom displacement U_n.

here k_F is the Fermi wavevector, i.e. the wavevector of the electrons at the Fermi level,

$$k_F = \pi/2a \qquad (7.10)$$

The Fermi level describes the occupied state of the system with highest energy, and the Fermi wavevector is that state's momentum.

Solving for the PA chain in this dimerized case yields a more complex dispersion relationship

$$E^2(k) = 4t_0^2 \cos^2 ka + \Delta^2 \sin^2 ka \qquad (7.11)$$

with the gap parameter $\Delta = 4\alpha U$.

This dispersion relationship is plotted schematically in Fig. 7.5b. The gap parameter $\Delta = 4\alpha U = 2t_0$ is a measure of the size of the energy gap, with $4t_0 = W$ being the total width of the π band, i.e. the energy differences between the highest unoccupied molecular orbital (LUMO) and the lowest occupied orbital (HOMO). For PA, $W \approx 8$–$10\,\mathrm{eV}$.

From the dispersion relationship we can see that with dimerization comes the opening of a band gap. As the system will be half filled, all the negative states (valence band) are full and the positive states (conduction band) are empty, thus the PA will be intrinsically semiconducting with a band gap found experimentally to be $1.4\,\mathrm{eV}$.[4] The density of states of the system $N(E)$ which is a measure of the number of states per unit energy of the band can be found from the tight binding model with

$$N(E) = \frac{N/\pi}{(4t^2 - (E - E_0)^2)^{1/2}} \tag{7.12}$$

This band filling is depicted in Fig. 7.5 for both the undimerized and dimerized cases. From these diagrams we can begin to see where the driving force for such a dimerization comes from. The transition was in fact described many years before for such 1-dimensional systems by Peierls.[13] Peierls' theory states, 'A 1-dimensional metal is unstable to a distortion that opens up a gap at the Fermi level, since the total energy of the occupied band states is reduced by the presence of this gap'. This can clearly be seen in Fig. 7.5, with the Fermi level of the dimerized system being at lower energy than that of the undimerized case. Thus, as the system reduces its energy by distorting and as long as the reduction energy is greater than that elastic energy required to produce the distortion, the system will dimerize. The energy 'gain' is related to the magnitude of the displacement from uniform lattice position, U_0, as in the case of PA,

$$E_{gain} = -U^2 \ln(t_0/\alpha U_0) \tag{7.13}$$

the elastic energy required to produce the distortion will be proportional to U^2 for small U_0, i.e. Hooke's law. Thus the total energy gained per carbon atom for the system will be

$$E(U) = 2K_0 U^2 - U^2 \ln(t_0/\alpha U_0) \tag{7.14}$$

In the weak coupling limit, this function can be analytically minimized[14] to yield

$$E_g = 16t_0 \exp[-(1 + \tfrac{1}{2}\Gamma)] \tag{7.15}$$

where

$$\Gamma = 2\alpha^2/\pi t_0 K_0 \tag{7.16}$$

Γ being the dimensionless electron-phonon coupling constant.

It should be pointed out that other theories have been put forward which do not implicitly require bond alternation to yield an energy gap. Ovchinnikov et al.,[15] working with the tight binding model but including electron-phonon coupling and electron-electron interactions, found that a band gap forms in their dispersion relationship. Similarly, Ashkenazi et al.[16] have shown that on-bond Coulomb correlation and 3-D interchain coupling lead to the formation of a band gap even in the case where $U = 0$, i.e. no bond alternation. This can be pictured in a simple way. When a π electron hops to an adjacent site it will form a carbon atom containing two π electrons. There must be some energy required to overcome the Coulomb repulsion between these two π electrons. This energy is equivalent to the band gap. Hence, to hop, the π electrons require kinetic energy to overcome the Coulomb repulsion, i.e. an energy barrier (energy gap). Thus, we can now understand the intrinsic semiconducting nature of PA and conjugation. Next we shall explore the subtleties of this system and the exotic physics such a theoretical model gives us.

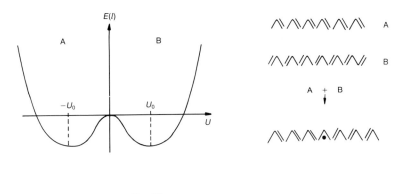

[ECl energy per unit length]

Fig. 7.7 Geometric potential energy minima for the two phases of polyacetylene, and how these two phases can combine to generate a bond alteration defect or soliton.

7.4 Solitons

In the previous section, we came to the general conclusion that PA must exist as a dimerized lattice, and that there exists a band gap between the valence and conduction bands. Now consider in more depth the expression of the total energy per carbon atom. If we plot this site energy versus U, the site displacement, we find that at $U = 0$, the undimerized case, the site energy goes through a local maximum,[17] Fig. 7.7. This again shows that the PA system is unstable with respect to the Peierls' transition. Further we see that two energy minima occur at $\pm U_0$,

$$U_0 = 4 \exp[-(1 + \tfrac{1}{2}\Gamma)] \tag{7.17}$$

We can reconcile these two minima quite easily as two phases of PA, A and B, see Fig. 7.7. Both phases are identical in terms of site energy, but phase A can be considered as having its double bonds pointing to the right whereas phase B has them pointing to the left. These two phases are identical and thus are degenerate ground states. The difference between these two phases becomes apparent when you join a piece of chain A with a piece of chain B. In this instance we end up with a bond mismatch, e.g. a carbon site with one too few π electrons so that it cannot form a double bond. This defect results in one unpaired π electron, but as a whole the system is neutral, the same number of protons as electrons. For all cases where a PA chain contains an odd number of carbon atoms, one of these bond mismatches must occur. Further, as the system has one unpaired electron it will have spin $\pm\tfrac{1}{2}$, i.e. a charge neutral, spin $\tfrac{1}{2}$ entity. These topological defects are called solitons, due in part to their nondispersive nature and the mathematical formalism from which they derive. The nondispersive wave or soliton is used throughout all of physics as it is a solution to many nonlinear differential equations; it is also an easy way to establish a close connection between pure mathematics and real physics.[18]

An everyday manifestation of the solitary wave or soliton is the bow wave of a boat, and this was in fact the source of their discovery in 1834 by Scott-Russel.[19] Scott-Russel noticed that when barges stopped in the Caledonian canal, their bow wave carried on for great distances 'without change of form or diminution of speed'.

The soliton in PA can be considered as an excitation from one geometric energy minimum (phase A) to the other, (phase B) see Fig. 7.7, the mediating pseudo particle for this excitation being the soliton. When the soliton moves up or down the chain it will convert phase A to B and vice versa at little energy cost. Thus what happens to the unpaired π electron in the soliton? How big is the soliton spatially and how does the chain perturbation affect the electronic structure of the PA?

Assuming the case of an infinite PA chain (i.e. ignoring effects due to chain ends), the domain wall separating phase A from phase B could occur at site n along the chain. However, such an abrupt discontinuity would have a high kinetic energy due to the uncertainty principal. Thus we must assume that the change over occurs over several carbon atom sites and hence will have some finite length. Rice[20] proposed that the wavefunction of such a domain wall or soliton should be of the form

$$\phi_n = U_0 \tanh\left(\frac{na}{\xi}\right) \tag{7.18}$$

with ξ being a fitting parameter governing the length (in carbon sites) of the soliton. Su et al.[21] investigated the soliton numerically and calculated that in the static limit the lattice (chain) distortion is given by

$$U_n = U_0 \tanh[(n - n_0)a/\xi] \cos n\pi \tag{7.19}$$

n_0 being the centre of the soliton with the zero energy bound state for the electron being described by a wavefunction

$$\phi_0(n) = a/\xi \operatorname{sech}[(n - n_0)a/\xi] \cos\frac{n\pi}{2} \tag{7.20}$$

Minimizing these equations yields the width of the soliton and its lowest energy level.[22] It is found that

$$\xi = \frac{aW}{\Delta} \approx 7a \tag{7.21}$$

Here W is the π electron band width and Δ is the gap parameter with the creation energy of the soliton state being $2\Delta/\pi$. Hence the soliton is more stable than the creation of an electron in the conduction band and a hole in the valence band, with the width of the soliton being 14 carbon atoms. As the SSH Hamiltonian possesses charge conjugation symmetry, i.e. changes sign under the transformation

$$C_n \rightarrow (-1)^n C_n \tag{7.22}$$

such that the energy spectrum of the system is symmetric about zero, the wavefunction of the soliton is made up of half a conduction band orbital

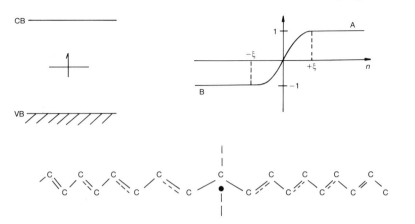

Fig. 7.8 Schematic representation of neutral soliton showing the associated distorted chain structure, width parameter and band structure of the soliton.

and half a valence band orbital, and lies symmetrically in the middle of the band gap, see Fig. 7.8.

We can see that this behaviour is very different from that observed in inorganic semiconductors. Because of the 'springy' nature of the carbon–carbon σ bonds, the excess, unbound electron within the soliton is localized due to the distortion of the polymer chain (lattice). The system changes to produce a new localized energy level for the unbound π electron at mid gap. One should also note that overall the system is charge neutral, but has spin $\pm\frac{1}{2}$, i.e. spin charge reversal.

This last point is of great significance and the reader should consider it in context with all other systems in solid state physics. For solitons the total charge and spin of the system is given by:

$$Q = e(N_+ + N_- - 1) \quad s = \tfrac{1}{2}(N_+ - N_-) \tag{7.23}$$

here $N_\pm$ is the occupation number, hence $Q = 0$ and $s = \frac{1}{2}$ for the neutral soliton and $Q = \pm 1$, $s = 0$ for an empty or doubly occupied soliton ($N_+ = N_- = 0, 1$). The latter will arise as a consequence of doping which we will turn to shortly. Solitons in more realistic systems, i.e. finite PA chains, etc., will be formed in soliton anti-soliton (S_0, S_0) pairs. Where these are separated by large numbers of carbon sites they will be non-interacting. Knowing this one can understand how a neutral soliton can have spin $\pm\frac{1}{2}$, as its anti-particle will be of $s\mp\frac{1}{2}$ and so in total spin is conserved. For further details on solitons, and their physics the reader may like to consult the excellent article by W. P. Su.[17]

7.5 Doping

As noted in the introduction, real interest in the conductive polymers arose when it was found that they could be made to conduct an electrical current after they had been doped.[5] Doping here is a misnomer as it tends to imply the use of minute quantities, e.g. parts per million or less, of impurities introduced into a crystal lattice. This could not be further from the truth. In the case of the conductive polymers, typically 1% to 50% by weight of chemically oxidizing (electron withdrawing) or reducing (electron donating) agents are used to physically alter the number of π electrons on the polymer backbone, leaving an oppositely charged counter ion alongside the polymer chain. These processes are true redox chemistry. This can be achieved because a π electron can be removed (added) without destroying the σ backbone of the polymer so that the charged polymer remains intact. This doping has a marked effect on the physical properties of the conductive polymer, usually increasing the conductivity by as much as 11 orders of magnitude, even approaching that of copper in some cases,[23] see Fig. 7.9. Therefore, what happens when the electronic structure is perturbed in such a fashion: where do the holes (electrons) go to, and what role do soliton defects play?

First, there are many ways to achieve doping in conductive polymers. The simplest is to expose the polymer to a vapour of an oxidizing (or reducing) agent. Typically, to oxidize the polymer, halogens are used, e.g. Cl, Br and I. This will form a polymer salt, e.g. $(CH^+(I_3^-)_{0.1})_x$ for 10% iodine doping of PA. Other common oxidants are $FeCl_3$ and AsF_5. Reductive doping is most commonly achieved using the alkali metals Li, Na and K. This is achieved using electrochemical reduction.[5] Electrochemistry is a very powerful way of doping conductive polymers and simul-

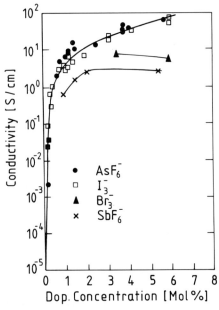

Fig. 7.9   Effect of oxidixing agents on the conductivity of polyacetylene, after Roth.[56]

taneously allows the redox chemistry of doping to be measured. Further, many of the conductive polymers can be synthesized electrochemically. For further reading on this topic, see the review of Diaz and Bargon.[24] There are two further ways to introduce charge carriers onto the polymer chain, namely photoinduced charge generation, i.e. optically exciting carriers from the valence band into the conduction band, and electron injection via the field effect within a FET structure. Both of these more esoteric methods will be discussed later.

7.6 Polarons, Bipolarons and Charged Solitons

During chemical oxidation or reduction of a conjugated polymer chain, electrons are removed (*p*-type doping) or added (*n*-type doping). We have already seen that in the case of PA, altering the electronic structure of the backbone causes charges in the carbon–carbon bond lengths at the charged site. This also occurs in the nondegenerate ground state polymers. However, in these materials, solitons are not stable, so when an electron is removed or added, the resultant change in bond lengths must be such that the phase of the polymer chain remains identical on either side of the defect. This means that bond length alternation is greatest in the middle of the defect. Such a species is termed a polaron if it is singularly charged or a bipolaron if doubly charged. In effect a polaron can be thought of as a charged and a neutral soliton bound pair.

In terms more familiar to the chemist, a polaron is simply a radical-ion associated with a lattice distortion. Thus, it is not necessary for the chain to be degenerate to support polarons and bipolarons, so that the charge carriers in all of the conjugated polymers, including polyacetylene, could take this form. The theoretical description of both the polaron and bipolaron can be made from an extension of SSH theory, describing them in terms of confined soliton pairs, but is beyond the scope of this chapter. However, if we consider those polymers containing benzene rings, using polyparaphenylene, PPP, as a model system, we can discuss qualitatively the consequences of such polaronic charge carriers. Further, for simplicity we shall consider *p*-type doping, i.e. removal of electrons.

In small, ring-containing organic molecules, such as biphenyl,[25] it is found that upon oxidation the rings change conformation markedly, from benzenoid to quinoid, see Fig. 7.10, i.e. changes in bond length occur because of a change in the electronic structure of the molecule.

In the case of the polymer, i.e. PPP, similar processes occur upon *p*-type doping. Again such localized species result, as opposed to the hole produced being free to move throughout the crystal lattice as holes in silicon do. This is because of the inextricable link between the electronic and geometric structure, i.e. strong electron-phonon coupling. When an electron is removed, the C–C bonds around the hole site lengthen, forming new electronic levels which localize the hole, yielding a charge, trapped at a localized chain distortion. The combination of this chain distortion and trapped hole (or electron in the case of *n*-type doping) is the polaron. In

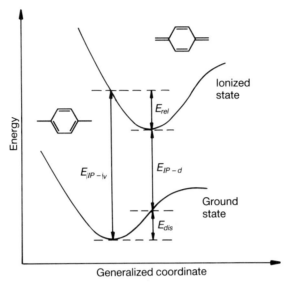

Fig. 7.10 Geometric and ionization energies of a typical organic molecule, after Baughman et al.[25]

Fig. 7.10 this is depicted schematically, exploiting the simple but elegant Frank–Condon principle of vertical electronic excitations in which the electron's momentum does not change during an excitation. The key to understanding this localization process is to remember that organic polymer chains are flexible. In the ionized state, the molecular structure changes, i.e. going from benzenoid to quinoid. This can be thought of as a shift of the centre of mass of the molecule or shift in molecular co-ordinate.

Thus from Fig. 7.10 it can be seen that the ionization energy, E_{IP-v}, of the ground state is much greater than the ionization energy, E_{IP-d}, of the molecule if it were in the distorted quinoid structure. It will take an amount of energy E_{dis} to distort the molecular structure from benzenoid to quinoid. If this occurs we will gain energy back because of the relaxation to the minimum of the ionized state, e.g. E_{rel}. Hence it is easy to see that if the energy gained E_{rel} by such a geometric distortion is greater than the energy required to drive that distortion the distorted state will be the preferred state, with the charge being localized at the distortion.

Defining the energy gained as:

$$\Delta\varepsilon = E_{IP-v} - E_{IP-d} \tag{7.24}$$

it can be seen that the overall LUMO will be lowered by $\Delta\varepsilon$ and the HOMO raised by $\Delta\varepsilon$ with respect to the Fermi level locally, and that the new charge resides in these new levels. The binding energy of the polaron is also given by:

$$\Delta\varepsilon^{pol} = \Delta\varepsilon - E_{dis} (= E_{rel}) \tag{7.25}$$

for PA, $\Delta\varepsilon^{pol} = 0.05\,\text{eV}$ and for PPP, $0.03\,\text{eV}$.

In the case of further oxidation or reduction the question arises: does the second charge form its own defect, or does it pair off with the first and so dig a bigger distortion for the pair? If the bipolaron is more stable than forming two polarons, the distortion created by the pair of charges must be large enough to overcome the Coulomb repulsion between the closely spaced like charges. Some of this Coulomb repulsion will be screened by the presence of the two counter ions, and thermodynamically a bipolaron should be more stable than two polarons.

In the special case of *trans* PA, with its degenerate ground state, the two charges forming a would-be bipolaron find it very easy to separate. Since there is no energy difference between phase A and B of the PA chain, no increase in distortion energy results from splitting the bipolaron, i.e. the chain structure produced between the two separated charges has the same geometric structures as that on either side of the charges. Thus the bipolaron will quite happily split to form two charged solitons. As an exercise, readers should prove to themselves that charge solitons are always produced in pairs.

Since the formation of these charged defects leads to new localized electronic energy levels being introduced into the gap, they will have very characteristic optical signatures, and hence optical absorption is an ideal tool to study these charged species. Figure 7.11 shows the possible optical transitions in the case of n and p-type polarons and bipolarons. In both cases polarons should exhibit three optical transitions below the gap, although ω_3 will be very weak, and for bipolarons, two new transitions should arise. Note, for positive bipolarons, there are no occupied mid-gap states so the ω_3 transition is not possible, and in the negative bipolaron case, both mid-gap levels are doubly occupied, i.e. full, hence Pauli's exclusion principle rules out the ω_3 transition. For polarons, it can be seen

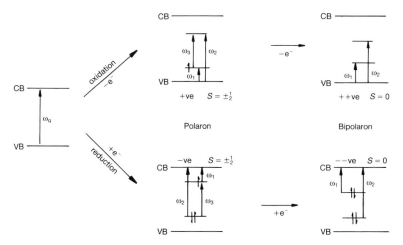

Fig. 7.11 Energy level diagrams with associated allowed optical transitions for polarons and bipolarons.

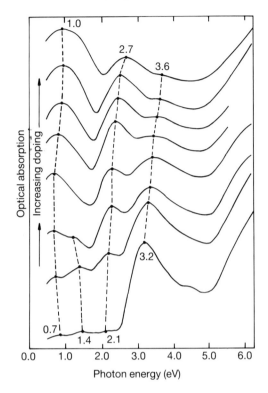

Fig. 7.12 Optical absorption of polypyrrole as a function of oxidation level, showing first polaron formation followed by bipolaron coalescence at high doping levels, after Yakushi et al.[57]

that a simple addition rule for the energies of the optical transitions holds

$$\omega_2 = \omega_1 + \omega_3 \qquad (7.26)$$

Such behaviour has been seen in non-degenerate ground state polymers such as polypyrrole, Fig. 7.12. In this case, it is believed that at low doping concentrations polarons are the stable defect but as the doping level is increased the polarons are forced closer together such that they interact, leading to stabilization of bipolarons, hence the spectra progress from three new transitions below the band gap to two, i.e. polarons going over to bipolarons. The reader is reminded that the two types of carrier, polaron and bipolaron, have different spin signatures. Thus to identify the carrier type fully electron spin resonance (ESR) measurements should also be made.

In the case of polyacetylene, the nature of the charge carriers as a function of doping level is still a matter of academic debate. At levels below about 1% doping, most theoreticians agree that charged solitons are the stable carrier, as evident by ESR data.[26] Between 1 and 7%, the ESR signal substantially decreases, leading to the concept of weakly interacting polarons. These weak interactions cause a splitting of the polarons to form

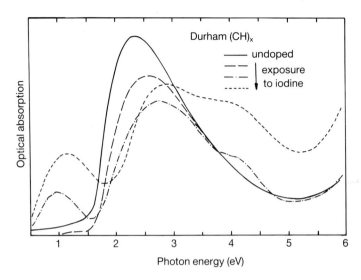

Fig. 7.13 Optical absorption of polyacetylene as a function of doping level. As doping increases the doping-induced mid-gap soliton state forms below the band edge. After Friend *et al.*[58]

charged solitons, which themselves form a new energy band in the band gap Fig. 7.13. Such charged solitons will be spinless, accounting for high conductivity with no associated Pauli susceptibility (free spins as in a metal). Above 7% doping level there are two main schools of thought. Stafstrom[27] postulates that the soliton band has grown wide enough to merge with the valence band, thus allowing electrons from the valence band to take over the role of charge carrier. This yields high conductivity with an associated (and measured) Pauli susceptibility. The electrons move from soliton site to soliton site, in effect the solitons become fixed defects, or a regular soliton 'lattice' with the electrons moving through this soliton lattice. Conversely, Drecksler *et al.*[28] suggest that instead of forming soliton lattice, a polaron lattice forms, again yielding the correct spin charge correlation measured in experiment.

7.7   Chain Alignment

Chemical doping, especially in the case of PA, can raise the electrical conductivity of the polymer up to quite high values, e.g. $1 \times 10^3 \, \text{Scm}^{-1}$. However, the conductivity of films can be enhanced further if the individual polymer chains in the film are aligned. This can be achieved in those polymers which are processible, e.g. PA, PTV and PPV (which are synthesized via a precursor route in which the precursor polymer is soluble[29]), and PANi where the 'conjugated' polymer is soluble in select solvents.[30] In these cases, the polymer films are subjected to uniaxial stress, causing the polymer chains to disentangle to form well ordered, aligned polymer crystallites. When such ordered films are doped, a further one or two orders of magnitude increase in conductivity can be produced. To date the highest recorded value for the conductivity of a PA sample is $>1 \times 10^5$ Scm^{-1}.[31] This should be compared to the room temperature conductivity of copper, 6.5×10^5 Scm^{-1}.

The above phenomenon can be understood simply. An amorphous polymer is like a plate of spaghetti. To measure bulk conductivity, two electrical contacts are made to the sample (plate of spaghetti). The charge carrier has to find its way across this network to travel from one electrode to the other. No single polymer chain will extend completely across, and make contact to both electrodes. Hence carriers have to make many jumps between polymer chains. As in the spaghetti, there are many points throughout the sample where chains come into close contact, but the distance over which chains closely approach each other is very short. This makes it very unlikely that carriers can cross from one chain to the other, limiting the macroscopic conductance of the polymer. In the aligned case the overlap distance of neighbouring chains is large. Therefore, it is much easier for carriers to flow from one chain to the next. Even though no one chain will stretch across the gap between the electrodes, with aligned chains the macroscopic conductance increases substantially, due to the greatly enhanced interchain overlap.

7.8   Photoinduced Absorption

This is one of the most powerful tools of the spectroscopist, allowing carriers to be created on chains without recourse to oxidative chemistry and its unavoidable consequence of counter ion insertion. The photoinduced carriers can then be studied in their intrinsic environment, without having to worry about the effects which charged counter ions adjacent to the polymer chain will have on them, e.g. counter ion pinning and chain deformation.[32] The complete topic of photoinduced spectroscopy requires a book on its own. The interested reader is pointed to the review article by Orenstein.[33]

As its name implies, photoinduced (PI) absorption is the induced optical absorption of light caused by photoexcitation of the sample. This can take the form either of increased absorption, i.e. the photocreation of chromophores, or bleaching, i.e. the loss of chromophores due to the photoexcitation. In practice, the sample is irradiated with photons (usually by a laser) which have energy greater than the band gap of the polymer. One then probes the optically excited polymer with white light in a similar fashion to the measurement of a normal absorption spectrum. Most experiments take the form of mechanically chopping the pump beam (excitation light source) and measuring the inphase signal from the probe beam using a lock-in amplifier.[34] This signal represents the induced absorption which is then normalized to the samples linear absorption to yield the final photoinduced absorption spectrum, Fig. 7.14.

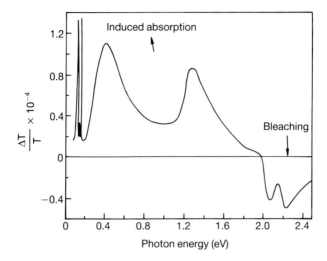

Fig. 7.14 Photoinduced spectrum of polythiophene, bandgap bleaching is seen above 2.0 eV. New induced features at 1.3 eV and 0.5 eV are ascribed to bipolaron formation. The sharp induced features in the IR are pinning modes showing that the bipolarons possess charge. After Vardeny et al.[37]

In the case of PA one observes that, upon photoexcitation across the band gap, a new optical absorption is created lying below the band edge. This new absorption is centred at 0.6 eV, i.e. close to mid gap, and so must be a signature of solitons. The sharp bands also seen in the IR are due to the soliton being charged, and this charge enhances the oscillator strength of the vibrational modes of the C–C bonds associated with the soliton chain distortion.[35] Thus, we can see that carriers generated by photoexcitation (even though given enough energy to cross over to the conduction band) relax back from the conduction band to form new mid-gap states or solitons. Hence, the self-localization phenomenon which gives rise to soliton formation is intrinsic and does not depend on the effects of counter ion (charge) pinning etc. By studying the time evolution of such photoinduced absorption using femtosecond lasers,[36] the formation and dynamics of such carriers can be observed. From the photobleaching signal it is also possible to determine from where the carriers came to produce the new mid-gap states. In PA, these can be seen to be from the valence band as optical bleaching occurs at the 1.4 eV absorption, i.e. the band gap transition.

In the case of polythiophene, two new induced absorptions are observed, at 0.6 eV and 0.3 eV. These are ascribed to the photogeneration of bipolarons,[37] showing that even in nondegenerate ground state systems carrier localization is an intrinsic mechanism. However, as with all the experimental results discussed here, the reader should be aware that these organic polymers are by no means perfect. Samples must include many chain ends, defects and chain mislinkages. These are in practice very difficult to measure and quantify[38] but could still be the major cause of carrier localization. Recent results by Heun et al.[39] may be pointing us in this direction, and so the field of conductive polymers may well be travelling down the same path as molecular crystals. Readers wishing to know more on these systems should consult Pope and Svenberg's book on the subject.[40]

7.9   Polyaniline

From Fig. 7.3 it can be seen that those polymers containing heteroatoms (sulphur and nitrogen) such as polythiophene and polypyrrole are really analogues of cis-PA with pendant groups containing the heteroatoms hanging from the cis PA backbone. However, polyaniline, PANi, can be seen as the only polymer where the heteroatom lies in the conjugation path along the polymer backbone. This leads to many physical consequences. First, the polymer can exist in three oxidation states: fully reduced leucoemeraldine which has no quinoidimine units, oxidized emeraldine and doubly (or fully) oxidized pernigraniline. In all three cases the polymer is insulating and it can be seen that only the pernigraniline form is strictly conjugated. However, the emeraldine form has the unique property that upon protonation of the imine sites, i.e. placing the emeraldine base into an acidic solution of pH < 3–4, the conductivity increases by up to 11 orders of magnitude. Values as high as 350 Scm^{-1} have been reported.[30]

The emeraldine form of PANi is also unique in that both the base form[41] and the salt (conductive) form[42] are soluble and can be processed. Further such films can be aligned by uniaxial stress to yield films with electrical anisotropies of up to 25:1.[43] Such processible, stable material has many applications, from conductive composites to electromagnetic radiation shielding in cables.

The charge carriers in PANi, and the role of lattice distortions, are more complex than the other nondegenerate ground state polymers. First, only protons are required to 'dope' emeraldine. Addition of a proton to an imine nitrogen causes the number of π electrons on the backbone to remain constant; however, the conductivity is seen to increase by many orders of magnitude. Furthermore, thermoelectric power measurements[42] reveal that, even in the highly conductive state, the charge carriers in the films have negative charge, i.e. electrons not holes. This seems strange as, upon protonation, holes are added to the backbone. This has been resolved with the idea that protonation removes one of the imine nitrogens lone pair electrons to form the N–H bond the remaining unpaired electron can then hop between the vacancies left at these sites, giving rise to negative charge carriers.[43] Protonation also cause the imine rings to change their geometry. This not only alters bond lengths but causes the rings to rotate to a more planar configuration, increasing the wave function overlap to yield a highly conductive state. This new degree of freedom, i.e. ring rotations, plays a dominant role on the physical properties of the polyaniline family. Thus, the emeraldine form displays thermochromic properties attributed to rotation of the rings,[46] and the fully oxidized form, pernigraniline, has been shown to exhibit photoinduced absorption which lasts for months.[47] The latter phenomenon has been attributed to ring rotations which cause self-localization of the photoinduced species. The reader may note that pernigraniline is a degenerate ground state polymer and hence joins polyacetylene in this unique class of materials. Pernigraniline should support soliton excitations, but these will be massive compared to those in the case of polyacetylene.[48]

Polyaniline (or the family of aniline polymers) is thus an extremely complex system. The nature of charge carriers, their dynamics and photogeneration mechanisms are still subjects of debate. Much work is presently being undertaken by many groups, and so to appreciate PANi and its physics fully the reader should refer to the current literature on the subject.

7.10   Device Applications

As mentioned earlier, charge carriers can be directly injected into PA, PPV and PTV using the field effect, as is exploited in a field effect transistor (FET). As these polymers can be solution processed (in their precursor form), thin films of conductive polymer, typically 0.1 μm thick, can easily be spun onto device substrates using analogous processes used in silicon chip fabrication. This new and potentially very exciting area of the field arose from the pioneering work of Burroughs et al.[49] in the Cavendish Laboratories. This group discovered that PA could be incorporated as the semiconducting layer in a metal-insulator–semiconductor (MIS) structure, Fig. 7.15. The 'device' can be characterized in an identical fashion to silicon MISFETs. The reader is referred to Sze[50] for the background to these devices. The group also found that the PA layer device could be switched between accumulation and depletion and could even be driven into inversion, i.e. both positive and negative carriers could be supported within the PA.

To investigate the nature of the carriers in these device structures, experiments analogous to photoinduced absorption (PI) can be performed. Whereas in PI experiments, a laser is used to generate carriers, in the FET devices the electric field generates the carriers but as in PI, these charges can be observed using differential absorption. By fabricating ultra thin gate devices, it is possible to measure the optical transmission of the PA layer below the silicon band gap. By doing this the Cavendish group determined that upon switching on the electric field, a new optical absorption is seen at 0.6 eV, e.g. around mid gap for PA, suggesting soliton states are being formed. Further, this field induced optical absorption was found to correlate extremely well to the total amount of charge injected into the PA layer found from differential capacitance measurements. Thus the group concluded that the charge carriers injected by the electric field into the PA semiconductor layer were charged solitons as opposed to electrons or holes. Because of this, one can see that these structures could give rise to

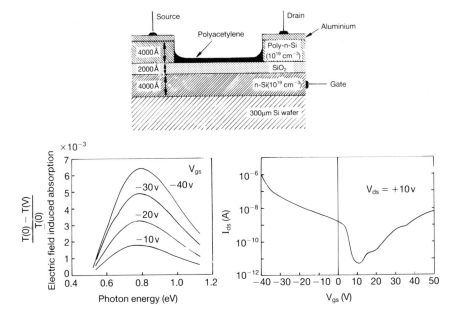

Fig. 7.15 Field effect transistor employing a thin polyacetylene layer as the semiconductor. Device characteristic, including induced absorption are also shown. After Burroughs et al.[52]

the possibility of producing electrooptic switches, i.e. the absorption can be controlled by an applied electric field. This at present remains simply an idea as the switching speed achievable is rather slow, the quantum efficiency of the device is small and the PA is too unstable.

From this work, many other groups have followed on. Horowitz et al.[51] have worked on the FET idea, trying to manufacture simple, cheap organic FETs to rival amorphous silicon devices. Instead of using PA, they started working with polythiophene (PT). This they found to be of little use, as both the carrier density and mobility were far too low to make useful devices (remember that the PT has to be in the semiconductive state to be useful as a field effect material). However, from this work on the polymeric form of thiophene they discovered that processible oligomers of thiophene (molecules contain from 3 to 8 repeat units) do make good FETs, especially sexithienyl which demonstrates a high field effect mobility. What is more, the FETs are almost entirely organic (except the metal contact layers and indium tin oxide (ITO) electrode) so are very easy to fabricate with no high temperature processing required. They are also flexible.

Following on from their initial investigations of PA FETs, Burroughs et al.[52] also discovered that good quality polymers, especially those made from a precursor route polymer synthesis, which fluoresce, when made into simple charge injection devices exhibit electroluminescence. This phenomenon is especially strong in the PPVs and PTVs, and arises from the polymers' ability to support both positive and negative polarons (or excitons). When electrons are injected into a polymer film from one side, and holes on the other, the two types of carrier can migrate through the polymer, meet and annihilate, leading to emission of light. At the time of writing this chapter a great deal of work is under way to perfect these light emitting diodes (LEDs). One great advantage that such polymeric LEDs offer over InP or GaAs devices, is that the band gap of the semiconductor, i.e. polymer, can be tuned via organic chemistry, thus yielding the potential of any colour LED being possible. Again, manufacture of these polymeric LEDs is cheap, and virtually the entire device can be made from polymeric materials. To gain an idea of the flexibility of polymeric LEDs, see the article by Gustaffsson et al.[53] At present PPV LEDs show very high efficiencies and look set to break through into the market place in the next few years.

Further uses of conductive polymers, more directly linked to their conductive nature, arise in the area of sensing. As the conductivity of all these polymers is directly controlled by doping, i.e. oxidation or reduction, the measured conductivity (or conversely resistance) will change in the presence of oxidizing or reducing ambients. Hence they can be used very effectively as sensing elements. In the area of biosensors, Fouldes and Lowe[54] have shown that polypyrrole (Ppy) is an effective sensing element in the measurement of glucose. When glucose is metabolized by glucose oxidase, one of the by-products is hydrogen peroxide. This will readily oxidize Ppy, changing its resistivity which can be measured and directly related back to the initial glucose concentration. In a similar fashion PANi can be used to measure gases such as NO_x, SO_2 and H_2S.[55] Here, the gas directly interacts with the polymer changing its resistance. Such polymeric sensors can measure down to very low levels of gas (down to about 1 ppm) and are reversible at ambient temperature.

7.11 Conclusions

Conductive polymers form a unique niche in materials science, chemistry and physics. They give the scientist a unique range of properties to explore and a number of ways to investigate them, be they synthetic chemistry, spectroscopy, polymer processing or solid state physics. In conductive polymers, electronic properties are controlled by chain geometry which in turn is intrinsically linked to the electronic energy levels. These properties are summed up well in the π orbital overlap, i.e. the transfer integral t. By studying what happens to t, much of the ensuing physics of the system can be understood.

References

1. H. Kuhn, *J. Chem. Phys.*, **16**, 840 (1948).
2. G. Natta, G. Mazzanti and P. Corrandini, *Att. Acad. Naz. Lincei. A. Sci. Fis. Mat. Nat. Rend.*, **25**, 2 (1958).
3. T. Ito, H. Shirakawa and S. Ikeda, *J. Polym. Sci. Chem.*, **12**, 11 (1974).
4. H. Shirakawa, T. Ito and S. Ikeda, *Polym. J.*, **4**, 460 (1973).
5. C. K. Chiang, C. Z. Fincher, Y. W. Park *et al.*, *Phys. Rev. Lett.*, **39C**, 1098 (1977).
6. D. J. Berets and D. S. Smith, *Trans. Faraday Soc.*, **64**, 823 (1968).
7. C. Fincher, C. E. Chen, A. J. Heeger, A. G. MacDiarmid and J. B. Hastings, *Phys. Rev. Lett.*, **48**, 100 (1982).
8. C. S. Vannoni and T. C. Clarke, *Phys. Rev. Lett.*, **51**, 1191 (1983).
9. J. A. Pople and S. H. Walmsley, *Mol. Phys.*, **5**, 15 (1962).
10. L. Salem, *Molecular Orbital Theory of Conjugated Systems*, (Benjamin, New York, 1966).
11. W. P. Su, J. R. Schrieffer and A. J. Heeger, *Phys. Rev. Lett.*, **42**, 1698 (1979).
12. M. J. Rice, *Phys. Lett. A.*, **71**, 152 (1979).
13. R. E. Peierls, *Quantum Theory of Solids*, (Oxford University Press, London, 1955), p. 108.
14. D. Baeriswyl, G. Harbeke, H. Kiess and W. Meyer, in *Electronic Properties of Polymers*, Eds J. Mort and G. Pfister, (Wiley and Sons, 1982), p. 268.
15. A. A. Ovchinnikov, I. I. Ukranski and G. V. Kventsel, *Sov. Phys. Usp.*, **15**, 575 (1973).
16. J. Ashkenazi, W. E. Picket, B. M. Klein, H. Krakaner and C. S. Wang. *Synth. Met.*, **21**, 503 (1987).
17. W. P. Su, *Handbook of Conducting Polymers*, (Marcel Dekker, New York, 1986), p. 760.
18. A. C. Scott, F. Y. F. Chu and D. W. McLaughlin, *Proc. IEEE*, **61**, 1443 (1973).
19. J. Scott-Russel, *Proc. Roy. Soc. Edin.*, 319 (1844).
20. M. J. Rice and E. J. Mele, *Solid State Comm.*, **35**, 48 (1980).
21. W. P. Su, J. R. Schrieffer and A. J. Heeger, *Phys. Rev.*, **B28**, 1138 (1983).
22. Y. Lu, *Solitons and Polarons in Conducting Polymers*, (World Scientific, Singapore, 1988), p.19.
23. H. Naarmann and N. Theophilou, *Synth. Met.*, **22**, 1 (1987).
24. A. F. Diaz and J. Bargon, *Handbook of Conducting Polymers*, (Dekker, New York, 1986), p. 81.
25. R. H. Baughman, J. L. Bredas, R. R. Chance, R. L. Elsenbaumer and L. W. Shacklette, *Chem. Rev.*, **82**, 209 (1982).
26. J. M. Pochan, J. Harbour and H. W. Gibson, *Bull. Am. Phys. Soc.*, **26**, 397 (1981).
27. S. Stafstrom, *Electronic Properties of Polymers*, (Springer Verlag, Berlin, 1992), p. 11.
28. S. L. Drecksler, J. Malek and M. Springborg, *Electronic Properties of Polymers*, Springer Verlag, Solid State Sciences 107, Berlin 1992, p. 38.
29. D. D. C. Bradley, *J. Phys. Appl. Phys.*, **20**, 1389 (1987).
30. A. P. Monkman and P. Adams, *Synth. Met.*, **40**, 87 (1991).
31. T. Schimmel, W. Riess, J. Gmeiner *et al.*, *Solid State Comm.*, **65**, 1311 (1988).
32. S. Kivelson and A. J. Heeger, *Synth. Met.*, **22**, 371 (1988).
33. J. Orenstein, *Handbook of Conducting Polymers*, (Dekker, New York, 1986), p. 1297.

34. P. O'Connor and J. Tauc, *Phys. Rev.*, **B25**, 2748 (1982).
35. B. Horovitz, *Solid State Commun.*, **41**, 729 (1982).
36. B. I. Green, J. Orenstein, R. R. Millard and L. R. Williams, *Phys. Rev. Lett.*, **58**, 2750 (1987).
37. Z. Vardeny, E. Ehrenfreund, O. Brafuncin *et al.*, *Phys. Rev. Lett.*, **56**, 671 (1986).
38. A. Kenwright, E. W. Feast, A. Adams, A. Milton, A. P. Monkman and J. Say, *Polymer*, **33**, 4292 (1992).
39. S. Heun, R. F. Mahrt, A. Greiner *et al.*, *J. Phys.: Con. Matter*, **5**, 247 (1993).
40. M. Pope and C. E. Svenberg, *Electronic Processes in Organic Crystals*, (Oxford University Press, 1982).
41. M. Angelopoulos, G. E. Austurias, S. P. Emer *et al.*, *Mol. Cryst. Liq. Cryst.*, **160**, 151 (1988).
42. Y. Cao, P. Smith and A. J. Heeger, *Synth. Met.*, **48**, 91 (1992).
43. A. P. Monkman and P. Adams, *Solid State Comm.*, **78**, 29 (1991).
44. A. P. Monkman, F. Hampson and A. Milton, *Electronic Properties of Polymers*, Springer Verlag Solid State Science 107, 255 (1992).
45. Y. W. Park, V. S. Lee, C. Park, L. W. S. Lacklette and R. H. Baughman, *Solid State Comm.*, **63**, 1063 (1987).
46. A. P. Monkman, P. Adams, A. Milton, M. Scully and S. Pomfret, *Mol. Cryst. Liq. Cryst.*, **236**, 189 (1993).
47. A. J. Epstein, R. P. McCall, J. M. Ginder and A. G. MacDiarmid, *Spectroscopy of Advanced Materials*, (John Wiley and Sons, 1991), p. 355.
48. M. dos Santos and J. L. Bredas, *Phys. Rev. Lett.*, **64**, 1185 (1990).
49. J. H. Burroughs, C. A. Jones and R. H. Friend, *Nature*, **335**, 137 (1988).
50. S. M. Sze, *Physics of Semiconductor Devices*, (John Wiley and Sons, 1981).
51. G. Horowitz, X. Peng, D. Fichou and F. Garnier, *J. Appl. Phys.*, **67**, 528 (1990).
52. J. H. Burroughs, D. D. C. Bradley, A. R. Brown *et al.*, *Nature*, **347**, 539 (1990).
53. G. Gustaffson, Y. Cos, G. M. Treacy, F. Klauetter, N. Colaneri and A. J. Heeger, *Nature*, **357**, 477 (1992).
54. N. C. Foulds and C. R. Lowe, *J. Chem Soc., Faraday Trans.*, **82**, 1259 (1986).
55. N. Agbor, M. C. Petty and A. P. Monkman, *Synth. Met.*, **55–57**, 3789 (1993).
56. S. Roth, in *Electronic Properties of Polymers and Related Compounds*, Eds H. Kuzmany, M. Mehring and S. Roth, Springer Series in Solid State Sciences 63, (Springer-Verlag, 1985), p. 2.
57. K. Yakushi, L. J. Lauchlan, T. C. Clarke and G. B. Street, *J. Chem. Phys.*, **79**, 4774 (1983).
58. R. H. Friend, D. C. Bott, D. D. C. Bradley *et al.*, *Phy. Trans. R. Soc. Lond. A.*, **314**, 37 (1985).

8

Conductive Charge-transfer Complexes

M R Bryce

8.1 Introduction

This chapter will consider organic molecular systems which possess electrical conductivity. Intermolecular charge-transfer interactions between the constituent molecules of the salt or complex are the source of the unpaired electrons which give rise to the conductivity. Polymeric organic conductors are dealt with in the previous chapter.

The vast majority of organic materials in their pure form are electrical insulators with room temperature conductivity values, $\sigma_{rt} = $ $<10^{-12}\,\text{Scm}^{-1}$. It is, therefore, intriguing that very high conductivity values, and even superconductivity, have been discovered in several specific families of organic solids. The first experimental observation that molecular compounds containing no metal atoms could exhibit interesting electrical properties was reported by Akamatsu *et al.* in 1954.[1] An unstable bromine salt of the polycyclic aromatic hydrocarbon perylene was shown to be a conductor, $\sigma_{rt} = 1\,\text{Scm}^{-1}$, whereas perylene itself is an insulator, $\sigma_{rt} = 10^{-15}\,\text{Scm}^{-1}$! This observation was a verification of predictions made as early as 1911 by McCoy and Moore that 'composite metallic substances' might be synthesized from nonmetallic elements.[2]

The field of molecular conductors has expanded dramatically in recent years. The materials which are at the forefront of attention today can be divided into four generic classes based on the chemical structure of the constituent molecules (Fig. 8.1). These are:

    (i) TCNQ and TTF systems;
   (ii) metal dithiolate complexes;
  (iii) metallomacrocycles;
  (iv) fullerenes.

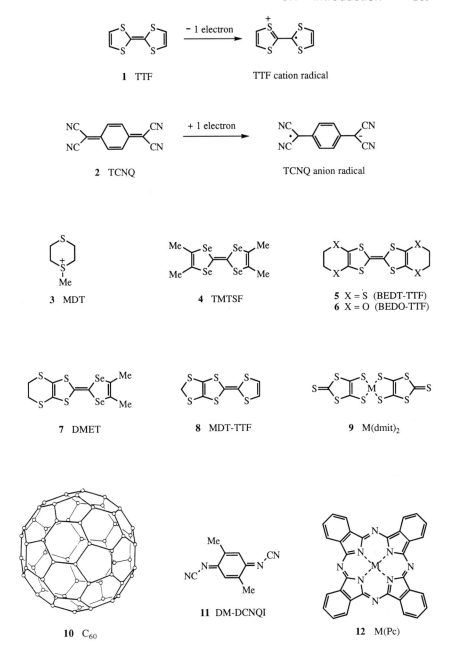

Fig. 8.1 Formulae of molecules that are components of molecular metals and super-conductors.

Compounds which fall into the first class have been the most extensively investigated, and they will, therefore, be given most coverage here. Detailed reviews of the field have been published regularly.[3-6] Before considering these classes of compounds in some detail, we will discuss the basic principles that give rise to the electronic conductivity they display.

8.2   Basic Concepts

A full understanding of how intermolecular charge-transfer can result in a material being an electronic conductor requires a detailed knowledge of the quantum theory of solids. None the less a simple picture can be used to explain the fundamental aspects. All metals (organic and inorganic) are conductors due to the presence of mobile electrons (conduction electrons) that transport electrical charge through the solid. These electrons are not usually present in organic solids, because the electrons are strongly involved in the bonding forces which hold the constituent atoms together. Hence, the vast majority of organic solids are electrical insulators.

However, consider the situation where an organic solid is formed by the combination of two (or more) types of neutral molecules, one of which is an electron donor (i.e. has a low ionization potential and can be easily oxidized) and the other is an electron acceptor (i.e. has a high electron affinity and can be easily reduced). An electron can be transferred from the donor molecule to the acceptor molecule; this leaves behind an organic cation that has a 'free' electron, in other words, an electron that is not strongly involved in the bonding. The corollary is that the acceptor molecule will have gained an electron to become an anion radical. If other strict electronic and structural criteria within the solid are met, then these electrons can become conduction electrons, just like those of traditional metals. Clearly this charge-transfer process depends upon a delicate balance between the ionization potential and the electron affinity of the constituent molecules: this is difficult to control, presenting a major challenge in the synthesis of molecular conductors.

For all four families of molecular conductor identified above, the shape of the molecules and the specific atoms they contain play a profound role in determining how the molecules will pack together and interact in the solid. These factors are also very important in regulating the number of electrons that are available to conduct electricity throughout the system. Typically, either the donor molecule, or the acceptor molecule (or both species) are planar (or nearly planar) in shape. This favours crystallization in the form of highly-ordered stacks throughout which the conduction electrons are delocalized.

At a more sophisticated level, the transport properties of molecular conductors are explained by band theory (see Chapter 7). The presence of a supermolecular orbital provides a mechanism for the metallic delocalization of electrons, with the width of the conduction band dependent upon the strength of interactions between molecular orbitals on neighbouring molecules. The electronic occupancy of these energy bands is critically important: when the energy gap between the valence band and the conduction band is large, the material will be an insulator. As the band gap decreases, thermal excitation of electrons across the band gap is possible and the material is an intrinsic semiconductor. Metallic behaviour requires partially filled bands in which it is possible for a large number of electrons to move easily into higher energy states within the band. For the complex formed by the donor molecule tetrathiafulvalene (TTF) **1** (Fig. 8.1) and the acceptor molecule tetracyano-*p*-quinodimethane (TCNQ) **2** (see below) the charge on each stack is non-integral (namely 0.59; in other words, there are 59 electrons delocalized over 100 molecules in each TCNQ stack) so, although the donor to acceptor ratio is 1:1, a mixed valence state is achieved. This requires a delicate balance between the ionization potential of the donor and the electron affinity of the acceptor.

A more common way of obtaining a mixed valence system is to have complete charge transfer in a complex where the donor to acceptor stoichiometry is other than 1:1. This is the case in conducting $(Cation^+)_1(TCNQ_2)^{-\cdot}$ salts, e.g. where the cation is 1-methyl-1,4-dithianium (MDT) **3**. (It is notable that analogous 1:1 TCNQ salts, e.g. K^+-$TCNQ^-$, are insulators). The conducting and superconducting salts formed by the donors tetramethyltetraselenafulvalene (TMTSF) **4** and bis(ethylenedithio)-TTF (BEDT-TTF) **5**, which have the general formulae $(Donor)_2^{+\cdot}(Anion^-)_1$, are prime examples of mixed valence organic radical ion salts. The anion is a closed shell species (I_3^-, PF_6^-, ClO_4^-, etc.) with the free electrons responsible for conductivity being provided solely by the fulvalene donor.

Accurate X-ray crystal structure determinations play a pivotal role in explaining the properties of charge-transfer complexes: several X-ray crystal structures are, therefore, shown in this chapter. Variable temperature conductivity values at ambient pressure for a range of highly conducting materials are shown in Fig. 8.2.

8.3   TCNQ and TTF Systems[3]

The first *stable* highly-conducting organic solids were reported by workers at DuPont laboratories in 1962. The key component of these materials was the new, powerful π-electron acceptor TCNQ **2**,[7] which in combination with many monovalent cations (e.g. caesium, morpholinium, quinolinium) formed salts of formula $(Cation^+)_1(TCNQ_2)^{-\cdot}$. There followed intensive activity from chemists and physicists in the search for new organic conductors, and a wide range of TCNQ salts were found to display semiconducting properties (i.e. to have thermally-activated electrical conduction) with activation energies of 50–200 meV.

X-ray structural analysis of many of these salts revealed that the planar TCNQ radical anions form segregated molecular stacks; this enables the π-electronic systems (the p_z orbitals) above and below the molecular plane of each TCNQ unit to interact, resulting in extensive electronic delocalization along the TCNQ stacks. The conductivities of TCNQ salts are

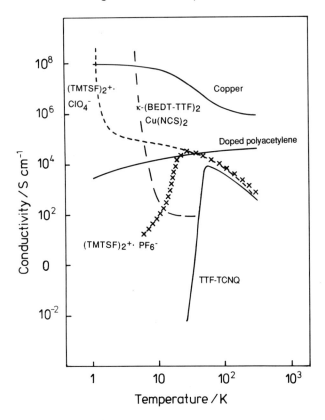

Fig. 8.2 Variable temperature conductivity values at ambient pressure for a range of highly conducting materials.

anisotropic, typically being 500–1000 times greater along the stacking axis than in the perpendicular direction, where there is very little interstack electronic interaction. The X-ray crystal structure of a typical conducting TCNQ salt, $(\text{1-methyl-1,4-dithianium})_{\overline{1}}^{+}\text{-}(\text{TCNQ}_2)^{-\cdot}$ is shown in Fig. 8.3.[8] The room temperature conductivity value for single crystals of this salt is $\sigma_{rt} = 2.0\,\text{Scm}^{-1}$. Within the crystal structure the TCNQ molecules form discrete pair (dimers) with intra-dimer separation between the mean planes of the TCNQ molecules of 3.2 Å. The cavities formed by this structure are occupied by the methyldithianium cations.

Molecular charge-transfer compounds with mixed stacks (i.e. the donor and acceptor molecules alternate within the same stack) are poor conductors because of electron localization on the acceptor species.

During the same period as the development of TCNQ salts as a major class of conducting solids, the understanding of superconductivity in elemental metals progressed dramatically due to the work of Bardeen, Cooper and Schrieffer (the BCS theory). The vibrations of atoms in a solid (termed 'phonons') usually scatter the mobile conduction electrons at a

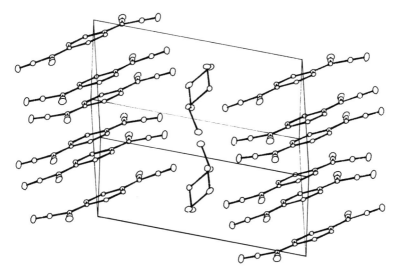

Fig. 8.3   X-ray crystal structure of the salt formed by the MDT cation **3** and TCNQ **2**: formula (MDT)$_1$ (TCNQ$_2$)$^-$ (from data presented in Bryce *et al.*[8]).

particular temperature, and so reduce the conductivity. However, these researchers demonstrated that a state could exist in which phonons could cooperate and allow mobile electrons to form pairs ('Cooper pairs') that could transport electrical charge without resistance. (This theory won the three workers the Nobel prize in 1972.) In the light of this new understanding, Little proposed that a high transition temperature might be obtainable in *molecular (organic)* materials where the high frequencies of internal vibrations might play the same role as phonons in elemental materials.[9] These advances provided an added impetus to the synthesis and study of new conducting charge-transfer systems, and although the model system proposed by Little has yet to be synthesized, organic superconductors are now well known (see below).

The charge-transfer complex formed by the donor TTF **1** and the acceptor TCNQ **2** is the archetypal organic metal.[10] The room temperature conductivity value of single crystals of TTF-TCNQ is about 500 Scm^{-1} and between the temperatures of 298 and 54 K, the complex displays genuine metallic behaviour (decreasing electrical resistance with decreasing temperature) with the conductivity rising to a maximum value about 15 times that at room temperature. Below 54 K, structural transitions (Peierls transitions) occur, and the material becomes a semiconductor due to a gap opening in the energy band associated with the metallic properties. The TTF and TCNQ molecules form independent molecular stacks in the complex and, like the cation-TCNQ salts mentioned above, the material is a one-dimensional conductor. Extensive chemical modification of the TTF and TCNQ units, notably by attachment of electron-withdrawing or electron-donating substituents and fused rings, has yielded a vast number of highly-conducting charge-transfer complexes.

An important difference between the cation-TCNQ salts and the TTF-TCNQ charge-transfer complexes is that the former are single-stack conductors (unpaired electrons are found only on the TCNQ stacks), while TTF-TCNQ is a two-stack conductor. Upon one-electron oxidation TTF yields a cation radical, so both components of the complex are open shell species. Both the TTF cation radical and the TCNQ anion radical species are stabilized by six π-electron aromaticity.

Organic superconductivity was first observed in 1979, by Bechgaard, Jérome and co-workers, in salts of the organoselenium donor TMTSF **4**.[11] The initial compound, $(TMTSF)_2^{+\cdot} PF_6^-$, has a transition temperature, T_c, of 0.9 K under about 9 kbar pressure; the compound was found to exhibit a Meissner effect, which is a true test of volume superconductivity. Several examples of superconducting salts of this donor with inorganic anions are now known at temperatures of <4 K. All except the perchlorate salt are superconducting only upon the application of hydrostatic pressure (typically about 10 kbar) which suppresses Peierls distortions.

More recently, emphasis has shifted to the sulphur-based donor molecule BEDT-TTF **5** which has yielded the largest family of organic superconductors, with transition temperatures as high as 12 K when the counter-anion is $Cu[N(CN)_2]Br$.[3g] The evidence of superconducting transitions for crystals of $\kappa\text{-}(BEDT\text{-}TTF)_2Cu[N(CN)_2]Br$ and $\kappa\text{-}(BEDT\text{-}TTF)_2Cu(NCS)_2$ is shown in Fig. 8.4. The X-ray crystal structures of a TMTSF and a BEDT-TTF salt are shown in Figs. 8.5[12] and 8.6,[13] respectively. These salts are two-dimensional materials (not one-dimensional like TTF-TCNQ) with a strong network of inter-stack, as well as intra-stack, chalcogen–chalcogen interactions, playing a key role in eliminating the

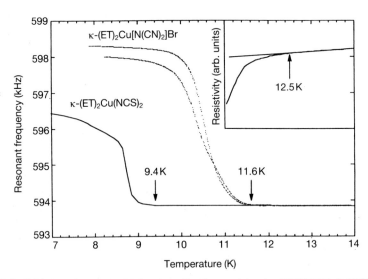

Fig. 8.4 Evidence of superconducting transition for crystals of $\kappa\text{-}(BEDT\text{-}TTF)_2Cu[N(CN)_2]Br$ and $\kappa\text{-}(BEDT\text{-}TTF)_2Cu(NCS)_2$ from Kini *et al.*[13]

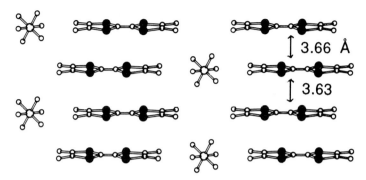

Fig. 8.5 X-ray crystal structure of $(TMTSF)_2^{+\cdot} PF_6^-$ which is a superconductor under pressure (redrawn from Thorup et al.[12]). Selenium atoms are shaded.

Peierls transition. Conduction electrons are mobile throughout the resulting extended sulphur or selenium networks.

All the $(TMTSF)_2^{+\cdot} X^-$ salts possess the same triclinic crystal structure at room temperature, with the TMTSF cations aligned in a zigzag column along the crystallographic a axis of the unit cell. The cation columns form infinite two-dimensional sheets in the ab plane; the sheets are separated from one another along the c axis by the anions. This infinite sheet network of short Se–Se contacts expands and contracts depending upon the size of the anions.

While TMTSF salts do form donor stacks, the structures of many $(BEDT-TTF)_2^{+\cdot} X^-$ salts possess little, or no, columnar stacking, and are characterized instead by short *inter*-stack S–S interactions. This is, at least in part, a consequence of the structure of the BEDT-TTF molecule. X-ray analysis reveals that although the central fulvalene unit is essentially planar (like TMTSF) the peripheral ethylene bridges of BEDT-TTF deviate considerably from planarity. This steric constraint reduces face-to-face stacking interactions, and consequently lateral interactions become dominant. For example, dimer pairs, not stacks, are seen in Fig. 8.6. This crystal packing motif, with orthogonal dimers, is termed a *kappa*-phase structure.

Despite the anisotropy observed in the metallic state of charge-transfer conductors, in the superconducting state electrical resistance is zero in all directions. However, the low-dimensional character of these materials has not been lost, and it is observed in other experimental measurements on the superconducting states: for example, the ability to remain superconducting in an applied magnetic field varies along the different crystal axes.

Other organic π-electron donors which yield superconducting salts are compounds **6–8**; the only acceptors which form superconductors are metal(dmit)$_2$ anions **9** and buckminsterfullerene, C_{60}, **10**. The acceptor molecule **11** is a close analogue of TCNQ **2**, and it is notable as a component of some of the most highly conducting materials known[14] [e.g. the copper salt, $Cu(11)_2$, has a conductivity value at $3.5\,K$ of $\sigma = 5 \times 10^5\,Scm^{-1}$) although **11** has not been reported to form any superconducting salts. The X-ray crystal structure of $Cu(11)_2$ reveals that

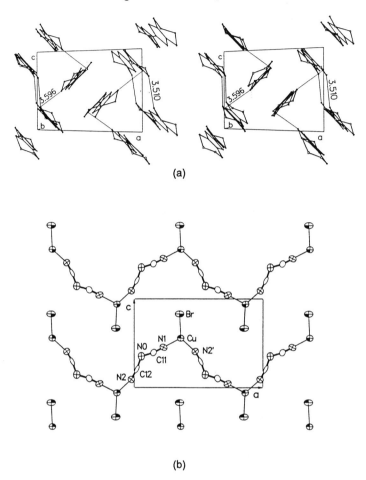

(a)

(b)

Fig. 8.6 X-ray crystal structure of the superconductor κ-(BEDT-TTF)₂Cu[N(CN)₂]Br. (a) Stereoview of the BEDT-TTF donor layer showing the kappa arrangement of orthogonal dimers, and intermolecular S–S contacts shorter than 3.60 Å represented as lines between the molecules; (b) polymeric Cu[N(CN)₂]Br anion layer (from Kini et al.[13]).

each metal atom is tetrahedrally coordinated to four organic ligands by strong Cu–N interactions and the acceptor forms one-dimensional columns (just like TCNQ) with an intrastack distance between the planes of the acceptors of 3.2 Å (Fig. 8.7).[15] The distance between the metal cations (3.8–3.9 Å) is too large for electron transport to occur along a metal ion chain: conductivity is, therefore, ascribed to a partially filled band formed by the acceptor molecules. The copper atoms are present in a mixed valence state, namely $Cu^{1.3+}$, which could arise from admixture of the $3d$ orbitals on copper with the π-electron conduction band of the DM-DCNQI ligands.

To produce single crystals of charge-transfer complexes, e.g. TTF-

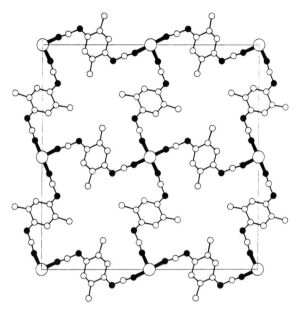

Fig. 8.7 X-ray crystal structure of the copper salt of the anion of acceptor molecule **11**: formula Cu(DM-DCNQI)$_2$. The nitrogen atoms are shaded (from Hünig and Erk[14]).

TCNQ, the two neutral components are dissolved, with heating, in a suitable solvent that has been rigorously dried and purified, such as acetonitrile; upon cooling, and perhaps partial evaporation of the solvent, the complex is precipitated. Similarly, to obtain organic cation-TCNQ salts, a solution of the cation iodide and TCNQ (or Li$^+$ TCNQ$^-$) is prepared. For organic superconductors the technique of electrocrystallization is needed, whereby the electron donor is oxidized electrochemically at a constant current in a solution containing the charge-balancing anions. An H-shaped cell is typically used consisting of two compartments separated by a glass frit with each compartment fitted with a platinum wire electrode.[16] A solution of the organic donor and a tetraalkylammonium salt of the anion (usually the tetrabutylammonium salt for solubility) is placed in the anode compartment, and a solution containing only the anion salt is placed in the cathode compartment. Typical solvents are dry tetrahydrofuran or 1,2,2-trichloroethane. A constant current (1–5 μA) or a constant voltage is applied to the electrodes. Crystals of the salt slowly grow on the anode over a period of about one week. The procedure is complicated for some materials where mixtures of salts having different crystal structures may be obtained in the same crystallization experiment. The donor BEDT-TTF **5** is notorious in this respect.

Magnetic susceptibility studies on organic conductors are also of considerable interest.[3c] In particular, these data provide a means of distinguishing between the various electronic states (charge-density waves and spin-density waves) which conductivity measurements cannot establish.

The close proximity of superconducting and antiferromagnetic ground states is a notable feature of these charge-transfer systems.

8.4   Metal-dithiolate Systems[4]

Metal complexes of the 1,2-dithiolene ligand have emerged as a molecular unit that yields a range of highly-conducting organic metals, and a few superconducting systems. There are three notable features of the metal(dmit)$_2$ system **9**: (i) it is nearly planar (ii) there are ten peripheral sulphur atoms which can engage in intra- and inter-stack interactions, and (iii) the redox properties can be tuned by varying the central metal atom (M = Ni, Pd, Pt etc.). The Ni(dmit)$_2$ compounds have been the most widely studied.

The salts of Ni(dmit)$_2$ **9** (M = Ni) can be conveniently classified into three types, based on the structures of the counter cations, which can be:

  (i) open-shell organic cations, e.g. TTF$^{+\cdot}$;
  (ii) closed-shell organic cations, e.g. Me$_4$N$^+$;
  (iii) or inorganic cations, e.g. Na$^+$.

A variety of salts of open-shell organic cations have been prepared. However, the salt of TTF is the only one that behaves as an organic metal and maintains the metallic state down to low temperatures (1.5 K). The X-ray crystal structure of this salt consists of segregated stacks of donor (TTF) and acceptor [Ni(dmit)$_2$] molecules (Fig. 8.8).[17] There are a large

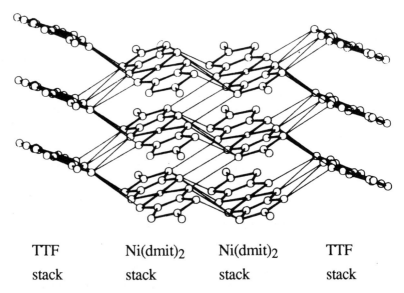

TTF	Ni(dmit)$_2$	Ni(dmit)$_2$	TTF
stack	stack	stack	stack

Fig. 8.8   X-ray crystal structure of TTF[Ni(dmit)$_2$]$_2$ viewed in the plane of the TTF ring. Thin lines indicate intermolecular S–S distances shorter than 3.70 Å (from Bousseau *et al.*[17]).

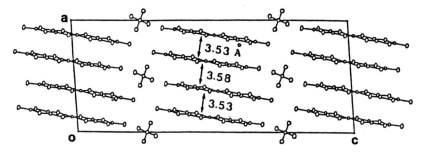

Fig. 8.9   X-ray crystal structure of the salt $Me_4N^+[Ni(dmit)_2]_2^-$ (from Kobayashi *et al.*[18]).

number of intermolecular, interstack S–S interactions which have distances shorter than the sum of the van der Waals' radii. Under a pressure of 7 kbar, the salt exhibits superconductivity with $T_c = 1.6$ K. This was the first superconductor containing a transition metal complex to be discovered. The partially-oxidized TTF stacks, as well as the anion stacks, play an important role in the conduction process in this salt, which is, therefore, a two-chain conductor (like TTF-TCNQ).

Salts of $Ni(dmit)_2$ anions with closed shell cations have also provided very interesting materials: the mixed valence salt $NMe_4[Ni(dmit)_2]$ is an anion stack superconductor ($T_c = 5$ K at 7 kbar). This compound provides the only example to date of superconductivity arising solely through interactions in an anion stack. The X-ray crystal structure of the tetramethylammonium salt (Fig. 8.9) reveals stacks of planar $Ni(dmit)_2$ anions separated by channels which contain the cations.[18] This is, in essence, the reverse situation to that of the TMTSF salts (cf. Fig. 8.5).

8.5   Metallomacrocycles[5]

Since the mid-1970s interest has developed in this class of conductive molecular solids. These materials become conductive upon oxidation (e.g. by doping with iodine or bromine) which produces lattices containing one-dimensional arrays of planar metallomacrocycles (Mmac) in which the metal cation radical is in a partial oxidation state. A representative structure of an Mmac species is the phthalocyanine system, (abbreviated to Pc) shown in formula **12**. A variety of transition metals are found to coordinate into these structures, e.g. M = Fe, Co, Ni, Cu, Zn and Pt. The iodine- or bromine-containing counterion is usually present as a polyhalide species (e.g. I_3^-); alternatively, anions such as BF_4^-, PF_6^- or SbF_6^- may be used.

In many of these Pc charge-transfer systems the metal retains a +2 oxidation state and plays only a spectator role in the conduction process. The mobile electrons (the charge carriers) are associated with delocalized π-orbitals on the macrocyclic ligand, from which electrons are removed. There are instances, however, where the metal- and ligand-oxidized states

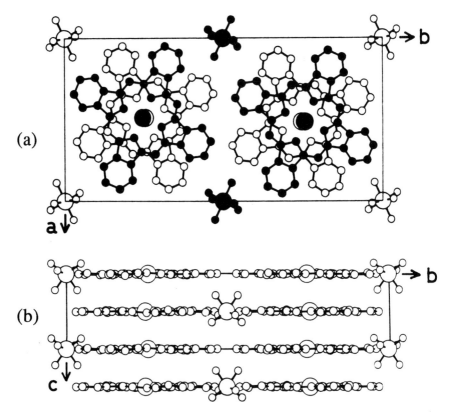

Fig. 8.10 Two views of the X-ray crystal structure of the nickel phthalocyanine salt $(NiPc)_2SbF_6$: (a) projection down the NiPc stacking axis, i.e. along the crystallographic c axis; (b) projection in the plane of the NiPc molecules, i.e. along the crystallographic a axis (from Yakushi et al.[19]).

are almost degenerate, and are interconvertible depending upon conditions. The X-ray crystal structure of $(NiPc)_2 SbF_6$ is shown in Fig. 8.10.[19] A typical NiPc columnar packing motif is seen, with the planar Pc units staggered by about 40° with respect to the nearest neighbours, and with the anions occupying channels between the Pc columns. The room temperature conductivity value of this material is as high as $\sigma_{rt} = 200$ Scm^{-1} with metallic behaviour observed between room temperature and 150 K. Ni(Pc)I was the first molecular material to remain conductive at very low temperatures (100 mK) without containing sulphur or selenium atoms to mediate interstack interactions.

Conclusive proof that a central metal ion is not needed to impart the metallic properties to stacked Pc complexes has been obtained by studies on a system in which the metal atom is replaced by two hydrogen atoms, and the counter ion is iodide: i.e. the complex H_2Pc I. The complex is a true molecular metal ($\sigma_{rt} = 700$ Scm^{-1} increasing to a maximum value at 15 K of 4000 Scm^{-1}).[20]

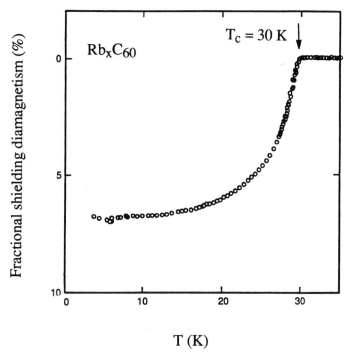

Fig. 8.11 Magnetic susceptibility curves for a powdered sample of Rb_xC_{60} showing a superconducting transition at $T_c = 30$ K (from Hebard *et al.*[21]).

8.6 Fullerenes[6]

The most recent family of organic superconductors to be discovered are salts of the electron acceptor molecule buckminsterfullerene, C_{60}, **10**. Clearly the structural properties of salts of this spherical molecule will be completely different from those of the flat systems described above. In 1991 a potassium fulleride salt, K_xC_{60}, was the first superconductor of this class to be reported $(T_c = 18\text{ K})$.[21] Shortly afterwards $(K^+)_3\ C_{60}^{3-}$ was identified as the superconducting phase.[22] The analogous rubidium salt was found to have a dramatically higher T_c value, namely 30 K (Fig. 8.11).[21] A combination of three techniques—microwave absorption and magnetic susceptibility of powders, and dc resistivity of films—asserted the occurrence of superconductivity in these salts (Fig. 8.12).[22] Shielding diamagnetism curves were used to assess the volume fraction of the superconducting phase as 74% for K_3C_{60} and approximately 7% for Rb_3C_{60}.

A variety of techniques have been employed to synthesize alkali metal fulleride salts, A_xC_{60}; many involve the direct reaction of C_{60} and alkali metal vapour at 200–400°C, with the precise conditions depending upon both A and x.[24] These fulleride salts are air-sensitive and must be kept in an inert atmosphere at all times.

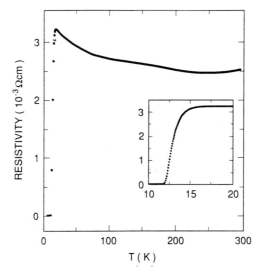

Fig. 8.12 Temperature dependence of the electrical resistivity of a 1600 Å thin film of K_3C_{60} between 300 and 4 K. The inset shows an expanded scale of the superconducting transition between 20 and 10 K (from Palstra et al.[22a]).

The alkali metal fulleride salts adopt face-centred cubic structures in which the alkali metal cations fit neatly into the spaces between the bulky C_{60}^{3-} anions. The structure of K_3C_{60}, obtained from synchrotron X-ray powder diffraction data, is shown in Fig. 8.13.[24] Potassium ions occupy all the available tetrahedral and octahedral vacancies in the host lattice. Intercalation of potassium expands the unit cell dimension from 14.11 Å in pure C_{60} to 14.24 Å in K_3C_{60}. These materials behave as three-dimensional synthetic metals.

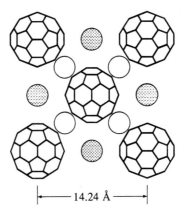

Fig. 8.13 X-ray structure of the fulleride salt K_3C_{60}. The open and hatched spheres represent the potassium ions at tetrahedral and octahedral sites, respectively (from Stephens et al.[24]).

8.7 Conclusions

Our understanding of superconductivity in materials in general was changed with the spectacular discovery in 1986 by Bednorz and Müller of superconductivity at about 30 K in the metal oxide $LaBa_2CuO_4$. Subsequently an oxide of thallium, calcium, barium and copper, $Tl_2Ca_2Ba_2Cu_3O_{10}$, was shown to have a transition temperature as high as 125 K. These findings, coupled with the advent of fullerene-based superconductors, suggest that entirely new families of conducting systems will be synthesized in due course. The bulk nature of the superconductivity in the organic systems has been rigorously established: they exhibit both zero resistance and expel an external magnetic field (the Meissner effect). A question that continues to motivate much research on organic charge-transfer complexes is 'How does the mechanism for electron pairing that causes superconductivity in these systems compare with that of conventional metallic superconductors or that of the new high T_c ceramic oxide superconductors?'

Much work remains to be done on the chemical analysis of new charge transfer systems. The rational design of new molecular conductors seems to be limited to a very restricted range of compounds, e.g. salts of BEDT-TTF **5** in which only minor changes to the overall structure results. The control of dimensionality is a fascinating challenge: only where two-dimensional interstack interactions become predominant is superconductivity observed. X-ray crystallographic studies at liquid helium temperatures, as well as at high pressures, are needed to shed light on the low temperature transitions that occur in many charge transfer organic conductors.

The technological impact of molecular conductors has, so far, been minimal. However, the preparation of conducting thin films (rather than single crystals) of these materials is a growing topic. It has recently been shown that vapour deposition affords superconducting thin films of BEDT-TTF iodide salt.[25] Good quality Langmuir–Blodgett films of some amphiphilic tetrathiafulvalene salts are now available, although their conductivity values are still relatively low (typically between 10^{-2} and $1.0 \, Scm^{-1}$).[26] Nevertheless, novel properties and technological application may result, especially from the controlled fabrication of superlattice structures.

References

1. H. Akamatsu, H. Inokuchi and Y. Matsunaga, *Nature*, **173**, 168 (1954).
2. H. H. McCoy and W. C. Moore, *J. Am. Chem. Soc.*, **33**, 1273 (1911).
3. For reviews on TCNQ and TTF conductors, including organic superconductors, see: (a) M. R. Bryce and L. C. Murphy, *Nature*, **309**, 119 (1984); (b) F. Wudl, *Accts. Chem. Res.*, **17**, 227 (1984); (c) P. M. Chaikin and R. L. Greene, *Physics Today*, **39(5)**, 24 (1986); (d) J. R. Ferraro and J. M. Williams, *Introduction to Synthetic Electrical Conductors*, (Academic Press, London,

1987); (e) T. Ishiguro and K. Yamaji, *Organic Superconductors*, (Springer-Verlag, Berlin, 1990); (f) M. R. Bryce, *Chem. Soc. Rev.*, **20**, 355 (1991); (g) J. M. Williams, A. J. Schultz, U. Geiser *et al.*, *Science*, **252**, 1501 (1991).

4. For reviews on metal dithiolate complexes, see: (a) reference 3e; (b) P. Cassoux, L. Valade, H. Kobayashi, A. Kobayashi, R. A. Clark and A. E. Underhill, *Coord. Chem. Rev.*, **110**, 115 (1991).

5. For reviews on metallomacrocycles see: (a) reference 3d; (b) B. M. Hoffman and J. A. Ibers, *Accts. Chem. Res.*, **16**, 15 (1983); (c) T. J. Marks, *Science*, **227**, 881 (1985).

6. For a review on fullerenes see Eds J. E. Fischer and D. E. Cox, a special issue of *J. Phys. Chem. Solids*, **53(11)** (1992).

7. L. R. Melby, R. J. Harder, W. R. Hertler, W. Mahler, R. E. Benson and W. E. Mochel, *J. Am. Chem. Soc.*, **84**, 3374 (1962).

8. M. R. Bryce, A. J. Moore, P. A. Bates, M. B. Hursthouse, Z-X. Liu and M. J. Nowak, *J. Chem. Soc. Chem. Comm.*, 1441 (1988).

9. W. A. Little, *Phys. Rev. A.*, **134**, 1416 (1964).

10. J. Ferraris, D. O. Cowan, V. V. Walatka and J. H. Perlstein, *J. Am. Chem. Soc.*, **95**, 948 (1973).

11. (a) K. Bechgaard, C. S. Jacobsen, K. Mortensen, H. J. Pedersen and N. Thorup, *Solid State Comm.*, **33**, 1119 (1980); (b) K. Bechgaard and D. Jérome, *Sci. Am.*, **247**, 52 (1982).

12. N. Thorup, G. Rindorf, H. Soling and K. Bechgaard, *Acta Cryst.*, **B37**, 1236 (1981).

13. A. M. Kini, U. Geiser, H. H. Wang *et al.*, *Inorg. Chem.*, **29**, 2555 (1990).

14. S. Hünig and P. Erk, *Adv. Mater.*, **3**, 225 (1991).

15. A. Aumüller, P. Erk, G. Klebe, S. Hünig, J. U. von Schütz and H.-P. Werner, *Angew. Chem. Int. Ed. Engl.*, **25**, 740 (1986).

16. J. M. Williams, *Prog. Inorg. Chem.*, **33**, 183 (1985).

17. M. Bousseau, L. Valade, J.-P. Legros, P. Cassoux, M. Garbauskas and L. V. Interrante, *J. Amer. Chem. Soc.*, **108**, 1908 (1986).

18. A. Kobayashi, H. Kim, Y. Sasaki *et al.*, *Chem. Lett.*, 1919 (1987).

19. K. Yakushi, M. Sakuda, I. Hamada *et al.*, *Synth. Metals*, **19**, 769 (1987).

20. T. Inabe, T. J. Marks, R. L. Burton *et al.*, *Solid State Comm.*, **54**, 501 (1985).

21. A. F. Hebard, M. J. Rosseinsky, R. C. Haddon *et al.*, *Nature*, **350**, 600 (1991).

22. (a) T. T. M. Palstra, R. C. Haddon, A. F. Hebard and J. Zaanen, *Phys. Rev. Lett.*, **68**, 1054 (1992); (b) K. Holczer, O. Klein, S-M. Huang *et al.*, *Science*, **252**, 1154 (1991).

23. D. W. Murphy, M. J. Rosseinsky, R. M. Fleming *et al.*, *J. Phys. Chem. Solids*, **53**, 1321 (1992).

24. P. W. Stephens, L. Mihaly, P. L. Lee *et al.*, *Nature*, **351**, 632 (1991).

25. K. Kawabata, K. Tanaka and M. Mizutani, *Adv. Mater.*, **3**, 157 (1991).

26. A. S. Dhindsa, Y-P. Song, J. P. Badyal *et al.*, *Chem. Mater.*, **4**, 724 (1992).

9

Liquid Crystals and Devices

D Lacey

9.1   Introduction

The subject of liquid crystals is vast, and this can be clearly seen from the liquid crystal family tree given in Fig. 9.1. Since the major topic of this book is molecular electronics, we shall confine this chapter to thermotropic liquid crystalline materials that exhibit liquid crystal properties, either on heating the material above its melting point or on cooling the material from its isotropic melt. As can be seen from Fig. 9.1, thermotropic liquid crystals can be further subdivided into two main classes: high molecular (or molar) mass materials, which include both main and side chain polymers; and low molecular (or molar) mass materials, which include calamitic (rod-like or lath-like molecules) and discotic (disc-like molecules) liquid crystalline materials. Although we shall be saying something about all these different classes of liquid crystals the main emphasis will be on thermotropic, low molar mass, calamitic liquid crystals and their application to electro-optic devices.

The distinction between calamitic and discotic liquid crystals is mainly due to the shape of the molecule. In calamitic liquid crystals the molecules can be described as being lath-like or rod-like, examples of which are given in Fig. 9.2(a). However disc-like molecules, similar to those given in Fig. 9.2(b), can also exhibit liquid crystalline behaviour on heating or cooling, either by arranging themselves into columns (columnar phases) or into an arrangement which is very reminiscent of a pile of coins (nematic phase) which have been pushed over. Schematic representations of both these discotic phases are given in Fig. 9.3. A detailed account of these intriguing phases is beyond the scope of this chapter and, therefore, the reader is referred to two very good articles on the subject of discotic liquid crystals.[1,2]

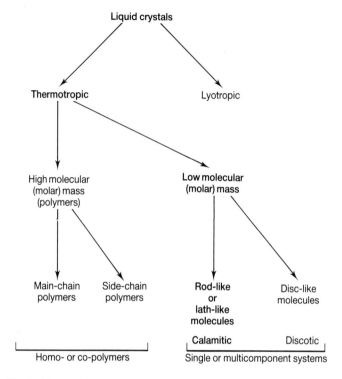

Fig. 9.1   The liquid crystal family tree.

9.2   Basic Concepts

What are liquid crystals? The simplest way to answer this question is first to consider the different common states of matter and then to see how these relate to liquid crystals. Normally matter can exist in three different states.

1   **The solid state**: where molecules (or atoms or ions) execute small vibrations around firmly fixed lattice positions but the molecules can rotate.
2   **The liquid state**: characterized by relatively unhindered rotation and translation of the molecules, resulting in the absence of any long-range order.
3   **The gaseous state**: where molecules can move freely through the entire volume of the container, under almost no constraint.

The sequence of events that we are most interested in is the one that occurs when an organic compound is heated and melts to form the liquid state or isotropic liquid. This process is normally called the melting point. The melting of normal organic compounds involves, in a single process, the abrupt collapse of the positional order of the lattice array. It marks the onset of essentially free rotation and translation of the molecules i.e.,

(a) Calamitic Liquid Crystals

(b) Discotic Liquid Crystals

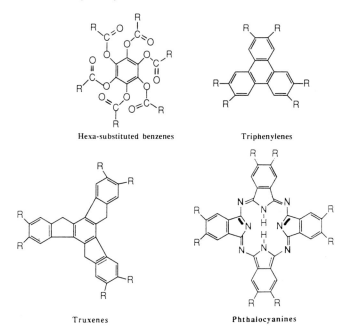

Hexa-substituted benzenes

Triphenylenes

Truxenes

Phthalocyanines

Fig. 9.2 The structure of molecules which form (a) calamitic and (b) discotic liquid crystalline phases.

the system passes from the 3-dimensional order of the solid state to the zero order of the isotropic liquid.

However, this very disruptive, single-step process from the crystal to the isotropic liquid is not universal and in certain compounds where the structure of the molecules is rod-like or lath-like (see Fig. 9.2(a)), this single-step process is not always observed. Instead, on heating the crystal-

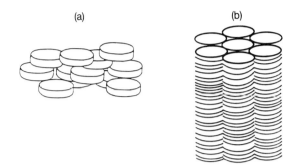

Fig. 9.3 The structure of (a) nematic and (b) columnar discotic phases.

line solid, a multiple step process can be observed whereby the crystal solid produces a new intermediate phase called the Mesophase, in which varying degrees of molecular order are retained prior to the onset of the isotropic liquid. Compounds exhibiting a mesophase or liquid crystal phase are called Mesogens.

SOLID	$\longrightarrow$	MESOPHASE	$\longrightarrow$	ISOTROPIC LIQUID
3D order		1 or 2D order		ZERO order
POSITION ORIENTATION		ORIENTATION (POSITION)		NONE

The term Liquid Crystal stems from the fact that the mesophase not only exists between the solid state and isotropic liquid, but it also exhibits two fundamental properties associated with these states of matter. The mesophase is both fluid to some degree, like the isotropic liquid, but also maintains some degree of molecular order, a property which is characteristic of the solid (crystal) state. Hence the term **liquid** (because it has fluidity) and **crystal** (because it maintains some degree of order). These two unique properties of the mesophase, fluidity and molecular order, coupled with its anisotropic nature are the essential ingredients that allow liquid crystals to be used in electro-optic display devices—a topic which we shall discuss in detail later.

9.3 The Mesophases

There are three distinct types of mesophase, namely the smectic, nematic and cholesteric (chiral nematic) phases. We shall now give a simplified description of the three different types of mesophase. A more detailed account of these mesophases are given elsewhere.[3,4]

Smectic Phase   This is a viscous opaque phase. On heating a layered crystal lattice the melting process evidently disrupts the end-to-end molecular cohesions of the solid state, but the temperature at which the mesophase is stable is not sufficient to weaken the lateral associations sufficiently to allow transitions in the direction of their long axes. The layers therefore remain essentially intact, and these are sometimes free to move over each other (with greater or lesser ease). Thus in this type of mesophase we have a lamellar structure. At present twelve different types of smectic phases have been identified and these are designated with alphabetical subscripts i.e. S_A, S_B, S_C etc. up to S_K. In fact this only gives eleven smectic phases, but there are two smectic S_B phases; the crystal B and hexatic S_B phases.

The true smectic phases (S_A, S_C, S_F, S_I and S_B hexatic) possess no, or very limited, positional correlation between layers. The other smectic phases (crystal B, E, G, H, J, and K) have very long range correlation of position over many layers and are more akin to crystals. They are indeed 'soft crystals' and layer flow does not occur. The S_D phase is not a lamellar phase and is optically extinct when viewed between crossed polarizers (all other smectic/crystal phases are optically birefringent—see later in Section 9.6).

The major structural differences between the smectic phases are due to the different arrangement of molecules within the layers, and also from the presence or absence of correlation from one layer to the next. This is best illustrated by considering the structure of a selected number of smectic phases, as shown in Fig. 9.4.

In the S_A phase (Fig. 9.4(a)), we have no ordering of the molecules within the layers and no positional correlation from one layer to the next. This is the least ordered smectic phase. The layer arrangement is relatively weak and diffuse, the centres of gravity of the molecules approximating on average to density planes of the molecules. Such smectic phases can be viewed as 1-dimensional systems.

In the hexatic S_B phase (Fig. 9.4(b)), the phase is more ordered in that we can now see a regular hexagonal ordering of the molecules within the layers, but there is still no positional correlation from one layer to the next. Such a phase can be said to be based on a weakly-coupled system of ordered layers, i.e., it is a 2-dimensional system.

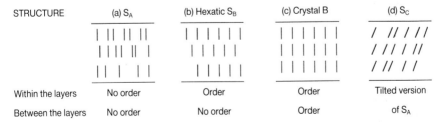

STRUCTURE	(a) S_A	(b) Hexatic S_B	(c) Crystal B	(d) S_C
Within the layers	No order	Order	Order	Tilted version
Between the layers	No order	No order	Order	of S_A

Fig. 9.4   Simple representation of a selected number of smectic phases.

However, in the very ordered crystal B phase (Fig. 9.4(c)), order prevails both within the layers (again hexagonal) and also between and over many layers. Smectic phases of this type can now be viewed as 3-dimensional ordered systems (a term normally used to describe crystal phases). However, the ordering in these smectic phases is less perfect than that found in normal crystal phases, and so, as mentioned earlier, these types of smectic phases are regarded as *disordered crystal phases* or *soft crystals*. Such smectic phases are denoted by the prefix crystal, i.e., crystal B phase.

So far we have discussed smectic phases where the molecules lie with their long axes orthogonal to the layer planes. However, there are a number of smectic phases, like the S_C phase (Fig. 9.4(d)) where the molecules within the layers are tilted. In this case, because the S_C phase has no ordering of the molecules within the layers and there is no positional correlation from one layer to the next, the S_C phase can be regarded as a tilted version of the S_A phase.

It is only prudent at this stage to point out to the reader once again that this is a very simplistic view of the structure of the smectic phases, and a more detailed account of the different smectic phases can be found in Demus[2] and Collings.[3] It must be remembered that these smectic phases are formed by the application of heat (energy) and so the molecules that form these smectic phases will have thermal motion. In the case of the smectic phases depicted in Fig. 9.4, the thermal motions of the molecules will not allow the molecules within the layer planes to be so well-defined as shown, nor will there be such a sharp boundary between the layers themselves. Indeed, the molecules can gain sufficient energy to 'hop' from one layer to the next. Although these diagrams are 'idealized' with a certain amount of 'artistic licence', they do convey the general features of the smectic/crystal phases. By considering (1) the ordering of the molecules within the layers, (2) the possibility of correlation from one layer to the next and (3) whether the molecules are tilted or orthogonal with respect to

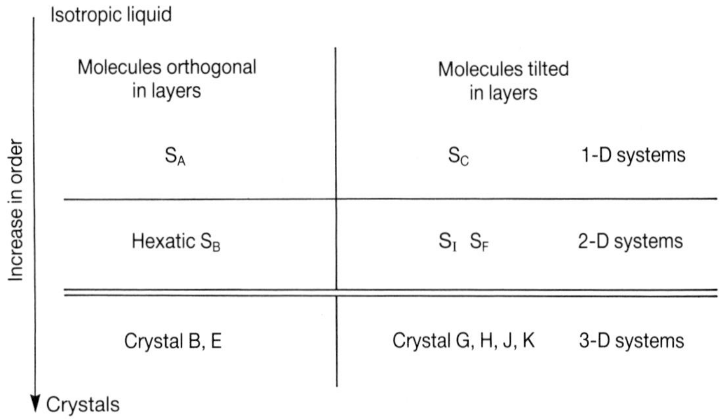

Isotropic liquid

Molecules orthogonal in layers		Molecules tilted in layers	
S_A		S_C	1-D systems
Hexatic S_B		S_I S_F	2-D systems
Crystal B, E		Crystal G, H, J, K	3-D systems

Increase in order ↓

▼ Crystals

Fig. 9.5 The different types of smectic phases.

the layer planes, the following picture (Fig. 9.5) of smectic/crystal phases can be built up.

Further structural features of the phases are necessary to differentiate between S_I and S_F phases or between crystal G, H, J, and K phases. Such a detailed account of smectic phases are beyond the scope of this chapter.[4]

Although theoretically it is possible for a liquid crystal compound to exhibit all 12 types of smectic and crystal phases, the highest number of these phases exhibited by any one compound is five, e.g., terphthalylidene-bis-4-n-pentylaniline exhibits five smectic phases and one nematic phase.

$$C_5H_{11}\!-\!\!\langle\bigcirc\rangle\!-\!N{=}CH\!-\!\!\langle\bigcirc\rangle\!-\!CH{=}N\!-\!\!\langle\bigcirc\rangle\!-\!C_5H_{11}$$

$$crystal \longleftarrow S_H \overset{61°C}{\rightleftarrows} S_G \overset{140°C}{\rightleftarrows} S_F \overset{149°C}{\rightleftarrows} S_C \overset{179°C}{\rightleftarrows} S_A \overset{212°C}{\rightleftarrows} N \overset{233°C}{\rightleftarrows} I$$

$$68°C$$

N = nematic phase (see later).

I = isotropic liquid.

The S_H phase is monotropic — only observed on cooling below the melting point.

On heating, a sequence of smectic/crystal phases can terminate in one of two ways: either the smectic phase will pass directly to the isotropic liquid, in which case the liquid crystal compound has exhibited smectic and crystal phases only and is called a Smectogen, or the molecules may translate out of the layered structure to form the second of the two types of mesophases, the nematic phase.

Nematic Phase   This opaque phase is very much more fluid than the smectic/crystal phase because the nematic phase is not lamellar and the only molecular order that this phase possesses, with respect to the isotropic liquid, is that the long molecular axes remain statistically parallel or near parallel to a preferred direction, called the Director (*n*).

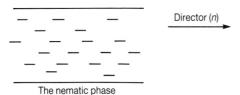

The nematic phase

The more fluid nature of the nematic phase makes it an ideal medium for electro-optic display devices but, as we shall see later, devices using smectic liquid crystals have also been developed.

There is only one nematic phase and so on heating this phase will finally yield the isotropic liquid. If a liquid crystalline material exhibits a nematic phase only, then this material is called a Nematogen.

So far we have considered two distinct mesophases, one lamellar (smectic/crystal) the other non-lamellar (nematic), but there is a third mesophase which is closely related to the nematic phase. This phase is called a Cholesteric or Chiral Nematic phase.

Cholesteric Phase (Chiral Nematic)   This fluid, opaque phase derives its name from the fact that the first compounds that gave this type of phase were derivatives of cholesterol, e.g. cholesteryl benzoate, acetate etc. However, the important feature of the molecules that distinguishes them from those which give rise to the nematic phase is that the molecules that form the cholesteric (chiral nematic) phase are optically active.

Gives a nematic phase   Gives a cholesteric phase
                                            * Chiral centre

The optical centre associated with such molecules gives rise to the unique helical structure of the cholesteric phase and this is shown diagramatically in Fig. 9.6.

Each slice (the cholesteric phase is *not* lamellar) taken from the cholesteric phase has a molecular arrangement very similar to that found in the nematic phase. In other words, the long axes of the molecules remain parallel or near parallel to the local director (this is shown in Fig. 9.6 as a bold arrow). As we travel from one successive slice to another along the optic axis, we find that the director gradually rotates, giving rise to the unique helical structure of the cholesteric phase.

Associated with the helical structure of the cholesteric phase is a physical parameter called the pitch length (p). This is simply the distance along the optic axis to observe a 360° rotation of the director (in Fig. 9.6, we see one pitch length). The wavelength of the light (λ) reflected by the cholesteric phase is governed by the product of the pitch length (p) and the refractive index (n) of the phase, according to the equation.

$$\lambda = np \tag{9.1}$$

For certain values of np, the wavelength (λ) of the light reflected by the cholesteric phase will be in the visible region of the electromagnetic

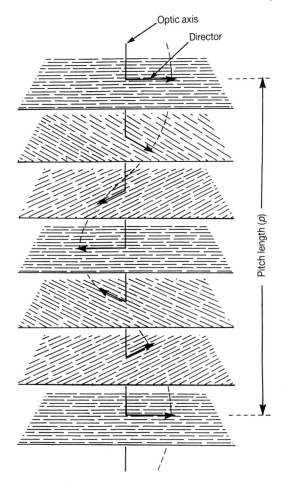

Fig. 9.6 The helical structure of the cholesteric (chiral nematic) phase.

spectrum, and the phase will then show iridescent colours. One important property of the pitch length is that it is sensitive to changes in temperature. If visible light is being reflected by the cholesteric phase, then changes in temperature will alter the pitch length. This in turn will cause a change in the colour being reflected by the cholesteric phase. Thus cholesteric liquid crystals can be used as temperature sensors, either in the form of a spray, a paint or an encapsulated polymer film.

The temperature dependence of selective reflection from a cholesteric phase is shown in Fig. 9.7. A particularly strong thermochromic effect is induced by the presence of a smectic phase. The colour play is the temperature range over which colours can be observed for the cholesteric phase.

Unfortunately, as the temperature is decreased so the wavelength of the

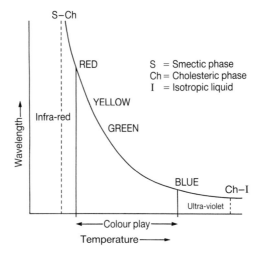

Fig. 9.7 The variation of the wavelength of reflected light with temperature for a cholesteric phase.

light being reflected by the cholesteric phase increases, thus reversing our perception of the association between colour and temperature. We are told that red is hot and blue is cold but fortunately no-one told the cholesteric phase this fact, and we observe blue colours when hot, red colours when cold.

These materials are used in a number of applications, some of which are given below.

1   Thermometers, both clinical and domestic.
2   Diagnostic thermal mapping of objects to detect structural flaws or hot spots, e.g. in electrical circuitry and machinery.
3   Decorative fabrics and jewellery, which use the visual appeal of colour change with temperature as their primary feature.

Recent applications of cholesteric liquid crystals can be found in cosmetics, paints and inks. There is on sale at the moment a clear, eye-moisturizing gel which contains a cholesteric liquid crystal. The liquid crystal appears as a continuous spiralling iridescent streak throughout the clear gel. Whether the liquid crystal has any cosmetic property or not is debatable, but the iridescence of the cholesteric streak makes the gel look very attractive. Paints can also use the iridescent property of the cholesteric phase to make the colours look more warm and homely.

Cholesteric liquid crystals are the main active ingredient in inks which can now be screen-printed onto fabrics. These fabrics are usually black, since the reflected colours are more pronounced on a black background than on a white one. T-shirts can be made out of this black fabric and when a person puts on the T-shirt, it just looks black—not too exciting. However, the heat from the person's body affects the cholesteric liquid crystal and soon vivid coloured patterns appear on the T-shirt.

The vivid colours associated with the cholesteric phase is a reminder of an incident which happened some years ago. Our research group at Hull was approached by a furniture manufacturer who wanted to use a flexible polymer material into which had been capsulated the cholesteric liquid crystal. The idea was that as one entered the room, the settee or chair made out of this material would simply look black, but when it was sat upon, body heat would affect the cholesteric liquid crystal and so the furniture would turn into a vivid cascade of colour. This would certainly give people a more 'colourful' life.

It must also be remembered that molecules containing a chiral centre may also form smectic phases, and chiral versions of S_A, S_C, S_I and S_F are now known. Indeed materials exhibiting the chiral smectic C phase are very important materials for the ferroelectric display device, which will be discussed later. The chiral smectic A is a very recent and fascinating discovery in which the formation of the helical structure of the phase is due to the layers of the smectic A phase forming blocks. These then twist gradually and continuously throughout the phase giving rise to the twisted smectic A* spiral layer order (the structure of this phase is given below).

TWISTED SMECTIC A*
MESOPHASE
SPIRAL LAYER ORDER

This is in sharp contrast to the cholesteric phase where the twisting of the chiral molecules manifest the helical structure of this phase. Like the cholesteric phase the chiral smectic phases also show reflected colours, but unlike the cholesteric phase they reflect red colours at high temperatures, blue colours at low temperatures.

Phase Sequences   We have seen that liquid crystals can exhibit three different types of mesophase, i.e. smectic (crystal), nematic and cholesteric (chiral nematic). The phase sequences observed on heating a liquid crystalline material are given below.

Solid ⟶ Nematic (N) ⟶ Isotropic Liquid (I)   (Nematogen)

Solid ⟶ Smectic/Crystal(s) ⟶ Isotropic Liquid   (Smectogen)

Solid ⟶ Cholesteric (Ch or N*) ⟶ Isotropic Liquid   (Cholestogen)

Solid ⟶ Smectic/Crystal(s) ⟶ Nematic ⟶ Isotropic liquid

Solid ⟶ Smectic/Crystal(s) ⟶ Cholesteric ⟶ Isotropic liquid

All these transitions, except the crystal-to-mesophase transition are precisely reversible and the transition temperatures, whether obtained during a cooling or heating cycle, should be within 1 or 2°C of each other if the material is pure.

Note that although smectic (crystal) phases can coexist with either the nematic or cholesteric phase, a liquid crystalline material cannot, for obvious reasons, exhibit both a nematic and a cholesteric phase. Well, this was true until very recently when members of our group synthesized a number of compounds which gave *twist inversion* of the chiral nematic phase. On cooling from the isotropic liquid the right-hand helical structure of the cholesteric phase inverts, by way of nematic phase where the pitch is infinite (no helical structure), to a cholesteric phase where the pitch is left-handed. Thus these compounds exhibit the unusual phase sequence,

cholesteric → nematic → cholesteric

in which the lower temperature cholesteric phase reflects light at higher wavelengths (red) on increasing the temperature. Slaney et al.[5] suggested that the helix inversion is due to changes in the population of the conformational isomers and that such a change is temperature dependent.

9.4   Identification of Phases

We have already seen that some liquid crystalline materials can give a complex sequence of phases, all of which have to be identified. In some cases this task can be quite a problem especially when the sequences involve a number of different smectic phases. Because of this it is quite common these days, when trying to identify phases exhibited by mesogens, to use a combination of the following techniques.

1   Thermal analysis.
2   Optical microscopy.
3   Miscibility studies.
4   X-ray diffraction and neutron scattering studies.

It is beyond the scope of this chapter to discuss all the above techniques but the most widely used are thermal analysis and optical microscopy. In thermal analysis, whether it is differential thermal analysis (DTA) or differential scanning calorimetry (DSC), the heat evolved or absorbed by the compound in going from one state of matter to another is detected and appears as a series of peaks called a Thermogram. A typical thermogram (DSC) for a liquid crystalline material is given in Fig. 9.8.

In Fig. 9.8 the first peak represents a transition from the crystal to a S_B phase and, in most cases, the first peak is usually the largest peak. The area under the peak is the enthalpy of the transition (ΔH) and the largest change in this value is always the crystal to mesophase transition. The

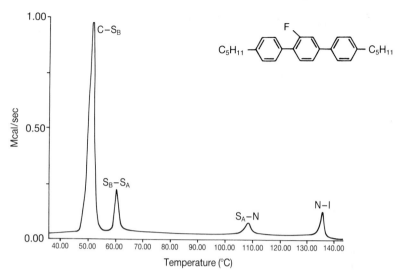

Fig. 9.8   DSC thermogram of a liquid crystalline material.

smaller peaks in the thermogram represent mesophase to mesophase/isotropic liquid transitions, in this case S_B to S_A, S_A to N and N to I. From the thermogram the number of mesophases exhibited by the liquid crystalline material and the temperature at which these transitions occur (and the enthalpy of transition) can be deduced. However, this technique will not give the type of mesophase associated with peaks in the thermogram. For this sort of information a complementary technique, optical microscopy, has to be used.

In optical microscopy a sample of the liquid crystal material, sandwiched between a microscopic slide and cover-slip, is placed into the hotstage assembly of a polarizing microscope; the temperature of the hotstage assembly can be controlled accurately. Many of the smectic/crystal, nematic and cholesteric phases exhibit birefringent patterns or textures, which, when viewed through a polarizing microscope (analyser and polarizer set at 90°), can be used to identify the different types of mesophases. A detailed description of the many types of texture exhibited by liquid crystalline materials is beyond the scope of this chapter, but this subject is covered in an excellent book by Gray and Goodby.[4]

A great deal of experience is needed to identify the phases using optical microscopy, and where there is uncertainty regarding phase identification then other techniques are employed, especially X-ray diffraction and neutron scattering studies which are the only techniques that examine the actual structure of the mesophase.

9.5   Liquid Crystal Polymers

Although this chapter is primarily concerned with thermotropic liquid crystals of the low molar mass type, it is important just to mention liquid crystal polymers, since this is one of the most rapidly expanding areas of liquid crystal research. There are two basic types of liquid crystal polymers, and these are shown in Fig. 9.9.

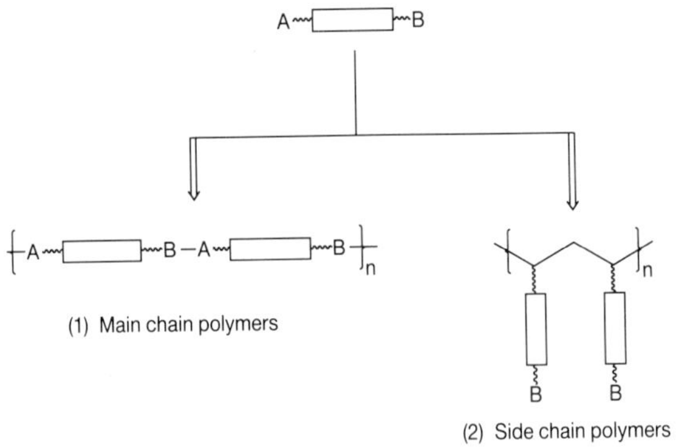

Fig. 9.9   The two basic types of liquid crystal polymers: (1) main chain, (2) side chain.

In Fig. 9.9(1), the mesogenic groups are connected in a head-to-tail fashion, forming the long polymer main chain. Polymers having this type of structure are called *main chain liquid crystal polymers*, an example of which is shown below.

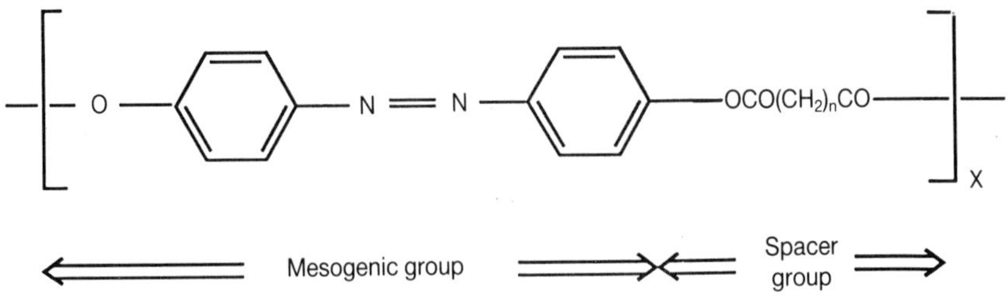

Research in this area of polymer science has led to the development of polymers such as Kevlar, PECK etc., which have outstanding mechanical properties, and are used as high tensile strength materials.

In Fig. 9.9(2), the mesogenic moieties are attached as side groups or chains to the polymer backbone, like teeth on a comb. These polymers are called side chain liquid crystal polymers, an example of which is shown later. Since *side chain liquid crystal* (SCLC) polymers are more relevant to molecular electronics than main chain liquid crystal polymers; we shall now briefly focus on these types of polymers.

Polymer
backbone

CH ——— COO(CH₂)ₙ ——— ⬡ ——— N = N ——— ⬡ ——— CN

CH₂

X

⇐ Spacer group ⇒ ✕ ══════ Mesogenic group ⟹

The unique feature of SCLC polymers is the bringing together in one material of two entities which have vastly different properties. On the one hand we have the mesogenic side chain. The molecules which make up this entity are lath-like or rod-like in structure and so will have the tendency to form well-ordered, anisotropic phases. On the other hand the polymer backbone is very long and flexible and will tend to form a random coil. The problem is how, in the SCLC polymer, do we bring these two entities together. It is a bit like putting a Tory and a Labour MP in the same room and telling them not to argue! Obviously, in SCLC polymers something has got to come between the backbone and side chain so that the random motion of the polymer backbone does not completely disrupt the ordering capability of the mesogenic side chain. Such a mediator in this case is the spacer group, which decouples (not completely) the motions of the polymer backbone from that of the side chain.

By careful design of the above three features of the SCLC polymers, i.e., polymer backbone, spacer group and the mesogenic side chain (see Fig. 9.10) one can control the thermal, rheological and physical properties of the SCLC polymer.[6–9] It must always be remembered that in SCLC polymers, polymer characteristics like degree of polymerization ($\overline{DP}$) and polydispersity (ratio of weight-to-number average molecular weights, M_w/M_n) play an important role in determining the properties of the SCLC polymer.[10–12]

Side chain liquid crystal polymers are used in both electrically and thermally addressed display devices,[12] but the mesogenic side chain and

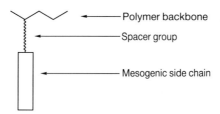

Polymer backbone

Spacer group

Mesogenic side chain

Fig. 9.10   The main features of SCLC polymers.

the polymer backbone need not be physically bound to produce such displays. A material developed for commercial display applications using electrically-induced index matching was that of Ferguson,[13] who applied encapsulation procedures to create micron-size liquid crystal capsules inside a polymer network. Such displays are called Nematic Curvilinear aligned phase (NCAP) display devices. On the application of an electrical field, this strongly light-scattering polymer composite becomes clear. By redesigning the liquid crystalline material and with the addition of dye, it is possible to produce a display device giving black information on a white background (positive contrast). Recently researchers at the Liquid Crystal Institute at Kent State University, USA, have developed phase separation procedures to obtain a dispersion of liquid crystal droplets inside a polymer network for electro-optic switching.[14] This type of system is called Polymer Dispersed Liquid Crystals (PDLC). Phase separation procedures not only greatly extend the types of polymer that can be used for display device purposes but they also simplify display fabrication methods.

9.6 Applications of Liquid Crystals

Already we have discussed two applications of liquid crystals; the use of the temperature dependence of the reflected light of cholesteric materials in temperature sensors, and the use of liquid crystalline materials in polymer dispersed liquid crystals devices such as NCAP. Although there are many applications that use liquid crystalline materials, the major application of these materials is in electro-optic display devices, where an electric field is applied to the liquid crystal to produce an optical effect. There are a number of features which make liquid crystals well suited for application to these types of display.

The mesophase envelops the properties of both the isotropic liquid and solid states of matter in that the mesophase is both fluid (mobile) and structured. This gives a medium in which the molecules can be orientated easily, but also have some degree of freedom of movement. In an electro-optic display device, construction of which is given in Fig. 9.11, the liquid crystal is vacuum filled through a small hole left in the seal and is later closed by epoxy resin.

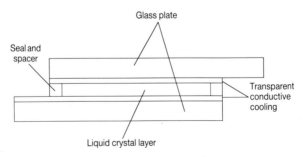

Fig. 9.11 Construction of an electro-optic display.

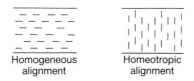

Homogeneous alignment   Homeotropic alignment

Fig. 9.12 Representation of the homogeneous and homotropic alignments.

If the inner surfaces of the supporting glass planes are suitably treated with an aligning agent, the liquid crystal will adopt one of two orientations: the homogeneous or the homeotropic alignments. A representation of these two alignments is given in Fig. 9.12. Thus in a thin cell ($<20\ \mu$m) the liquid crystal will adopt an orientation which will have been determined by the condition of the two supporting surfaces. One of the simplest ways to achieve the homogeneous alignment is to place a thin layer of a polymer, e.g. polyvinyl alcohol, nylon-66, polyimide etc., onto the surface of the glass support and then the preparation is rubbed in one direction only. For homeotropic alignment the glass support is treated with lecithin or trichlorosilanes which contain a long aliphatic chain, e.g. *n*-octyltrichlorosilane.

Orientational Elasticity As well as having the ability to align under the influence of surface alignment, the liquid crystal also has orientational elasticity. In a nematic liquid crystal the molecules are orientated parallel, or near parallel, to the director (n). However, this situation is not perfect and certain deformations are allowed, the free energy of which is given by the Frank–Oseen continuum theory

$$F = \tfrac{1}{2}K_{11}(\operatorname{div}n)^2 + \tfrac{1}{2}K_{22}(n.\operatorname{curl}n)^2 + \tfrac{1}{2}K_{33}(n.\operatorname{curl}n)^2 \qquad (9.2)$$

The orientational elastic constants K_{11}, K_{22} and K_{33} describe the splay, twist and bend deformation[15] and the system will try to resist these deformations (about 10^{-11} N). Thus the liquid crystal system can be aligned easily, but if the system has the required energy certain reorientations of the director are allowed, making the liquid crystal phases 'elastic'. The orientational elastic contrasts are very important in display devices, as we shall see later.

Anisotropic Properties The anisotropic physical properties of liquid crystals are also important in their application to display devices. The most obvious feature is the large anisotropy of refractive index, sometimes called Optical Anisotropy or Birefringence (Δn), which is due to the anisotropic nature of liquid crystalline materials which gives rise to double refraction. This means that these materials will have two refractive indices which, for a nematic phase, are related by the following equation,

$$\Delta n = n_{\parallel} - n_{\perp} \qquad (9.3)$$

where $n_{\parallel}$ and $n_{\perp}$ are the refractive indices measured parallel and perpendicular to the nematic director (n) respectively. In the case of the homogeneous alignment the preparation will appear bright or birefringent, but the homeotropic preparation will be dark or optically extinct because we are viewing the preparation along its optic axis, and double refraction cannot occur. Therefore, by changing the surface alignment we can produce two different optical states, dark and bright, i.e. black and white.

This is a good starting point for most optical effects in liquid crystals but we still need a mechanism by which we can switch electrically from one optical state to the other. The electric permittivity of the liquid crystal, which is also anisotropic, is the one most important factor in performing this task. The anisotropy of electric permittivity (dielectric anisotropy $\Delta\varepsilon$) is given by:

$$\Delta\varepsilon = \varepsilon_{\parallel} - \varepsilon_{\perp}, \tag{9.4}$$

where $\varepsilon_{\parallel}$ and $\varepsilon_{\perp}$ are the electric permittivities measured parallel and perpendicular to the nematic director (n) respectively. The anisotropy of electric permittivity is extremely important because it is the driving force in most display devices. The electric contribution to the free energy density (F_E) contains a term that depends on the angle between the director (n) and the applied electric field E and is given approximately by

$$F_E = -\tfrac{1}{2}\varepsilon_0\Delta\varepsilon(n.E)^2 \tag{9.5}$$

In the absence of any other constraints, the director will rotate to minimize this contribution to the free energy. There are therefore two possible orientations of the director (n) to the applied electric field: the director (n) orients parallel to the applied field for positive anisotropy ($\varepsilon_{\parallel} > \varepsilon_{\perp}$), and perpendicular to the applied field for negative anisotropy ($\varepsilon_{\parallel} < \varepsilon_{\perp}$).

Using the different orientations of the director manifested by both surface alignment and the application of an electric field, a simple electro-optic display device can be constructed. Such a display device is called the Freedericksz Cell (see Fig. 9.13) which clearly demonstrates the following three basic features of most liquid crystal display devices.

1   Surface alignment.
2   Reorientation of the director by applied electric field.
3   Optical effect.

In the top-half of Fig. 9.13, the nematogen must exhibit positive $\Delta\varepsilon$ ($\varepsilon_{\parallel} > \varepsilon_{\perp}$), whereas for the bottom illustration in Fig. 9.13, the nematogen must exhibit negative $\Delta\varepsilon$ ($\varepsilon_{\parallel} < \varepsilon_{\perp}$).

For optima contrast between the 'off' and 'on' states the cell is viewed through crossed polarizers.

Although there are many types of electro-optic display devices which use liquid crystalline materials, we shall only be discussing the following four major types

1   Twisted nematic.
2   Supertwist.
3   Active matrix (TFT).
4   Ferroelectric.

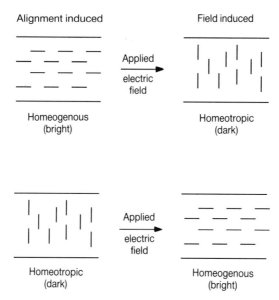

Fig. 9.13   The Freedericksz cell.

Twisted Nematic Display   The twisted nematic (TN) display device made its first appearance in electronic products some 15 years ago, giving a low-voltage, low-power consumption display for watches and calculators. Improvements both in liquid crystalline materials (stability, wider temperature ranges etc.) and device technologies (better uniformity and alignment layers, multiplex addressing schemes etc.), enabled the TN devices to expand into other areas including consumer electronics, automotive instrument panels and test and measurement equipment.

Although the vast majority of commercial liquid crystal display devices use the TN electro-optic effect, the TN device does have a number of shortcomings which supertwisted and active matrix display devices have overcome (see later).

(a) The Twisted Nematic Effect   The basic operation of a conventional twisted nematic display device is given in Fig. 9.14.

In the 'off' state of the TN device (left-hand side of Fig. 9.14) a thin film (6–10 μm) of a nematic liquid crystal is sandwiched between two glass plates (electrodes) whose inner surfaces carry a patterned, transparent conductive coating of indium–tin oxide (ITO). The inner surfaces of the supporting glass plates have also been coated with an alignment layer, usually polyimide, which is then unidirectionally rubbed to align the nematic director (optic axis) at the surfaces parallel to the rubbing direction. The lower glass plate is rubbed at right angles to the upper glass plate, inducing a 90° twist in the nematic director. Polarizing sheets are laminated to the outside of the glass supporting plates so that the linear polarized light entering and leaving the TN device is parallel to the rubbing

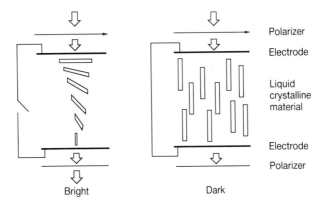

Fig. 9.14   The twisted nematic display device.

direction of the alignment layer at each of two glass plates, i.e. the polarizers are set at 90° to each other. Plane polarized light entering the device and vibrating parallel to the director at the upper surface of the device will now propagate through the nematic medium. Because the director is the optic axis, the plane of polarization of the light will rotate in tune with the twisted structure induced in the nematic phase, so that by the time the light reaches the bottom polarizer the plane of polarization of the emerging beam of polarized light is parallel to the transmission axis of the lower polarizer. In this configuration the TN device will appear bright.

On the application of a small electric field (2–5 volts) (right-hand side of Fig. 9.14), the director reorients and lies parallel to the applied electric field. Thus the twisted structure of the 'off' state disappears and the plane of polarization of the emerging beam of polarized light is now perpendicular to the transmission axis of the bottom polarizer. The light is absorbed by the bottom polarizer causing this part of the TN device to appear dark.

The threshold voltage, V_c, above which the director will reorientate is given by

$$V_c = \pi \sqrt{\frac{K_{11} + \frac{1}{4}(K_{33} - 2K_{22})}{\varepsilon_0 \Delta \varepsilon}} \tag{9.6}$$

This shows the importance not only of orientational elastic constants (K_{11}, K_{22} and K_{33}) but also the dielectric anisotropy ($\Delta \varepsilon$) in this type of display device.

It must be remembered that although many of the liquid crystal materials that we used in the TN display device are polar, as shown below; the applied electric field does *not* align the dipoles; only about 1–2% of the molecules align with the electric field due to their dipolar nature. The applied electric field aligns the *director* and because of this, an AC and not a DC electric field can be used in the TN device. The significance of this is that if a DC electric field was continually being applied to the TN device, because the liquid crystalline material is a dielectric medium, a permanent

C_nH_{2n+1}———CN

$C_nH_{2n+1}O$———CN

C_nH_{2n+1}———CN

charge would build up in the medium and, in time, render the device useless because the device would be permanently in the 'on' state. The use of an AC electric field would prevent this charge accumulation, and would also eliminate the likelihood of the liquid crystalline material being degraded by electrolysis.

The success of the TN device would now depend on whether or not the director could return to its original 90° configuration when the applied electric field is removed, i.e. can the 'off' state be regenerated? Under normal circumstances there is no reason why the 'off' state should be regenerated, but in the TN device the small voltages that are used to orientate the director are not sufficiently strong to affect the molecules very close to the upper and lower glass supporting surfaces. When the applied electric field is removed the molecules close to the glass-supporting surfaces act as 'markers' and the original 'off' state configuration is regenerated.

The information on a TN device can be displayed either as a series of dots or as a display using seven-segment arrays as shown in Fig. 9.15. In some TN devices a hybrid consisting of both dots and segments is used. Only in the dot or segment areas do we have the ITO layer, all around these areas the ITO has been etched away. Therefore, by carefully selecting the correct dots or segments (shaped areas) the required information can be displayed.

In today's ever increasing demand for display devices that can present complex information, such display devices contain a larger number of connections between the display and the electronic circuitry. The increase in the number of connections between the display device and the inte-

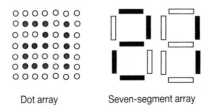

Dot array   Seven-segment array

Fig. 9.15   Information displayed as dots or using seven-segment arrays.

grated circuits becomes large and expensive, and eventually for very complex displays the necessary connections between the display and the electronic circuitry becomes physically impossible. Most of the display devices today—this includes all types of display device using liquid crystalline materials, and not just the TN display device—use multiplexed addressing schemes to overcome this problem.[16]

Commercially available mixtures for TN devices contain a small amount of a chiral dopant to ensure that the helix in the device is always of the same handedness, i.e. that it is either left- or right-handed. If this is not carried out, then disclination walls will appear in the display device which will give the device a patchy appearance.

An example of the use of a TN device in a calculator is given on Plate 9.1, picture 1.

(b) Material Requirements for TN Display   The success of the twisted nematic device (TN) as a display device has produced a good sound basis for the liquid crystal display industry, from which more complicated and faster switching display devices have been developed (see later). A great deal of the success of the TN device, and indeed of the whole of the liquid crystal display industry, has been due to the successful collaboration between chemists, physicists, electrical and electronic engineers. Without stable, good quality liquid crystal materials produced by the chemists, both the physicists and engineers could not have developed displays like the TN. Probably more importantly, they would not have gone on to develop the more complicated and faster switching devices like the supertwist and ferroelectric displays. Liquid crystals indeed represent a new high-technology arena for multidisciplinary collaboration.

The role of molecular structure on the thermal and physical properties of liquid crystalline materials has been studied by many research groups, and at present we have a good understanding of structure/property relationships. It is beyond the scope of this chapter to go into detail regarding structure/property relationships, but to show how the structure of the mesogen can alter both its thermal and physical properties, we have chosen two examples. A more detailed account of the relationship between the structure of the liquid crystalline materials and their physical properties can be found elsewhere.[17–20]

The first example, which relates to the role of molecular structure on the thermal properties of liquid crystalline materials, illustrates the importance of collaboration. In 1986 our research group at the University of Hull embarked on a research programme on the synthesis and evaluation of a small ring system, namely the cyclobutane ring,[21] since at that time very little was known about the effect of incorporating a small ring system on the properties of liquid crystalline materials. The cyclobutane ring can exist in one of two puckered (butterfly) conformations, the structure of which is given later.

A mixture containing the *cis*- and *trans*-isomers of the mesogenic ester was made, and by using high performance liquid chromatography and a considerable amount of patience, the two isomers were separated. We

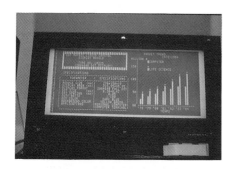

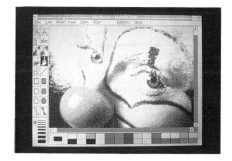

Plate 9.1 (1) Calculator by Sharp using a T/N·display device. (2) Laptop computer by Hitachi using a supertwist display device. (3) Supertwist display device (11.5″ diagonal) showing the large amount of information this type of device can display. (4) Colour TV by Panasonic which uses an active matrix (TFT) display device. (5) 3″ digital colour TV by Epson which uses an active matrix (TFT) display device. (6) 10.4″ diagonal ferroelectric display device developed by Thorn EMI.

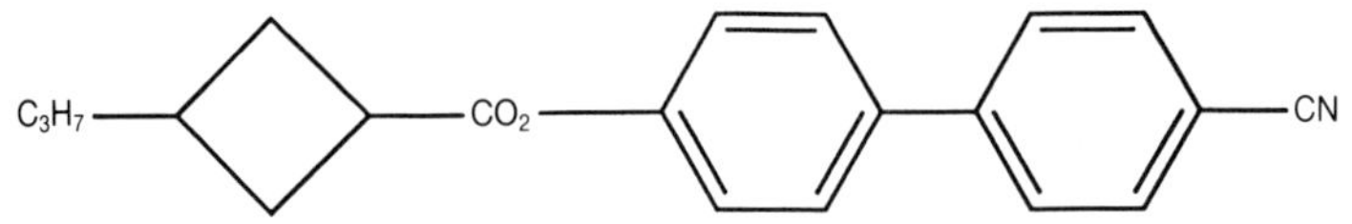

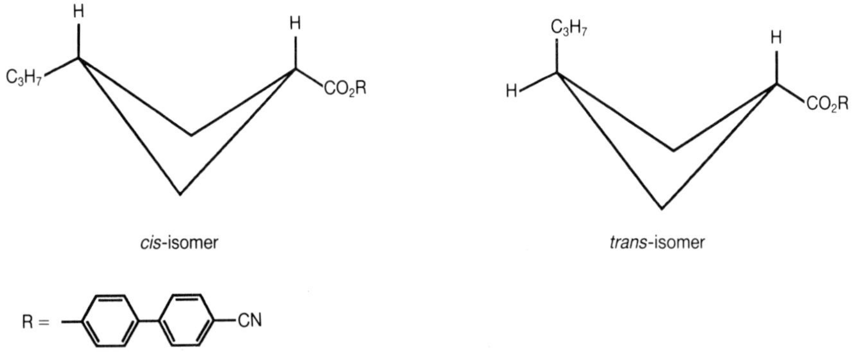

cis-isomer   trans-isomer

found to our astonishment a wide variation in their transition temperatures. The structure and transition temperatures for the *cis*- and *trans*-isomers are given in Fig. 9.16.

The literature was scanned for an explanation for this large difference in transition temperatures, and we found that for simple (non-liquid crystal-

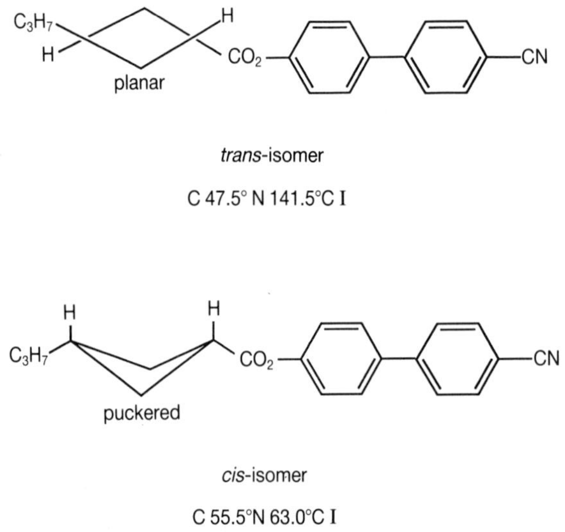

planar

trans-isomer

C 47.5° N 141.5°C I

puckered

cis-isomer

C 55.5°N 63.0°C I

Fig. 9.16 The transition temperatures and conformations for the cyclobutane compounds.

line) compounds incorporating the cyclobutane ring, the conformational structure of the *trans*-cyclobutane ring is temperature dependent. As the temperature is increased the puckered, non-planar conformation of the cyclobutane ring changes to a more planar (linear) form, which is certainly structurally more conducive to the formation of liquid crystal phases.

The *cis*-cyclobutane ring remains in a puckered structure over a wide temperature range. Thus in our particular materials, the more planar conformation of the cyclobutane ring could have significant enhancing effects upon the thermal persistence of the nematic phase found for the *trans*-isomer. At the time this seemed a reasonable explanation as to the wide variations in nematic thermal stability between the two isomers.

However, subsequent examination by X-ray diffraction of the *cis*- and *trans*-isomers by Dr R. M. Richardson at the University of Bristol gave us a much better explanation.[22] He had found different populations of overlapping core dimers for both the *cis*- and *trans*-isomers. Overlapping core dimers are quite common for nematogens incorporating a terminally positioned cyano-group. The structure of the overlapping core dimer is given below. By extending the linearity of the original monomer unit by dimer formation effectively increases the nematic thermal stability of the system.

Core-overlap

What Dr Richardson had found was that at low temperatures there was a low population of dimers for the *trans*-isomer, but as the temperature increased the dimer population for the *trans*-isomer also increased, and thus enhanced the thermal stability of the nematic phase (see Fig. 9.17). However, the converse was true for the *cis*-isomer, and the decrease in dimer population for the *cis*-isomer as the temperature increased would be detrimental to the stability of the nematic phase (see Fig. 9.17). This was

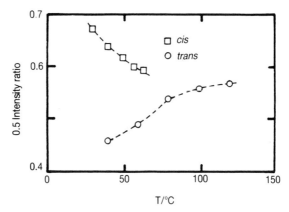

Fig. 9.17   The plot of intensity ratio against temperature (°C) for the *cis*- and *trans*-isomers.

the first time a temperature induced change in molecular conformation had been shown to influence the stability of the nematic phase directly. This clearly shows that one's first reasoned explanation to account for an unusual phenomenon may not always be the correct answer, no matter how well constructed the explanation may be, and that multidisciplinary collaborations can be very useful in this respect.

The second example to illustrate the role of molecular structure on the thermal and physical properties of liquid crystalline materials refers to the physical properties of these materials. In the pursuit of the role of molecular structure on the physical properties of liquid crystal materials, multidisciplinary collaboration is paramount and the success of liquid crystals in electro-optic display devices has sprung most notably from research conducted in a multidisciplinary environment.

Different manufacturers of electro-optic display devices may need specific physical properties from the liquid crystalline material, e.g., in the TN devices, a relatively high birefringence is needed (0.20), but for the supertwisted nematic display device (see later), a much lower birefringence is needed (0.12 to 0.15), and it is only the physicist and/or device engineer who can guide the chemist as to their needs. By manipulating the molecular structure of the liquid crystal material, the organic chemist can vary the birefringence of the material from a low value of 0.03 to a relatively high value of 0.35, as shown in Fig. 9.18. However, although the organic chemist can manipulate the molecular structure to accommodate a number of physical properties, e.g. viscosity, dielectric anisotropy, elastic constants etc., it is very important to realize that a change in molecular structure of a liquid crystalline material to obtain say a high birefringence, will and can adversely affect other physical properties of the material. A good example of this is shown in Fig. 9.19. To obtain a material with a high birefringence, it is important that ring A (Fig.9.19) is aromatic. However, a biphenylyl ring structure would only give a moderate viscosity (a low viscosity is needed in a display device to ensure a fast response time) and a

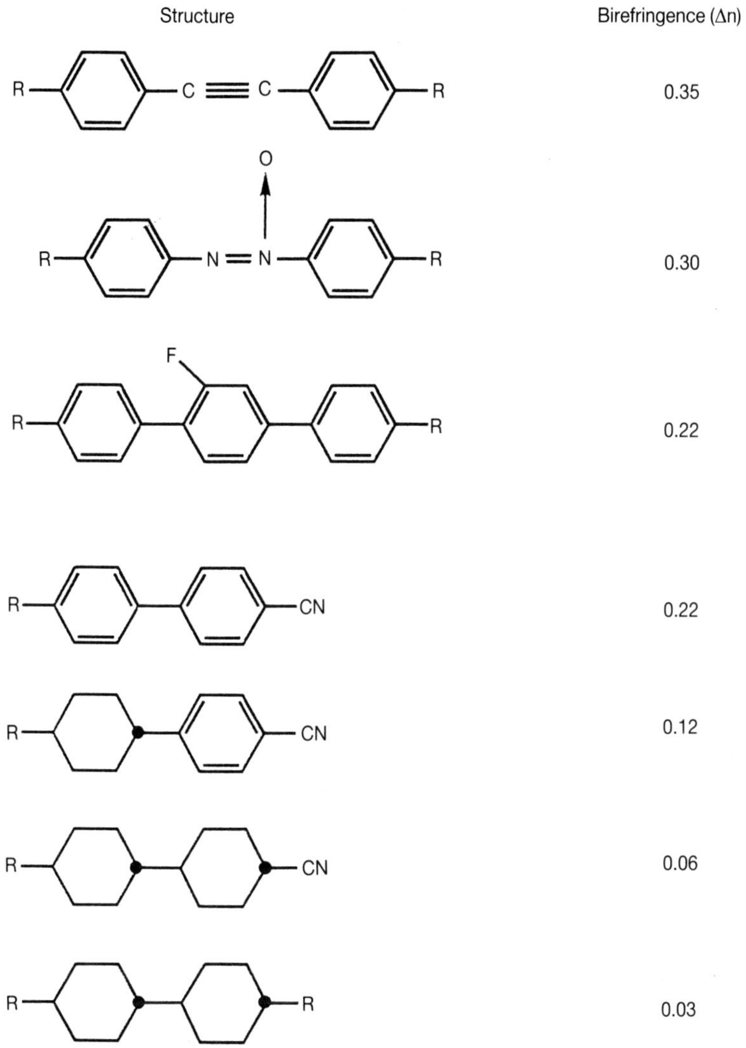

Structure	Birefringence (Δn)
	0.35
	0.30
	0.22
	0.22
	0.12
	0.06
	0.03

Fig. 9.18   The birefringence (Δn) for a selected number of liquid crystalline materials.

poor temperature dependence of the threshold voltage. A low $1/V_C$ ($\delta V_C/\delta t$) is needed to ensure that a change in temperature does not alter the status of the pixels, i.e. a pixel which is in the 'off' state does not change to the 'on' state. Many commercial liquid crystal mixtures are multi-component, to accommodate the vast array of thermal and physical properties required from the liquid crystalline material for optimum performance in electro-optic display devices. Many of these mixtures are complicated both from the viewpoint of their construction and performance, and once again the collaboration between chemists, physicists and

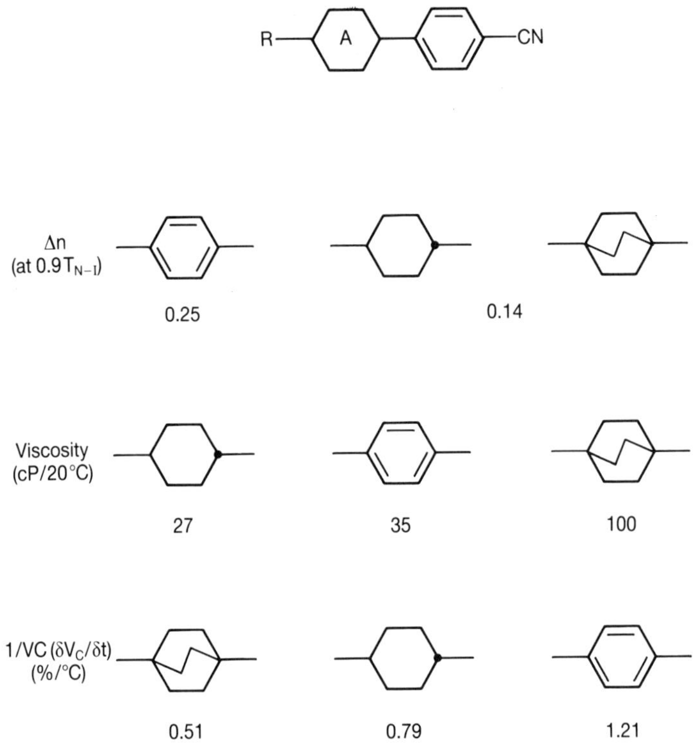

Fig. 9.19 The birefringence (Δn), viscosity and temperature dependence of the threshold voltage for a selected number of two ring compounds.

device engineers is important in the development of high performance, liquid crystal mixtures.

Supertwisted Nematic Display Although the twisted nematic display device is still one of the most commonly used display devices its shortcomings in size, response times and angle of view do not meet today's demands for large area displays, either for high density information display devices (laptop and desktop computers) or for very fast switching devices capable of video frame-rate switching (about 50–60 μs, for TV application). Over the last few years significant advances have been made in display technologies using liquid crystalline materials to overcome the limitations of twisted nematic device, and today we have a considerable range of display devices using liquid crystalline materials.

One such display device which overcomes many of the disadvantages of the TN display device is called the supertwisted (STN) display device.[23] The construction of the STN device is very similar to the TN device (see Fig. 9.20), except that the twist angle has been increased and lies in the range 180–270° (the twist angle varies from one manufacturer to another).

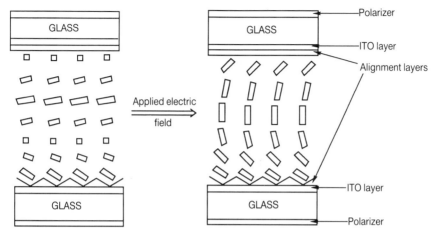

Fig. 9.20   Construction of the STN display device.

Also, one of the alignment layers on the glass supporting surface is so arranged to give the molecules on this surface a pretilt angle. In a STN device the polarizers are oriented 'off-axis' and depending on the Δnd value for the display device, where Δn is the birefringence of the liquid crystalline material used in STN and d is the cell thickness, they are adjusted to give optimum combinations of the contrast ratio and brightness. As with the TN display device, the two polarizers in the STN display device can be arranged to give either the positive contrast mode, with a dark 'on' state and a bright 'off' state, or the negative contrast mode (bright 'on' state, dark 'off' state), which can be achieved by rotating one of the polarizers through 90°. The STN display device gives improved contrast, angle of view and voltage characteristics over the TN display device, and various displays up to A4 in size (30×21 cm) have been produced by a number of Japanese manufacturers, e.g. Sharp, Seiko Epson, Asaki Glass, Hitachi etc.

However, because of the wavelength dependency of the transmitted light, most STN display devices are coloured, having a pale blue or a pale yellow/green background. The coloured background to these STN display devices is a disadvantage since many customers are used to black and white display devices and will not accept the colour generated unintentionally by the STN devices. Furthermore, black and white operation, i.e. no colour dependency, is a necessary condition if a STN device is to give true full-colour operation. To overcome this problem of colour, the double layer supertwisted display device (D-STN) was manufactured in 1987. Attached to a conventional STN layer is a second supertwisted device: a passive compensating layer, of identical thickness (the same Δnd value), but the twist sense of the two layers must be opposite. Thus, if the helix of the chiral nematic in the drive cell is right-handed, then the helix of the chiral nematic in the passive compensating cell must be left-handed. The D-STN display device gives complete compensation over the entire nematic temperature range. By using coloured filters on the pixels a multi-colour display device based on D-STN technology has now been manufactured.

Unfortunately, although giving complete compensation, the D-STN display device is both costly to make (it requires two STN devices to be put together very carefully) and heavy. In 1988 the first film-compensated STN display devices[23] were manufactured, in which the second STN device (i.e. the passive compensating cell) was replaced by a polymer film retardation layer, usually polycarbonate or stretched polyvinyl alcohol. Although this film-compensation layer does not give complete compensation over the entire temperature range, it is much lighter and the film-compensated STN display devices are easier to manufacture. Examples of displays using the STN device are given on Plate 9.1, pictures 2 and 3.

Active Matrix Display   The supertwisted nematic display devices, in all their forms, have now become the most commonly manufactured display devices using liquid crystalline materials, giving full-coloured, flat, large area display devices. Although the supertwisted nematic display device can be manufactured to give large area display devices, it is in fact much slower than the TN device in its response time, that is, in the time taken for the director to respond to the applied electric field. The response time for a typical STN display device is about 200 ms, whereas for the TN display device the response time is about 30–50 ms. Neither of these two display devices have response times fast enough for TV application, but in the active matrix display device—where a thin film transistor (TFT) forms an integral part of the pixel in a TN-type display device—such liquid crystal devices can be used in TV applications.

Such a matrix of these types of pixel can be considered as an array of pixel switches (see Fig. 9.21). When a pixel is selected, a given voltage is applied to that pixel alone and not to any unselected pixel. Because the voltage can be supplied to the pixel very quickly through a TFT switch the response times are very fast; fast enough for TV application. A number of small hand-held *colour* TV sets have now been developed by many

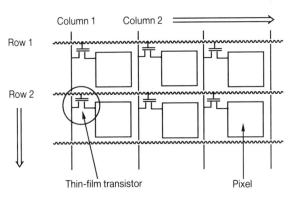

Fig. 9.21   Representation of the active matrix display device.

Japanese companies, e.g. Sharp, Toshiba, Seiko Instruments, Seiko Epson, Hitachi, Matsushita, etc. A number of European companies—e.g. GEC (Borehamwood), CNET and Letti (France), and Philips (The Netherlands)—have also joined the race to produce display devices using TFT technology.Although many televisions using a liquid crystal display are small in size, i.e. 4–6″ diagonal, much larger area screens, e.g. 12.5″ diagonal by Matsushita (480×640 pixels), 14″ diagonal by Sharp and Seiko (440×1950 pixels) and 14.3′ diagonal (1100×1440 pixels) by Toshiba/IBM, are currently being developed for colour TV. Indeed one company, NTT, are thinking of making an experimental 50″ diagonal TFT display for television. However, the active matrix display devices are very expensive to manufacture due to their low yield in manufacture. Examples of the use of active matrix devices are given on Plate 9.1, pictures 4 and 5.

Ferroelectric Display   We have seen so far a number of display devices using liquid crystalline materials which can produce both large area flat panel and fast switching display devices, fast enough for TV applications. In all these display devices, i.e. TN, supertwisted nematic and active matrix, it is the director of the nematic liquid crystalline medium which is being reorientated by the applied electric field.

A competing technology in the race to produce fast switching, large area, flat panel displays is based on ferroelectric *smectic* liquid crystals (FSLC). It may seem a little strange that we have to consider using the more viscous smectic phase to produce a display device which will have a faster response time than display devices based on the more mobile nematic/cholesteric phase. It is the way the molecules are switched to give an optical effect in the FSLC display device which is the key to their success (see later). Although many electronic companies, especially in Japan, have pinned their hopes on active matrix (TFT) to produce large area, flat, fast switching display devices, several companies have now developed ferro-electric display devices as the 'alternative' technology. In 1985 Seiko produced the first large area (12″ diagonal) display device based on FSLC technology, followed a year later by the first multicoloured display device from Toshiba. Canon have so far produced the largest ferroelectric display device, a 14″ diagonal high resolution display device. However, the first ferroelectric display device to work at video frame-rates was not Japanese but British. It has been developed by Thorn EMI (Hayes) from an Alvey/JOERS research programme sponsored by the SERC/DTI. This collaborative research programme involved STC Technology (Harlow), Thorn EMI (Hayes), Merck Ltd (Poole), DRA (Malvern) and the University of Hull's Liquid Crystal Group. This full colour 10.4″ diagonal display device works at TV frame-rates (64 μs per line) and has excellent contrast and angle of view.[24] A picture of this display device is given on Plate 9.1, picture 6.

Ferroelectricity in smectic liquid crystals was first realized by R. B. Meyer in 1974. In the chiral S_C phase (S_C^*), the molecules are arranged as for an ordinary S_C phase, but if the molecules possess a lateral dipole, then this dipole (black dot) will lie parallel to the layer plane but perpendicular

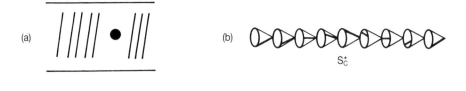

Fig. 9.22 Important features of ferroelectric smectic liquid crystalline materials.

to the tilt plane (Fig. 9.22(a)). However, on passing through a stack of layers (Fig. 9.22(b)), the tilt direction (black line) rotates, producing a helical arrangement and the dipole averages out to zero; the phase is not ferroelectric. If the helix is now unwound however, the dipoles align in one preferred direction and the phase becomes ferroelectric.

On the application of an electric field to the ferroelectric phase the molecules in the S_C^* phase move around the surface of a cone, a natural gyration movement of molecules in the S_C phase, and this, plus the fact that the molecular dipoles (spontaneous polarization) are coupled directly to the applied electric field, enables this display device to operate at video frame-rate. Thus in the 'off' state (Fig. 9.22(c) position 1), the molecular dipoles are facing upwards ('up' state), but on application of an electrical field the dipoles face downwards (position 2, 'down' state), moving through an angle of 2θ where θ is the tilt angle of the S_C^* phase. The dipoles can thus have two possible orientations within a layered structure, and these are shown in Fig. 9.22(d).

In a ferroelectric device (Fig. 9.23(d)) the molecules are homogeneously aligned with the director in the plane of the surface and the layer planes perpendicular to the plates, i.e. in Fig. 9.23, the device is viewed from above. In the 'off' state the molecules (represented by the line) have their dipoles (represented by the open circles) pointing upwards (the 'up' state), and the lower polarizer is aligned parallel to the director; the upper analyser is at 90° to the polarizer. Under these conditions no light can pass through the cell. On the application of an electric field the dipoles move to a downward pointing direction ('down' state; solid black circles). The new director now lies parallel to neither the polarizer nor the analyser and so

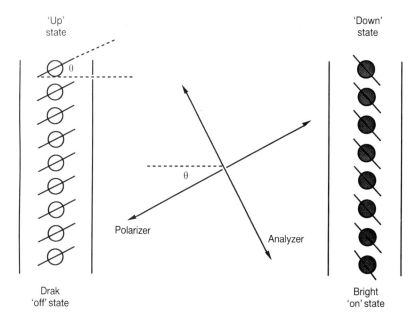

Fig. 9.23 Operation of the ferroelectric liquid crystal display device.

the device becomes bright, although only about 50% of the incident light is transmitted.

9.7 Conclusions

Although liquid crystals may be used in a variety of ways to produce large area displays, probably the most significant advance made recently in liquid crystal research has been the vast decrease in response time of such materials on the application of an electrical field. The development of ferroelectric liquid crystalline materials has pushed the speed of response of such materials to the application of an electric field from ms (10^{-3} second) to μs (10^{-6} second); and with the advent of the electroclinic effect, the best recorded response of a liquid crystalline material to an applied electric field is about 60 ns (10^{-9} second). This is at least five to six orders of magnitude faster than typical switching speeds of liquid crystal display devices of the 1970s and early 1980s.

At present screen sizes of 12 or 14″ diagonal can be commercially manufactured but these types of display devices will not have either the size or resolution for the next generation of television—high definition television or HDTV. Already a number of Japanese electronic companies, e.g. Seiko Epson, Sharp, Sanyo, Hitachi and Matsushita are developing projection systems[25] for HDTV application. The definition required for such an application can be gauged from the fact that the ferroelectric display device developed by Matsushita for their projection system (HDTV) requires *4 million* pixels on each of the three, 3.3″ diagonal display devices. The screen sizes for projection systems can vary from 25 to 100″ (diagonal).

It must be remembered that all these advances in liquid crystal technology, both in terms of materials and their devices, have been made possible due to the willingness of chemists, physicists and device engineers to collaborate. The multidisciplinary nature of liquid crystal research has clearly shown that when there is a willingness for scientists of many disciplines to sit down and truly communicate with one another in a 'common language', anything can be achieved.

Increasing interest in electro-optic and optical devices in telecommunications and computing has realized the potential application of liquid crystalline materials in optical signal processing, i.e. processes where instead of using electricity to perform certain functions in electronic circuitry, light itself is used. The concept of an 'optical computer' is no longer just a pipe-dream for the future, for with the advances now being made in materials for optical processing, such a dream could become a reality—computing at the speed of light.

References

1. S. Chandrasekhar, in *Liquid Crystals; their Physics, Chemistry and Application*, Eds G. Hilsum and E. P. Raynes, (The Royal Society, London, 1983), pp. 93–103.
2. D. Demus, in *Liquid Crystals, Applications and Uses*, Ed. B. Bahadur, (World Scientific, 1990), pp. 28–36.
3. P. J. Collings, *Liquid Crystals, Nature's Delicate Phase of Matter*, (Adam Hilger, Bristol, 1990).
4. G. W. Gray and J. W. Goodby, *Smectic Liquid Crystals, Textures and Structures*, (Leonard Hill, 1984).
5. A. J. Slaney, I. Nishiyama, P. Styring and J. W. Goodby, *J. Mater. Chem*, **2**, 805 (1992).
6. G. Nestor, G. W. Gray, D. Lacey and K. J. Toyne, *Liquid Crystals*, **7**, 669–81 (1990).
7. H. Ringsdorf and A. Schneller, *Makromol. Chem. Rapid Commun.*, **3**, 557 (1982).
8. H. Finkelmann, *Angen. Chem. Int. Engl.*, **26**, 816 (1987).
9. G. W. Gray, J. S. Hill and D. Lacey, *Makromol. Chem.*, **191**, 2227–35 (1990).
10. V. Percec and C. Pugh, in *Side Chain Liquid Crystal Polymers*, Ed C. B. McArdle, (Blackie, Glasgow and London, 1989), pp. 30–100.
11. H. Stevens, G. Reharge and H. Finkelmann, *Macromolecules*, **17**, 851 (1984).
12. H. Finkelmann, *Phil. Trans. R. Soc., London*, **A309**, 105–08 (1983).
13. J. L. Ferguson, *SID Digest of Technical Papers*, **16**, 68 (1985).
14. J. W. Doane, N. A. Vaz, B. G. Wu and S. Zumer, *Appl. Phys. Lett.*, **48**, 269–75 (1986).
15. P. G. de Gennes, in *Liquid Crystals and Plastic Crystals*, Eds G. W. Gray and P. A. Winsor, (Ellis Horwood, Chichester, 1974), pp. 67–91.
16. M. Schadt and W. Helfrich, *Appl. Phys. Lett.*, **18**, 127 (1971).

17. G. W. Gray, *Proc. R. Soc. London*, **A402**, 7 (1983).
18. K. J. Toyne, in *Thermotropic Liquid Crystals*, Ed. G. W. Gray, (John Wiley and Sons, Chichester, 1987), pp. 28–63.
19. D. Coates, in *Liquid Crystals Applications and Uses*, Ed. B. Bahadur, (World Scientific, Singapore, New Jersey, London and Hong Kong, 1990), pp. 92–132.
20. I. Pohl and U. Finkenzeller, in *Liquid Crystals Applications and Uses*, Ed. B. Bahadur, (World Scientific, Singapore, New Jersey, London and Hong Kong, 1990), pp. 140–69.
21. L. K. M. Chan, P. A. Gemmell, G. W. Gray, D. Lacey and K. J. Toyne, *Mol. Cryst. Liq. Cryst.*, **147**, 113–39 (1987).
22. J. M. Allman, R. M. Richardson, J. K. M. Chan, G. W. Gray and D. Lacey, *Liquid Crystals*, **6(1)**, 31–38 (1989).
23. T. Scheffer and J. Nehring, in *Liquid Crystals, Applications and Uses*, Ed. B. Bahadur, (World Scientific, Singapore, New Jersey, London and Hong Kong, 1990), pp. 251–74.
24. P. W. Ross, *Proc. 8th IRDC*, San Diego, 1988, 185.
25. W. P. Bletia and S. E. Shields, in *Liquid Crystals, Applications and Uses*, Ed. B. Bahadur, (World Scientific, Singapore, New Jersey, London and Hong Kong, 1990), pp. 437–91.

10

Langmuir–Blodgett Films

T Richardson

10.1 Introduction

A large part of this book is concerned with the properties of *individual* organic molecules, and how their chemical structure relates to the macroscopic behaviour of the materials observed when large numbers of molecules are grouped together in a solid. Much effort has been directed towards considering how specific molecules, possessing predesigned, functional characteristics may be manipulated and organized into molecular assemblies. Preferably, these assemblies would display a high degree of compatibility with the existing planar technologies associated with the microelectronics and optoelectronics industries. The preparation of organic *thin films* containing functional molecules provides one approach to satisfying this requirement.

One method of producing extremely thin organic films with a high degree of control over their thickness and molecular architecture is the Langmuir–Blodgett deposition technique described in this chapter. This technique enables insoluble molecular films, floating on a water surface, and usually only a single molecule in thickness, to be deposited onto solid substrates whilst retaining their orientational properties. The reader will learn in later sections how multilayer structures can be formed by sequentially transferring such monolayer films. In this context, each monolayer film can be considered as a fundamental building unit in the assembly of a complex organic multilayered structure.

Historically, the study of monomolecular films began in the eighteenth century with the astute observations of Benjamin Franklin, an American politician, during his visits to England. Here he carried out simple yet fruitful experiments using small quantities of oil at a pond in Clapham,

London.[1] He poured a little of the oil onto the water surface which was roughened by the wind, and was struck by the speed at which it spread out and produced a calming effect on the waves. His fascination grew on noticing that the oil film continued to expand so that interference colours could be seen. Eventually, it became so thin that it was no longer visible, although it continued to still the water ripples. As well as noting these facts, Franklin questioned why the spreading of oil on water was so remarkably different to its behaviour when dropped onto a glass surface where it remains drop-like. Although further experiments were carried out by Franklin, the subject was not strengthened scientifically until Rayleigh, in the 1880s, began investigating the lowering of the surface tension of water using oil films spread on its surface.[2] He succeeded in estimating that these films were between 10–20 Å thick, now known to be close to the thickness of monomolecular films of fatty acids. In 1891, Rayleigh received a letter from an amateur scientist, Agnes Pockels, whose work and ideas effectively laid the foundations for present day monolayer research. She described her design of a rectangular water basin filled to the brim with water with a strip of metal placed across its width which just made contact with the water surface.[3] She could move the strip of tin along the basin in order to clean the water surface or to compress a thin oil film floating on the water. In this way, she studied the effects of various oils on the surface tension, and helped Rayleigh to further his own measurements of the sizes of oil molecules.

However, Irving Langmuir is generally accredited as being responsible for the development of improved experimental apparatus, based on Pockels' trough, and for using it to determine accurately the molecular size and orientation of a wide range of materials.[4] He was able to conclude from these experiments that fatty acid type molecules were orientated with their polar functional group immersed in the water, and the long non-polar hydrocarbon chain directed nearly vertically from the surface. His early work concentrated on floating monomolecular layers (Langmuir films) but the arrival at his laboratory of an assistant, Katherine Blodgett, enabled him to turn towards developing the technique of transferring water-borne monolayers onto substrates such as glass slides. Blodgett succeeded in producing the first built-up multilayer films which are now termed Langmuir–Blodgett films.[5] After the Second World War, the subject was little studied until around 1965 when Hans Kuhn began work on the organization of monolayers and their spectroscopic properties.[6] Since this time, the number of related research groups has grown radically because of one important fact. The expertise of today's organic research chemists and physicists has facilitated the design of molecules which possess specific functional properties and which can be used in conjunction with the Langmuir–Blodgett technique. Thus, it is possible to prepare thin films possessing well-defined thickness and molecular infrastructure which display numerous exciting physical phenomena.

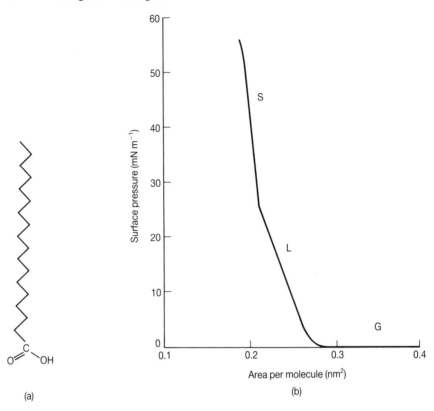

(a)   (b)

Fig. 10.1 (a) The chemical structure of octadecanoic acid, showing the hydrophilic and hydrophobic parts of the molecule. (b) The surface pressure–area isotherm of octadecanoic acid on a water subphase.

10.2 Basic concepts

Formation of Insoluble Monolayers   The most well-known compounds which are able to form monomolecular layers at the air–water interface are the carboxylic (fatty) acids and their metal salts. An example of one such material, octadecanoic acid, is depicted in Fig. 10.1(a). The molecule consists of a hydrophilic (attracted to water) carboxylic acid headgroup, $-CO_2H$, which is soluble in water and a long, hydrophobic (repels water) alkyl chain, $-(CH_2)_{16}CH_3$, which is insoluble in water. In order to obtain a monolayer film of octadecanoic acid, the solid must firstly be dissolved in a suitable solvent such as chloroform or toluene. The resulting solution is then applied drop-wise using a microlitre syringe onto the cleaned surface of an ultra-pure water bath. On contacting the surface, the solution droplets immediately spread outwards across the surface, distributing the solute molecules over the entire surface area of the water subphase. During this spreading process, the organic solvent evaporates, leaving an initially disordered distribution of octadecanoic acid molecules residing at the air–water interface. At this stage, the average displacement between nearest neighbour molecules is relatively large (several molecular lengths) and intermolecular interactions within the layer are weak. As a result, the surface pressure, Π, (that is the difference between the surface tension with γ' and without γ the floating layer, $\Pi = \gamma - \gamma'$) remains at zero. However, the surface pressure can be seen to change dramatically if the surface number density of molecules is increased by reducing the area available to the floating layer.

The specific apparatus required to perform the compression of the Langmuir film will be described later in this chapter. The measurement of Π as a function of the area occupied by the film is one of the most important experiments, allowing the size and molecular orientation of molecules at the air–water interface to be determined. These data are presented in the form of a surface pressure–area isotherm. A typical Π–A plot for octadecanoic acid is shown in Fig. 10.1(b). There are several distinct regions of the isotherm, each one of which corresponds to a particular type of phase within the film. At large area (G), the surface pressure remains at zero (more correctly a very low pressure of about $0.1\,\mathrm{mNm^{-1}}$ is observed if the pressure sensor is sensitive enough) since there is little interaction between the octadecanoic acid molecules. As the area is reduced (L), an abrupt transition occurs in which the surface pressure begins to rise linearly. In this region, the molecules are beginning to orientate on the water surface such that the alkyl chains protrude approximately normally to the water surface although the molecules are not yet close-packed. This phase can be thought of as corresponding to a two-dimensional liquid state.

Upon further compression (S), the surface pressure undergoes another transition in which it begins to rise even more steeply. This represents a phase-change to an ordered arrangement of the molecules within a two-dimensional, quasi-solid layer which is highly incompressible. If this zone of the isotherm is extrapolated to zero surface pressure, the area per molecule within the solid layer can be found. (This actually represents the hypothetical case of a close-packed structure which is not under compression). Clearly, for octadecanoic acid, it can be seen that the area occupied by each molecule is $0.22\,\mathrm{nm^2}$, which is in excellent agreement with the modelled value of the cross-sectional area of the molecule. It can thus be concluded (there are many other methods) that the floating layer is monomolecular in thickness.[7] If the area is reduced still further, eventually the monolayer will collapse or buckle.[8]

By introducing divalent metal chlorides into the water subphase, it is also possible to incorporate divalent cations into the monolayer of octadecanoic acid. The exact degree of cation uptake depends on the pH of the subphase as well as the stability of the aqueous ion. It is relatively straightforward to obtain mixed monolayers containing both octadecanoic acid, for example, and cadmium octadecanoate if the pH lies in the range 4.2–6.4. Several different ions can be incorporated[9] such as Ca^{2+}, Ba^{2+}, Mn^{2+}, Mg^{2+}, Cu^{2+} Al^{3+} and Fe^{3+}, although not all are straightforward to use. It is also possible to absorb dye molecules, dissolved initially in the subphase, onto a floating monolayer.[10] This is useful in cases where the dye materials of interest are soluble in water.

The fatty acids are not the only compounds which form monolayer films. There are many other different types of molecule which can be spread from organic solvents on a water subphase to produce such Langmuir layers. Amongst them are different kinds of dye molecules,[11] fluorescent compounds,[12] biological proteins/enzymes,[13] oligomeric and polymeric layers[14] and even the much discussed fullerenes.[15] It is beyond the scope of this chapter to detail each category and so only a few of the materials particularly useful in applied research areas will be mentioned (see Section 10.4).

Langmuir–Blodgett Deposition   Floating monomolecular layers can be transferred onto solid substrates in a number of different ways. By far the most common method used by the majority of research groups studying organic monolayers and multilayers is the Langmuir–Blodgett deposition technique.[16] This consists of inserting and withdrawing a substrate (a glass plate or silicon wafer for example) through the monolayer–water interface at a well controlled rate, whilst maintaining a constant surface pressure within the monolayer. The mechanisms which enable the Langmuir layer to bind to a solid substrate are not fully understood, and indeed may vary depending upon the nature of the molecules within the monolayer and of the substrate surface. However, the principle features associated with the LB process can be understood by considering the deposition of a simple fatty acid material such as the previous example, octadecanoic acid.

If the substrate is chemically treated so that its surface is hydrophilic, the first monolayer can be transferred as the substrate moves upwards through the water–monolayer interface. It is crucial that the surface pressure of the monolayer, which would normally be chosen at a value within the quasi-solid region of the isotherm (typically $30\,\mathrm{mNm}^{-1}$), should be maintained at a constant value throughout the deposition process. Otherwise, as soon as some material has been removed from the floating monolayer (by deposition onto the substrate) the remaining molecules reorganize and the surface pressure falls dramatically. A constant pressure is achieved by means of an electronic feedback mechanism which enables the area available to the film to be reduced until the preset pressure is regained. More will be said about Langmuir trough designs later in this section. The transfer of the first monolayer onto a hydrophilic substrate is depicted in Fig. 10.2(a). It can be seen that the hydrophilic (polar) carboxylic acid headgroups are bound to the substrate surface with the hydrocarbon chains orientated roughly normal to the substrate plane. If the substrate is subsequently lowered through the floating monolayer, a second layer will be deposited, this time with the hydrophobic alkyl chains of the second layer lying on top of those from the first, Fig. 10.2(b). A further upstroke deposits the third layer, Fig. 10.2(c) and so on. It is thus possible to transfer many monolayers sequentially by repeating this process. The resulting structure is known as a Y-type structure and is shown in Fig.

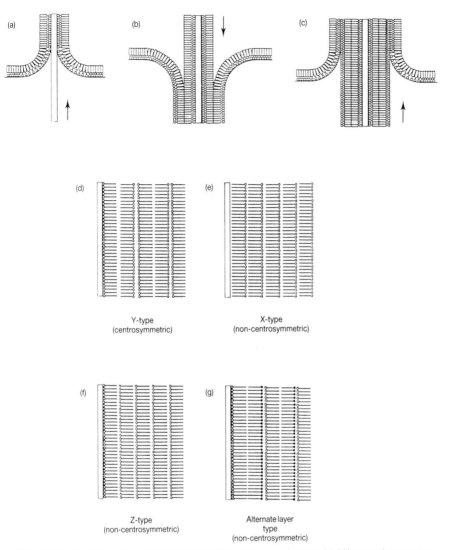

Fig. 10.2 The LB transfer process showing the deposition of an amphiphilic monolayer on to a hydrophilic substrate (a, b, c). The four possible structures achievable using the LB technique: (d) Y-type, (e) X-type, (f) Z-type and (g) the alternate-layer superlattice.

10.2(d). As the floating monolayer is depleted, it is obviously necessary to replenish the supply of material by respreading a further aliquot of solution.

An important parameter which aids characterization of the deposition process is the *transfer ratio*. This is simply given by τ where:

$$\tau = A_L/A_{LB} \tag{10.1}$$

where A_L is the reduction in the area occupied by the monolayer on the water surface and A_{LB} is the area of the solid substrate coated. Ideally, a transfer ratio of unity would always be obtained, but in practice values between 0.9 and 1.0 are achieved (for fatty acids). The mode of deposition outlined above is termed Y-type deposition, but there are other forms of transfer which lead to modified molecular architectures within the built-up LB multilayers. It is possible to deposit certain materials only as the substrate moves downwards through the monolayer, no transfer being achieved during each upstroke. The resulting layer structure is shown in Fig. 10.2(e), the process being referred to as X-type deposition. Alternatively, transfer is sometimes only successful on each upstroke; this mode is called Z-type deposition and the structure is depicted in Fig. 10.2(f).

Another deposition mode which, for many applications, is the most important is the alternate layer method.[17] This involves the codeposition of two *different* monolayers each confined within its own area of the water surface. The monolayers are deposited alternately such that the substrate may, for example, be inserted through the first monolayer and withdrawn from the second monolayer and so on. In this way, it is possible to build up ABABA . . . type structures. The structure of such assemblies is shown in Fig. 10.2(g). It is very similar to the Y-type structure (Fig. 10.2(d)) in that there are head-to-head and tail-to-tail interactions between adjacent layers. The usual difference, however, is that the headgroups (and sometimes the hydrophobic regions additionally) of the two materials are quite different in chemical structure. For example, a great deal of work has been done on the long chain alkanoic acid/alkylamine alternate layer system in which the hydrophobic alkyl chains of both materials are effectively identical but the headgroups are different chemically, being $-CO_2H$ and NH_2 respectively.[18] In this case and in other systems, the alternate layer method prevents the total cancellation of the dipole moments within each monolayer, thus allowing a macroscopic polarization to exist. As will be described later, such a polar assembly of molecules can give rise to several exciting physical phenomena.

The experimental instrumentation which is required to prepare both Langmuir and LB films is known as a Langmuir trough. Many different designs exist[19] and several types of trough are available commercially. However, all have several common features which will be outlined here. Firstly, there must be a means of containing the subphase (which is usually pure water) on which the spread monolayer will float. Normally this is a trough made out of a relatively inert material, such as PTFE, which is straightforward to clean in order to minimize contamination of the floating monolayer. The water which is held by the trough must be extremely pure. As mentioned previously, metal ions present in the subphase can be taken up by a monolayer. Therefore, significantly large amounts of ions or other impurities must be absent from the subphase unless added purposely in well-defined quantities. Ultra-pure water polishing systems are generally used although these are very expensive.

In order to confine the floating monolayer to a well-defined surface area, it is necessary to use some form of barrier. Pockels[3] and subsequently

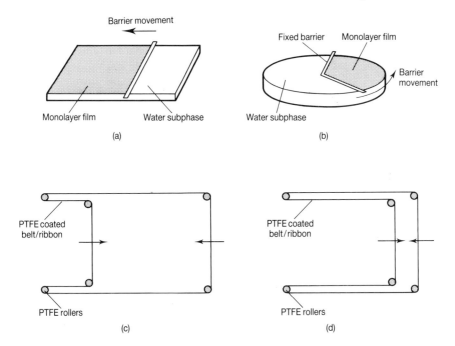

Fig. 10.3 Barrier systems used in Langmuir troughs. Single moveable barrier for (a) rectangular trough and (b) circular trough. The maximum (c) and minimum (d) area configurations for a constant perimeter barrier system.

Langmuir and Blodgett[4,5] used a single moveable barrier which could be swept across a tank filled to the brim with water, on which floated their monolayer film, as shown schematically in Fig. 10.3(a,b). Such a barrier can be used either with a rectangular or circular trough as indicated. Another means of confinement is to use a constant perimeter barrier. In this case the monolayer is enclosed within a continuous flexible belt which is semi-submerged in the water and held in place by small PTFE rollers. By changing the position of some of these rollers using sophisticated motor drives, the exact shape of the barrier can be modified so that the area it bounds can be changed continuously as shown in Fig. 10.3(c,d). Such a barrier avoids possible leakage problems associated with single barrier systems (although these occur infrequently in the well-engineered troughs available commercially). The moveable part of any of these types of barrier is usually connected to a high resolution potentiometer, such that the position of the barrier drive arm can be related to the area of the confined monolayer simply by measuring the voltage across the variable resistor.

Once able to compress a monolayer, a means of monitoring the surface pressure is required. Although there are several techniques, the most common is to use a Wilhelmy plate arrangement. This consists of suspending a small (typically 2 cm) length of filter paper from a very sensitive microbalance via a thin wire. A change in surface pressure within the monolayer, arising from the change in surface tension caused by the compression, is detected because of the change in the force experienced by the partially immersed filter paper. The sensitivity of this method is typically $10^{-2}\,\mathrm{mNm}^{-1}$ which is much greater than is required in most practical situations. The surface pressure measurement device and the barrier motor drives are coupled via a sensitive electronic feedback system; and this allows the area available to the floating monolayer to be modified automatically so that its surface pressure can be kept constant at all times during deposition. This coupling can, of course, be disabled when a surface pressure–area isotherm is recorded.

The substrate is moved through the monolayer–water interface normally by means of clamping to a motor-driven micrometer via a suitable gearbox, so that typical deposition rates of 0.5–$10\,\mathrm{mm\,min}^{-1}$ are achieved. The objective is to obtain a smooth movement of the substrate at a well-controlled, constant speed. A schematic diagram of a constant perimeter single layer Langmuir trough is shown in Fig. 10.4(a).

Finally in this section a few words will be said about the alternate-layer trough. It is sometimes desirable to prepare ABABA . . . type assemblies which contain two different materials. Without special equipment, this would be extremely time-consuming as it would be necessary to spread a monolayer of each compound alternately throughout the preparation of the multilayer. Consequently, troughs have been designed which contain two independent surface zones on which monolayers of different types can be manipulated. The subphase, however, is common to both monolayers. The surface pressure of each layer is independently controllable, because each surface zone is separated by a fixed central barrier on which the deposition mechanism is housed. A schematic diagram of one example of an alternate layer trough is shown in Fig. 10.4(b). In this case, the substrate is rotated about an axle which runs along the central barrier, firstly moving downwards through monolayer A followed by withdrawal through mono-layer B. Thus, alternating structures can be prepared.

10.3   Characterization of LB Assemblies

There are many techniques for characterizing and evaluating the properties of Langmuir monolayers and transferred LB films, only a few of which can be described in detail in this chapter. This section will be divided into two parts. Firstly, some of the principal methods associated with characterizing the deposition itself will be discussed, followed by a synopsis of the structural techniques used to gain insight into the arrangement of molecules within the multilayers.

General Methods   The most straightforward method of evaluating the deposition of a monolayer onto a solid substrate is simply to record the variation in the area occupied by the monolayer (at constant surface pressure) as a function of time. This leads to the quantification of the deposition process in that the transfer ratio may be derived. This technique

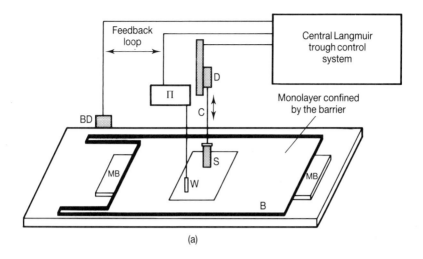

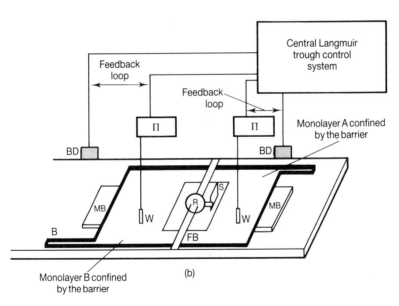

Fig. 10.4   (a) Constant perimeter single layer Langmuir trough showing the barrier ribbon (B), the barrier drive motor (BD), the moveable barrier (MB), the surface pressure monitor (W and Π), the substrate (S) and its holder (C), and the motor driven micrometer deposition head (D). (b) Constant perimeter alternate layer Langmuir trough showing the fixed central barrier (FB) and the rotary deposition mechanism (R). All other labels are the same as in (a).

is not realiable in cases where the monolayer is relatively unstable on the water surface, that is if the area diminishes slowly due to slight solubility in the subphase.

If an LB film absorbs ultraviolet or visible radiation in the region

280–1200 nm, then its absorption or transmission spectrum may be recorded at various stages of the sample preparation. For instance, Fig. 10.5(a) shows the optical absorbance spectrum for an organoruthenium complex LB film, indicating the linear variation in peak absorbance as a function of the number of layers transferred. This confirms that the sequential deposition of this monolayer film has occurred reproducibly. This is not always the case, however, and a deviation from linearity (at thicknesses $\ll \lambda/4$ where $\lambda =$ the wavelength of the light) alerts the researcher of a potential deposition problem. Consideration must be given to interference effects which arise when the film thickness approaches a multiple of the quarter wavelength of the incident light. Interference between beams reflected from different surfaces within the multilayer structure causes changes in the reflectivity of the multilayer, which can often show up as changes in the apparent absorbance of the film. This technique can also be used to study the uniformity of the transferred layers by scanning across the sample.

The quartz crystal microbalance technique[20] is also used to characterize the deposition of LB films. This uses a small (about 1 cm^2) piezoelectric quartz disc electroded on each face, which is forced to vibrate using a simple oscillator drive circuit, and this causes it to vibrate at its fundamental frequency. If a coating is applied to the crystal surfaces the frequency of oscillation decreases, rather like the mass-loading effect on a spring. Furthermore, the frequency change is directly proportional to the mass added to the system. It may be expected therefore that successive depositions of single monolayers would give rise to a total mass change, which would simply be the product of the number of layers and the mass per monolayer (per unit area). The total frequency change would be Δf_{tot} where:

$$\Delta f_{tot} = N \Delta f_L \qquad (10.2)$$

where $\Delta f_L =$ the frequency change per monolayer. Figure 10.5(b) shows a plot of the frequency change versus the number of bilayers of stearic acid transferred onto a 6 MHz quartz crystal. The frequency change per bilayer is approximately 120 Hz. It is straightforward to measure frequencies in the MHz region to a precision of around ± 1–5 Hz, so the technique is extremely sensitive. The frequency change per unit mass is also proportional to the nominal fundamental frequency squared, so that for an 18 MHz device a frequency change of ~1 kHz per bilayer is observed.

The last example of a general characterization method to be discussed here is electrical in nature and involves measuring the capacitance of a metal–insulator–metal structure in which the insulating dielectric material is the LB film assembly.[21] One of the main features of the LB technique is that, since single monolayers form the fundamental building blocks of the LB films, the total film thickness is well defined. The capacitance is simply related to the film thickness by C where:

$$C = A \varepsilon_0 \varepsilon_r / d \qquad (10.3)$$

where $A =$ the electrode area, $\varepsilon_0 =$ the permittivity of free space, $\varepsilon_r =$ the

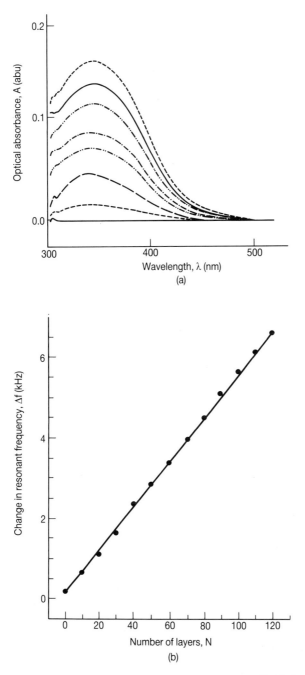

Fig. 10.5 (a) Optical absorbance spectra for LB layers of an organometallic LB film for 2, 6, 10, 16, 18 and 22 layers transferred on to a hydrophobic substrate; and (b) the change in resonant frequency of a quartz crystal oscillator versus number of transferred layers of stearic acid.

relative permittivity and d = the total film thickness which in principle is equal to the number of transferred monolayers, N, multiplied by the thickness per layer. A plot of reciprocal capacitance versus N should yield a straight line whose gradient will be the dielectric thickness, d/ε_r, provided that the deposition is reproducible from layer to layer. The thickness of the film may also be evaluated if the permittivity of the material (at the frequency of the measurement) is known.

Structural Characterization Techniques   There are a number of different techniques which can be used to obtain structural information regarding LB films, such as X-ray, electron and neutron diffraction, infra-red spectroscopy and Raman scattering to name only a selection. Most of the X-ray diffraction work has been aimed at long chain fatty acid multilayers containing divalent metal ions.[22] The presence of heavy metal ions gives rise to strong X-ray scattering; that from hydrogen and carbon alone is comparatively weak. Thus, most of the data collected from experimentalists refer to the interplanar distances associated with the planes of cadmium, barium or manganese ions. It can be seen from Fig. 10.6(a) that it is possible to observe many orders of Bragg diffraction peaks corresponding to the interplanar lattice spacings of the multilayer.[23] The inter-cadmium plane d-spacings yield monolayer thicknesses in excellent agreement with thickness measurements using interferometric techniques.

Electron diffraction has enabled the particular crystal structures to be identified successfully for the simple fatty acid type films.[24] Spot diffraction patterns have been obtained indicating that the material irradiated is in the form of a single crystal at least over dimensions corresponding to the beam width. Less work has been done using neutron irradiation although Bragg reflections have similarly been observed for fatty acid salts.

Infrared spectroscopy has been used increasingly in the last decade to study layer structure and the intra-layer molecular organization.[25] Within organic molecules, infrared radiation excites vibrational transitions associated with bond stretching, twisting or rocking. Each organic molecule possesses a specific infrared absorption spectrum, depending primarily upon the particular chemical groups present within the molecule and their environment. The technique has been developed so that data from very small numbers of layers can be obtained. A wide range of LB materials have been studied from the simple fatty acids to complex biological molecules such as the potassium ion binding valinomycin. The example of Fig. 10.6(b) shows how infrared spectroscopy can be used to gather information regarding changes in molecular structure and environment which occur when different surfaces are used as the LB film substrate. This is a powerful method of great importance to the study of molecular interactions within LB films.

In this section, only a very small proportion of the available techniques capable of providing useful information about the structure and properties of LB films has been presented. The interested reader can refer to the longer review articles[19] for a more in-depth account of specific methods.

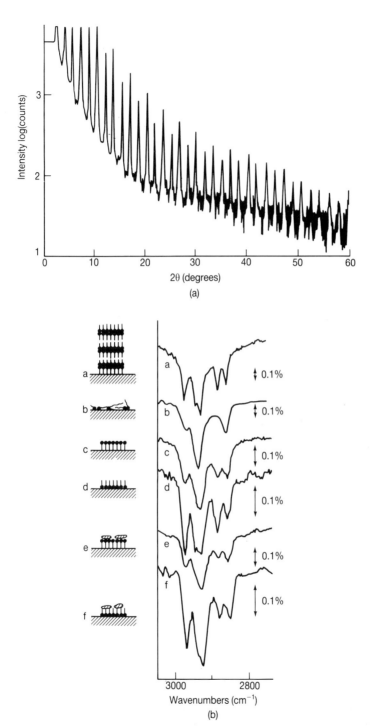

Fig. 10.6 (a) X-ray diffraction peaks of a 50 layer cadmium arachidate LB film deposited on a Nima trough (after Amm and Johnson[23]) and (b) infrared spectra obtained for cadmium arachidate monolayer assemblies deposited on to a range of surfaces (after Knoll et al.[38]).

10.4   Fundamental and Applied Research

This section provides a summary of the main subject areas in which LB films are currently being studied throughout the world. These can be divided into two major categories, the first concerned with the use of Langmuir and LB films as model systems for a broad range of scientific disciplines, and the second involving applied areas of research awaiting industrial and commercial exploitation. The applied areas will be given more attention, in order to give the reader a strong insight into the range of potential applications of LB films.

Fundamental Research   The LB technique has been used extensively in the photophysical and biological fields as a method of assembling artificial arrays of cooperating molecules on a solid substrate. On the photophysical side, Kuhn has studied the fluorescence from mixed layer films containing synthetic dye molecules in a matrix of fatty acid molecules.[6] One example of his work is depicted in Fig. 10.7(a) in which a block of monolayers of a compound X, which absorbs in the uv part of the spectrum and fluoresces in the blue, is separated from another block of monolayers containing a blue-absorbing dye, Y, which fluoresces in the yellow. The spacer layers between these two fluorescent blocks are monolayers of simple acids. Provided the separation between the regions X and Y is sufficiently wide, the fluorescence emitted is that of X as Y does not absorb in the uv region. However, below a threshold separation, the excitation energy of X is transferred to Y and the yellow fluorescence of Y appears (Fig. 10.7(b)). It is believed that the energy transfer occurs via quantum mechanical tunnelling through the spacer layers.

The main driving force for the use of LB films within the biological sphere has been the close resemblance of many classical LB materials to naturally occurring biological cell membranes.[26] It has been suggested that a two layer LB film might provide a suitable model for a cell membrane albeit in an artificial environment. Additionally, there has been great interest in immobilizing biological molecules such as enzymes and proteins which can actively bind specific ions or small molecules, with a resulting change in some physical or chemical property. Hence, biochemical or biophysical sensors can be envisaged possessing a highly selective response.

Applied Research   It is the diversity of potential applications of LB films which has maintained the enthusiasm and dedication of researchers working in this area. Although very few truly commercial products incorporating LB layers exist at present, it must be remembered that the field is still relatively young. The modern era of LB research is only 20 or so years old, although it has only been during the last decade that large numbers (200–300) of research groups have arisen. Only a small selection of potential applications can be included here.

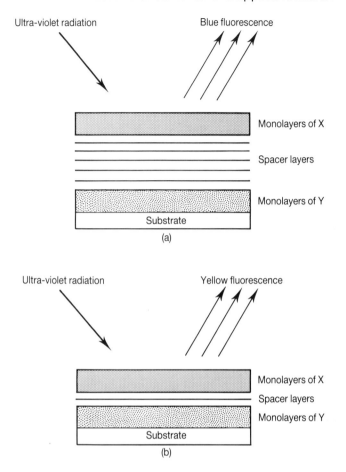

Fig. 10.7   Schematic diagram showing a block of dye monolayers X separated from a block of dye monolayers Y by fatty acid spacer monolayers. (a) The blue fluorescence of X appears, but in (b) the yellow fluorescence of Y is observed, even though Y does not absorb in the uv region.

Passive Uses of LB Multilayers

One of the passive applications of certain LB film materials which was recognized in the 1980s involves their use as microlithographic resists.[27] The need for faster and greater memory storage capacity within electronic integrated circuits has meant that the techniques for patterning electronic components on semiconductor wafers have undergone constant retuning in order to obtain increasingly smaller circuit elements. The main disadvantage with electron-beam lithography, which is used a great deal at present, is that although the beam can be focused to a spot of diameter of only 10 nm to polymerize (and therefore pattern) a resist layer, a large halo is formed around the point of impact due to scattering within the resist material itself. The resolution achievable is only improved by using extremely thin resists. By using LB films, extremely thin layers of precisely controlled thickness can be prepared, although this preparation is presently more time-consuming than other thin film coating methods. Remarkable resolution (better than 10 nm) has been achieved using simple fatty acid salts; these sublime when irradiated with strong electron beam radiation, therefore acting as positive resists. Even better performance can be obtained using negative resists, such as 22-tricosenoic acid which polymerizes when irradiated.

Two further passive applications of LB films are magnetic tape lubrication and the alignment of liquid crystals. Both rely on the change in the nature of a surface which the transferred LB film imparts. Lubrication is an important area in the magnetic tape industry; the application of a lubricant bound to the surface of magnetic tape reduces wear during passages of the tape across the tape-head. Seto et al.[28] showed that the use of only seven layers of barium stearate was sufficient to reduce the friction coefficient of an evaporated cobalt tape by roughly a factor of four after a hundred transits across the tape-head. Whilst this effect is dramatic, there still appears to be no systematic study of lubrication properties of LB films.

An industry as important as the magnetic tape industry is the liquid crystal device sector. A great deal of current research is being directed towards the properties of ferroelectric liquid crystals since they do not require addressing by a thin-film transistor in display applications. A fundamental requirement associated with any liquid crystal is that of alignment. Some effort previously has been given to investigating the alignment effect of particular LB layers[29] although again little systematic work has yet been carried out on entire families of LB molecules. The control of the LB film structure and thickness suggests that the mechanisms which cause liquid crystal alignment might be elucidated in the future.

Active Uses of LB Multilayers
There are several modern uses of LB films in which the monolayer or multilayer provides a more active function than in the cases mentioned above. A few examples from two broad categories will be discussed here, namely applied optics and sensing. A subject of intense current interest in the field of organic materials is electroluminescence. This is the emission of (preferably) visible radiation which occurs when an electric field is applied across a suitable material. The first report of electroluminescence in an LB film was made by Roberts et al.[30] who used a lightly substituted anthracene derivative, shown in Fig. 10.8(a). Although blue emission was achieved, the cells had a lifetime of only 2–3 minutes. This was attributed to chemical degradation of the anthracene molecules or crystallization brought about as a result of heating effects. As the films were extremely thin, only a relatively low voltage (6 V) was required to obtain a certain electroluminescence intensity, compared with about 200 V for an evaporated anthracene film. Unfortunately, work in this area quickly lost momentum, and only in the last 2–3 years have other groups, particularly in Japan, made further progress. Era et al.[31] have studied a range of evaporated organic layers and an LB film of a cyanine dye, shown in Fig. 10.8(b). The

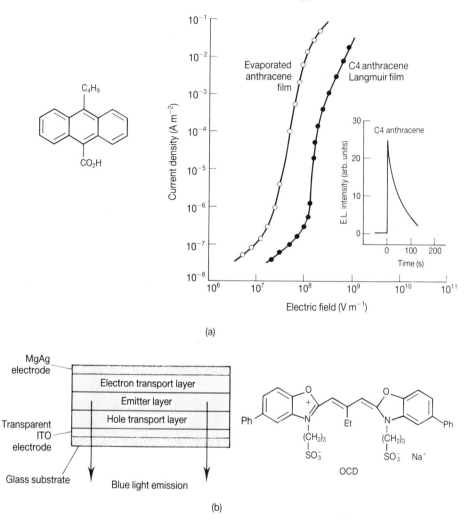

Fig. 10.8 (a) Lightly substituted anthracene molecule (left) and the corresponding emission data (right); and (b) the heterostructure (left) used by Adachi *et al.*[31], showing the cyanine dye (OCD) molecules used (right).

device is actually a three layer heterostructure with the active LB emitting layer sandwiched between charge transport layers, which improve the injection of charge into the device. It is thought that the emission process occurs as a consequence of electron-hole recombination within the emitter layer. The area of electroluminescence is important to the development of flat-panel colour displays; it will no doubt receive increasing attention during the next few years.

Another optical process which has been investigated in LB films for several years is optical second harmonic generation (SHG). Certain organic molecules possess highly conjugated, delocalized π-electron systems substituted with powerful electron-withdrawing or donating groups. Such materials can be polarized more easily in one direction than in the opposite direction. Thus when the molecules interact with the intense electric field associated with a laser beam, the electrical dipoles are set into nonlinear oscillation, which leads to the emission of some photons with different frequencies to the fundamental input frequency (see Chapter 5). The materials generate photons at twice the input frequency; this means that if the infrared (1.06 μm) of a Nd:YAG laser is incident upon these materials, some of the output will consist of visible radiation at 0.532 μm (green). For this conversion to occur, the molecules must be arranged noncentrosymmetrically within the LB film. Therefore the monolayers must be deposited either X or Z-type, or the alternate layer method must be used. Much of the early work was carried out on hemicyanine and nitrostilbene molecules[32] which possess dipole moments (more correctly molecular hyperpolarizabilities), whose directions occur in opposite senses with respect to the hydrophobic alkyl chains. When deposited in the ABABA... type structure, their individual dipoles add and a large macroscopic polarization is built up. Recently, Ashwell et al.[33] have designed and characterized a series of zwitterionic compounds which show extremely high second-order nonlinear optical coefficients. These molecules, one of which is shown in Fig. 10.9(a), can be deposited in the Z-type manner with little indication of the structural instability often shown by other Z-type dipolar molecules. The intensity of the second-harmonic radiation should theoretically increase quadratically with the thickness of the sample. Many researchers have found that this behaviour can be observed only for a relatively small number of layers, the subsequent saturation being attributed to the progressively worsening dipolar alignment within the LB film. However, this trend has been observed for more than 50 layers for the zwitterionic molecule discussed above, as is shown in Fig. 10.9(b). There are several applications for materials such as these, the principle one being electro-optic switching. Future developments in this field will undoubtedly attract further attention from industry.

The term 'sensors' is almost infinitely broad, and as such is appropriate for describing some of the potential applications of LB films in which the multilayer responds in some fashion to an external stimulus. From the many examples of LB sensors designed to date, only a small selection will be discussed here.

Firstly, pyroelectric detection offers a method of sensing temperature changes induced by the absorption of incident infrared photons. A pyroelectric material is one which possesses a temperature-dependent spontaneous polarization (see Chapter 3). Within LB film structures, the temperature dependence probably arises from ionic interactions between headgroup of molecules lying within adjacent monolayers, or via a realignment of polar chemical groups. The effect can be studied by short-circuiting a capacitor (metal-insulator LB-metal) device through a sensitive current-measuring system and by successively heating and cooling

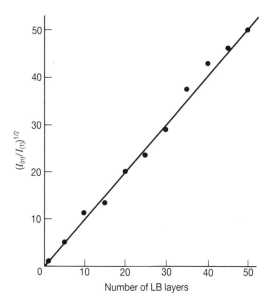

Fig. 10.9 (a) A zwitterionic compound developed by Ashwell[33] which shows a linear dependence of the square root of the normalized second harmonic intensity on the number of transferred layers as shown in (b).

the capacitor whilst simultaneously monitoring the current, i. The relationship between the current and the temperature is given by:

$$i = pA\,(\mathrm{d}T/\mathrm{d}t) \tag{10.4}$$

where p = the pyroelectric coefficient (the parameter which describes the sensitivity of the material to a unit change in the rate of temperature variation), A = the area of the electrodes and $(\mathrm{d}T/\mathrm{d}t)$ = the rate of change of temperature. Figure 10.10 shows a schematic diagram of the static pyroelectric measurement apparatus and the current and temperature traces for a pyroelectric LB film incorporating polar molecules.

Most of the work regarding the pyroelectric effect in LB films has been concerned with molecules possessing carboxylic acid and alkylamine groups. Work by Christie et al.[18] showed that pyroelectric coefficients

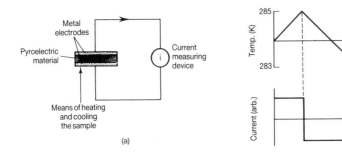

Fig. 10.10 (a) Schematic set-up for static pyroelectric measurement, and (b) the idealized temperature and current temporal profiles associated with such a measurement.

(dP/dT) for 22-tricosenoic acid/docosylamine alternate layer films were about $1.9 \, \mu \text{Cm}^{-2}\text{K}^{-1}$ (static measurement). Studies by others on a family of organo-ruthenium complexes[34] yielded a maximum coefficient of $2.7 \, \mu \text{Cm}^{-2}\text{K}^{-1}$ when mixed layers were used. Recently, researchers have concentrated on studying the mechanisms through which the pyroelectricity arises using a range of monomeric, oligomeric and polymeric compounds. Coefficients as high as $10 \, \mu \text{Cm}^{-2}\text{K}^{-1}$ have now been obtained.

Another kind of sensor which has attracted the attention of the LB researcher is the chemical or gas sensor. A successful sensor must be capable of providing an easily measurable signal in response to the adsorption, binding or reaction of the sensing agent with the ion or molecule to be detected. One of the early examples of such a device took the form of a chemiresistor incorporating a copper phthalocyanine LB film whose conductivity changes upon exposure to nitrogen dioxide gas.[36] Figure 10.11 shows the current–concentration response of the device. Although sensitive to a few volume parts per million, such materials are seldom selective to just one gas or chemical. Another approach is to deposit the LB detector onto a piezoelectric quartz crystal of the type discussed in Section 10.3. Upon exposure to gas, adsorption of the gaseous molecules occurs causing a slight increase in mass and consequently a reduction in oscillation frequency.[20] The use of quartz crystals in liquids has recently been reported.[37] This work involves a study of the uptake of certain anaesthetic drugs from aqueous solution by LB phospholipid membranes deposited onto the crystal surfaces. By using surface acoustic wave devices the sensitivity of the devices can be improved dramatically to levels typically around $75 \, \text{Hz ppm}^{-1}$.[38]

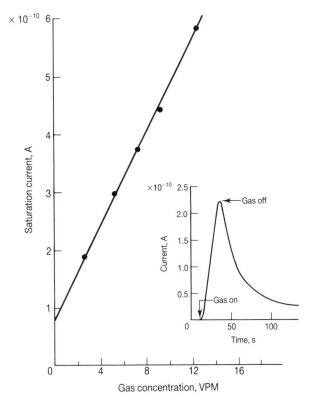

Fig. 10.11   The saturation current–gas concentration response of a copper phthalocyanine LB film chemiresistor. The inset shows the temporal response to 120 VPM nitrogen dioxide in nitrogen.

10.5   Conclusions

The subject of LB films has advanced greatly since the early experiments of Pockels, Langmuir and Blodgett. The number of different chemical compounds prepared as monomolecular films is enormous and grows each year, as does the number of potential applications of these assemblies. It can surely be only a question of time before commercial devices incorporating LB films become more common. The 1990s, more than any other time, present a wonderful opportunity to researchers contemplating entering this field. Anyone who has worked with LB films must have felt a surge of excitement on realizing that, with the help of this simple yet powerful technique, molecules can be organized into single monolayers which can subsequently be manipulated to form complex, artificial molecular arrays possessing predesigned physical and chemical properties. The future of this area of research holds great promise.

References

1. B. Franklin, *Philos. Trans. R. Soc. London*, **64**, 445 (1774).
2. Lord Rayleigh, *Proc. R. Soc. London*, **47**, 364 (1890).
3. A. Pockels, *Nature (London)*, **46**, 418 (1892).
4. I. Langmuir, *J. Am. Chem. Soc.*, **39**, 1848 (1917).
5. K. B. Blodgett, *J. Am. Chem. Soc.*, **57**, 1007 (1935).
6. H. Kuhn, *Thin Solid Films*, **99**, 1 (1983).
7. G. L. Gaines, Jr., *Insoluble Monolayers at Liquid–Gas Interfaces*, (Wiley-Interscience, New York, London, Sydney, 1966).
8. P. Joos, *Bull. Soc. Chim. Belg.*, **80**, 277 (1971).
9. R. D. Newman, *J. Colloid Interface Sci.*, **53**, 161 (1975).
10. C. Okazaki, *Japanese J. Appl. Phys.*, **29**, No. 11, 2506 (1990).
11. M. A. Schoondorp, A. J. Schouten, J. B. E. Hulshof and B. L. Feringa, *Thin Solid Films*, **210/211**, 166 (1992).
12. R. C. Ahuja and D. Mobius, *Thin Solid Films*, **210/211**, 228 (1992).
13. W. M. Heckl, *Thin Solid Films*, **133**, 73 (1985).
14. T. Richardson, W. H. Abd Majid, E. C. A. Cochrane, S. Holder and D. Lacey, *Thin Solid Films*, in press (1993).
15. G. Williams, *Thin Solid Films*, **209**, 150 (1992).
16. K. B. Blodgett, *J. Am. Chem. Soc.*, **56**, 495 (1934).
17. B. Holcroft, M. C. Petty, G. G. Roberts and G. J. Russell, *Thin Solid Films*, **134**, 83 (1985).
18. P. Christie, C. A. Jones, M. C. Petty and G. G. Roberts, *J. Phys. D*, **19**, L167 (1986).
19. G. Roberts (Ed.), *Langmuir Blodgett Films*, (Plenum Press, New York, 1990).
20. Y. Okahata, K. Ariga and K. Tanaka, *Thin Solid Films*, **210/211**, 702 (1992).
21. G. G. Roberts, P. S. Vincett and W. A. Barlow, *J. Phys. C*, **11**, 2077 (1978).
22. R. M. Nicklow, M. Pomerantz and A. Segmuller, *Phys. Res. B*, **23**, 1081 (1981).
23. D. Amm and D. Johnson (private communication).
24. I. R. Peterson and G. J. Russell, *Philos. Mag. A*, **49**, 463 (1984).
25. G. H. Davies and J. Yarwood, *Spectrochimica Acta*, **43A**, 1619 (1987).
26. S. Ohki and C. B. Ohki, *J. Theor. Biol.*, **62**, 389 (1976).
27. A. N. Broers and M. Pomerantz, *Thin Solid Films*, **99**, 323 (1983).
28. J. Seto, T. Nagai, C. Ishimoto and H. Watanabe, *Thin Solid Films*, **134**, 101 (1985).
29. Y. Nishikata, M. Kakimoto, A. Morikawa and Y. Imai, *Thin Solid Films*, **160**, 15 (1988).
30. G. G. Roberts, M. McGinnity, W. A. Barlow and P. S. Vincett, *Thin Solid Films*, **68**, 223 (1980).
31. M. Era, C. Adachi, T. Tsutsui and S. Saito, *Thin Solid Films*, **210/211**, 468 (1992).
32. D. B. Neal, M. C. Petty, G. G. Roberts, M. M. Ahmad, I. R. Girling, N. A. Cade, P. V. Kolinsky and I. R. Peterson, *Electron Lett.*, **22**, 460 (1986).
33. G. J. Ashwell *et al.*, *J. Chem. Soc. Faraday Trans.*, **86**, 1117 (1990).
34. R. Colbrook, G. G. Roberts and B. Holcroft, *Ferroelectrics*, **91**, 209 (1989).
35. T. Richardson, D. Lacey and S. Holder, *Supramolecular Science*, in press.
36. S. Baker, G. G. Roberts and M. C. Petty, *Proc. IEEE Part I*, **130**, No. 5, 260 (1983).
37. M. D. Ward and D. A. Buttry, *Science*, **249**, 1000 (1990).
38. M. Ohnishi, C. Ishimoto and J. Seto, *Thin Solid Films*, **210/211**, 455. (1992).

11

Organic Molecular Beam Epitaxy

M Hara and H Sasabe

11.1 Introduction

Molecular electronics has attracted a great number of investigators from both the fundamental and applied view points.[1–5] Many interesting molecules and/or molecular assemblies have been proposed for molecular electronic devices (MED), but in order to realize the true MED, there still remain many problems to be solved, e.g., how to handle and arrange functional molecules in the desired order or the patterned structure, how to communicate with single molecules, how to transmit the information through single strand of π-conjugated chain, and so forth. The key issues for MED are classified into three categories[3]: (1) materials—design and synthesis of functional molecules (including native molecules such as proteins and cells), (2) fabrication—control of molecular assemblies in low dimensions and direct molecular handling, and (3) lithography—circuitry design and/or patterning. Among these, issue (2) has seen remarkable progress in this decade along with the development of organic thin film technology and the STM (scanning tunnelling microscopy) technique. In this chapter we will first review the ultrathin film techniques for organic materials[6] and then introduce the technique of organic molecular beam epitaxy (OMBE) developed in our laboratory in detail.

Ultrathin film fabrication processes include techniques from both liquid and gas phases, i.e., Langmuir–Blodgett method, liquid phase epitaxy (adsorption), chemical vapour deposition (CVD) and physical vapour deposition (PVD). The Langmuir–Blodgett method is one of the most promising techniques for MED fabrication; amphiphilic molecules which have a hydrophilic head (e.g. –COOH, –OH, –NH$_2$) and a long hydrophobic tail (e.g., alkyl chain) can be spread over the air/water interface and by increasing the surface pressure (or reducing the area occupied by a molecule) they form a stable monolayer in the form of two dimensional crystal, which can then be transferred on to a solid substrate (see the previous chapter). Monolayers of different molecules can be stacked alternately in the designed manner, that is, a superlattice structure can be formed on the substrate. Various kinds of molecule with functional groups (electro- and/or optical-active groups) have been designed for this type of molecular heterostructure. The liquid phase epitaxy technique enables organic molecules to be adsorbed on the solid substrate through silane-coupling or disulphide interaction. This is quite useful for the site-specific adsorption of protein molecules. These two techniques are really chemical procedures, i.e. wet processes.

CVD and PVD, on the other hand, are the dry processes. The CVD process is based on the decomposition and/or radical generation of chemical species (monomers) by stimulating vapour with heat, plasma (discharge) or light (laser), followed by film formation on a solid substrate. In the case of radical generation, the films form a cross-linked polymer network, whereas in the case of decomposition of organometallics, the well organized monolayer of metal and/or semiconductor can be formed on a single crystal substrate (chemical vapour epitaxy). The PVD process includes both vacuum deposition and sputtering. In the vacuum deposition technique, the substances are heated and evaporated in a crucible (or boat) *in vacuo*, then condensed on a substrate. For organic substances, the first step of heating/evaporation may give the thermal decomposition, and hence particular care may be required. In the sputtering technique, an inert gas such as Ar and Xe is introduced in the vacuum chamber, ionized to form a plasma by applying a high frequency electric field and accelerated by a DC field. A target material is then bombarded with these ions and the resulting molecules fly off and reach the substrate where they form a film. When the accelerating field is too large, the decomposition of target substances occurs easily.

In the case of polymeric materials, we can design the molecular structure (chemical structure) of the main chain and/or the side chain with functional pendant groups, and control the higher order structure such as a random coil, α-helix, β-sheet, microphase separation and so forth. Table 11.1 indicates two approaches to make polymeric ultrathin films. The first corresponds to a thin film formation during the polymerization process and the other to the direct formation of thin films of polymers. Usually the CVD technique for a monomer gas gives high quality (pinhole free) ultrathin, but highly cross-linked polymer films. However, the control of molecular orientation on the solid substrate is difficult by this technique.

Among the PVD techniques, vacuum evaporation is most popular: a target material (polymer powder or film) is heated higher than its melting or sublimation point under a vacuum of 10^{-6} Torr, followed by deposition onto a solid substrate. This technique enables (1) control of higher order structure of the film by varying the evaporation speed and/or substrate temperature, and (2) creation of a hybrid film in which different types of molecules are intimately mixed by employing multiple evaporation sources

Table 11.1   *Techniques for polymeric ultrathin film formation*

During polymerization process	Direct process for film formation
Catalyst coating technique	Interfacial adsorption technique
	Gas phase adsorption
	Liquid phase epitaxy
Polymerizable LB film technique	LB technique
	Spin coating technique
Electro-deposition technique	Vacuum evaporation deposition technique
	Thermal evaporation
	Laser ablation
Chemical vapour deposition (CVD) technique	Molecular beam epitaxy
Thermal CVD	Clustered ion beam
Light (laser) CVD	Sputtering technique
Plasma CVD (plasma polymerization)	RF sputtering
	Ion beam sputtering

and simultaneous deposition. For example, a poly-*p*-phenylene sulphide (PPS) film can be formed on the glass substrate under the conditions of a crucible temperature of 400°C and a substrate temperature in the range of 180–200°C.[7] It should be be noted that the melting point, crystallization temperature and the glass transition temperature of PPS are 280, 135 and 92°C, respectively. The film thickness ranges from 7 nm to 500 nm without defects. The PPS film (10 nm thick) shows a highly crystalline structure (almost single crystal) with the chain axes perpendicular to the substrate. In the vacuum deposition of polymers, the scission of the main chains occurs often, lowering the molecular weight. The crystalline structure of deposited films, however, remains unchanged from that of the original polymer. The laser ablation technique is an alternative to vacuum evaporation; instead of heating in a crucible, the target material can be evaporated selectively by laser beam. This is available not only for the film formation but for the lithography of a polymer film.

Sputtering is based on the momentum exchange of accelerated ions incident on the target. Therefore, polymer chains of target material might be scissioned into small fragments with radicals, and then repolymerized on the substrate. The typical RF (radio-frequency: 13.6 MHz) sputtering apparatus has a large target electrode and hence a quite large number of sputtered molecules react on the substrate to form a cross-linked structure; this is similar to plasma polymerization. Although the film thickness can be controlled easily by adjusting the sputtering time, RF power, gas pressure and flow rate, the control of the molecular orientation on the substrate is difficult. To reduce the cross-linkage of films and orientate molecules on the substrate, an ion beam sputtering technique can be introduced. The fine ion beam sputters the target materials by scanning the surface, and the sputtered molecules aggregate epitaxially on the substrate.

Molecular beam epitaxy (MBE) is, in principle, similar to vacuum evaporation. However, this technique involves an ultrahigh vacuum to eliminate the scattering of target molecules (molecular beam) with residual gas molecules and multi-crucibles (Knudsen cells) to create a superlattice structure. This enables the control of molecular orientation and packing in

2-dimensions on an atomically flat substrate, such as cleaved single crystals and layered materials (graphite, transition metal dichalcogenides). We have developed an MBE apparatus particularly for organic materials. In the following section we describe the details of this equipment.

11.2   Basic Concepts

Figure 11.1 shows an ultra-high vacuum (UHV) organic MBE (OMBE) component system designed in our laboratory. This system is composed of three independent chambers, C1, C2 and C3, connected through gate valves, specifically designed for fabrication and characterization of organic molecular systems under a base pressure of less than 10^{-10} Torr.[8,9] The combination of diffraction (reflection high-energy electron diffraction (RHEED) in this system) with imaging by scanning tunnelling microscopy (STM) allows direct studies of epitaxial layers at the atomic level. The outside appearance of the OMBE system is shown in Fig. 11.2. The RHEED screen is about 1.6 m high and the main body spans about 3 m, including the magnetic feedthrough (transfer rod: TR) for the sample transfer.

C1, C2 and C3 are main growth chamber, substrate entry/prebaking chamber and portable transfer chamber, respectively. The smallest chamber, C3, can be isolated from the main body and used as a portable UHV chamber, which is maintained in a UHV state by an ion pump equipped with an independent power supply. Thus the C3 chamber can incorporate independent UHV systems, such as a remotely located STM chamber hanging on a vibration isolation table, into one system with an interchangeable sample holder assembly. Since a cigarette lighter port in automobiles can be used as a power supply for the C3 chamber, samples can be transferred over some distance without breaking the vacuum state.

For RHEED analysis, a charge-coupled device (CCD) camera was positioned in front of the screen for RHEED patterns. The diffraction patterns are stored in 600 megabytes magneto-optical (MO) disks through a microcomputer with two-dimensional (2-D) intensity profiles (16 bit) divided into 510×492 pixels, so that *in situ* RHEED patterns and 2-D intensity changes can be investigated at the same time. In addition, atomic or molecular layer phase locked epitaxy (PLE) can be managed by the feedback from the results of the intensity analysis.

The UHV STM system used was based on NanoScope II system (Digital Instruments, Inc., Santa Barbara, USA) with a modified STM head for the UHV use. The STM chamber is mounted on a vibration isolation table and kept at a vacuum pressure of less than 10^{-9} Torr during operation.

The main difficulties in developing UHV MBE methods for organic molecular systems are their inherently high vapour pressure, thermal instability, multiplicity of the condensed structures and impurity contents of other organic substances. A low-temperature Knudsen cell was specially designed for such organic samples in this UHV OMBE system. The temperature of the Knudsen cells can be controlled not only by electrical

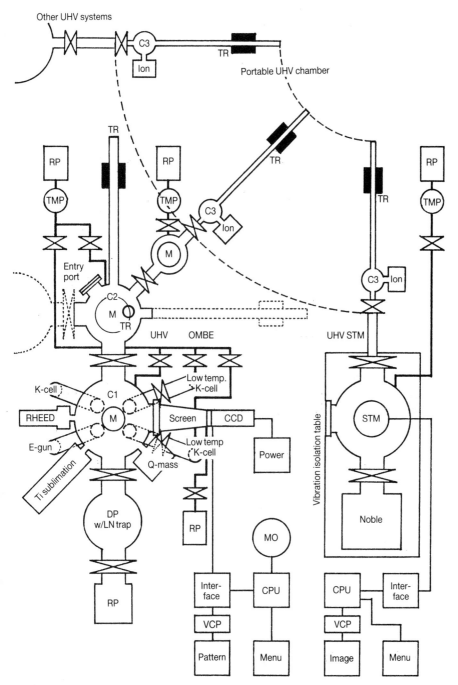

Fig. 11.1 Schematic diagram of UHV OMBE component system. RP: rotary pump, TMP: turbo molecular pump, DP: diffusion pump with liquid nitrogen trap, M: manipulator, VCP: video copy processor, TR: transfer rod, Q-Mass: mass spectrometer.

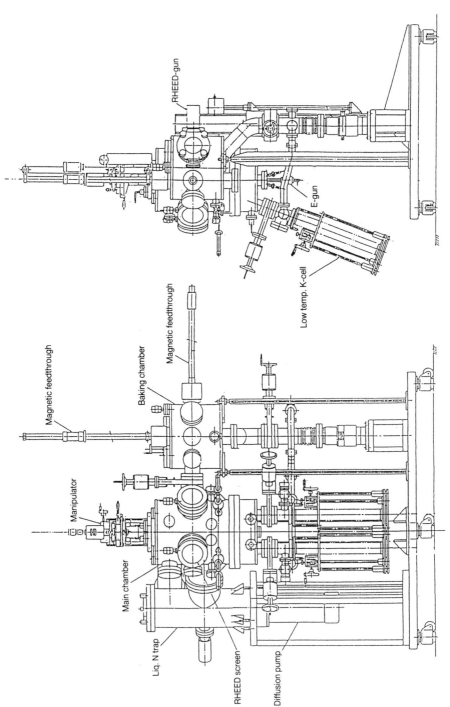

Fig. 11.2   The outside appearance of the OMBE system.

heating but also by circulating liquid nitrogen, resulting in a wide temperature range, from $-70°C$ to $400°C$. In addition to the usual precautions exercised during UHV growth of semiconductor structures, there are several additional barriers to obtaining fine structures of organic molecules on single crystal substrates. One of the most important procedures is the degassing of the organic source. During degassing of purified organic molecules under UHV conditions or during the baking process of the OMBE system itself, the low-temperature Knudsen cells can be drawn outside the growth chamber through load lock bellows, while the UHV condition in the cells is retained.

A further difficulty of UHV MBE methods for epitaxial growth of organic thin films is the lattice mismatch between organic overlayer and inorganic substrate. Compared to the lattice constant of an inorganic substrate, the lattice constant of an organic crystal can be quite large. Hence the simple concept of lattice mismatch is not sufficient to explain the orientational growth of organic molecules. Various orientational growth mechanisms for organic molecules have been proposed. In the case of the usual single substrates which have dangling bonds on the surface, nucleation sites are randomly located for large organic molecules, resulting in three-dimensional island structures which are generated by multicrystalline growth.

More recently, the van der Waals epitaxy method, which was proposed by Koma *et al.*, has been shown to provide outstanding inorganic heterostructures, even between materials having a large lattice mismatch.[10] In this heteroepitaxial technique, the film growth proceeds via van der Waals interactions between layered inorganic materials, such as a transition metal dichalcogenide (MX_2), which has no dangling bonds on its cleaved surface. The first *in situ* observation of van der Waals epitaxy for an organic thin film on a layered inorganic substrate was achieved for a combination of metallophthalocyanine (MPc) molecules on a MoS_2 substrate (a typical layered MX_2 single crystal[11]).

11.3 *In Situ* Observation of Film Growth— RHEED

For the first trial observations of the real-time epitaxial growth in organic molecular systems, MPc monolayers were grown heteroepitaxially on cleaved surfaces of MoS_2, HOPG (highly oriented pyrolized graphite) and alkali halide substrate under 2×10^{-10} Torr by the OMBE technique. *In situ* RHEED was used to monitor the actual film growth in the OMBE system. In order to avoid beam damage to the MPc film by the 15–25 keV electron beam, the irradiation time was usually shorter than 30 seconds to record one RHEED pattern. In the case of CCD camera imaging, the irradiation time was about a few seconds for one frame.

Figure 11.3(a) shows the RHEED pattern of the as-prepared MoS_2 surface. The electron beam is parallel to the $[11\bar{2}0]$ orientation of MoS_2, so

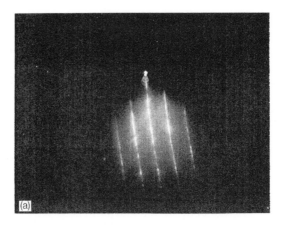

Fig. 11.3   (a) RHEED pattern of MoS$_2$ substrate; (b) RHEED pattern of CuPc ultrathin film on MoS$_2$ at the initial stage.

that the spacing between streaks in the pattern corresponds to the MoS$_2$ lattice spacing, about 2.74 Å, in agreement with an earlier report.[12] After opening the shutter to a copper phthalocyanine (CuPc) source, the pattern changed markedly. The change of the RHEED pattern at the initial growth stage is shown in Fig. 11.3(b), where the surface of the MoS$_2$ substrate is just slightly covered by less than 1/5 of a monolayer of CuPc molecules.[11] The very sharp and narrow streaks between the bright MoS$_2$ streaks originate from a growing CuPc layer that is molecularly flat. This was the first observation of epitaxial growth of an organic thin film confirmed by *in situ* RHEED during actual organic film growth.

The lattice constant of the CuPc layer is calculated from this pattern to

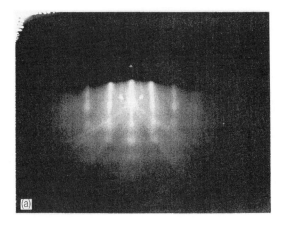

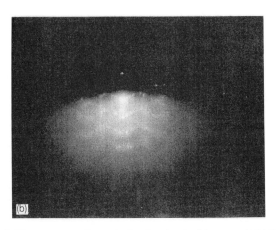

Fig. 11.4   (a) RHEED pattern of KCl at the [110] azimuth; (b) crossed RHEED pattern from CuPc layers on KCl at the [110] azimuth.

be nearly equal to a molecular diameter of CuPc, about 13 Å. This indicates that a molecularly flat CuPc 2-D crystal has been grown initially heteroepitaxially on MoS_2, with the organic film retaining its own lattice constant in a 2-D single-crystal lattice. When the substrate was rotated 60 degrees in the plane during film growth, the RHEED pattern was exactly the same as that at 0 degree for both MoS_2 and CuPc.

Figure 11.4(a) shows the RHEED pattern from an as-prepared KCl surface in the [110] azimuth. The spacing between streaks in the pattern corresponds to the characteristic KCl lattice spacing. The change in the pattern after opening the shutter to the CuPc source is shown in Fig. 11.4(b). Although this pattern shows rather spotty features due to diffraction from a three dimensional structure, the crossed RHEED patterns highlight the growth direction, especially the b-axis orientation. When the electron beam is parallel to the [100] orientation of KCl, the deduced angle of the b-axis from the crossed RHEED pattern is 32 degrees from the normal to the substrate surface. When the substrate was rotated 45 degrees in the plane, the angle changes to 24 degrees (Fig. 11.4(b)). This indicates that the RHEED pattern can also reveal the growth direction from the initial stage for organic 2-D films during the actual film growth. These data are in very good agreement with earlier electron microscopy results on large crystalline domains.[13]

In order to investigate the possibility of RHEED intensity oscillations during organic molecular layer growth, CCD camera imaging of the RHEED pattern in 2-dimensions was carried out. Figure 11.5 shows the intensity change as a function of time for a RHEED pattern from CuPc layers on MoS_2. Some periodic structure is apparent in the intensity profile corresponding to the layer growth. While further detailed investigations are required by changing incident angles of the electron beam and growth conditions, Fig. 11.5 is a definite indication that RHEED intensity oscillation occurs, even for organic molecular layer growth. If this sort of monitoring is established generally, molecular layer phase locked epitaxy (PLE) can be managed by the feedback of results from the intensity analysis. From this point of view, this is an encouraging first step towards eventually realizing novel 'nanoscopic' material structures such as organic or organic-inorganic molecular superlattices.

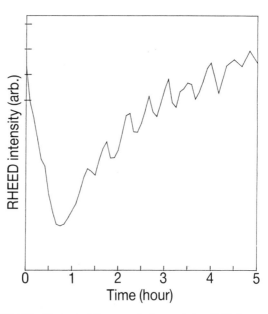

Fig. 11.5   RHEED intensity changes of the specular beam during CuPc layer growth.

11.4   Characterization of Assembled Molecules—STM

Since scanning tunnelling microscopy (STM) was introduced by Binnig and Rohrer,[14] the STM technique has been shown to provide outstanding results in structural analysis with ultrahigh resolution, principally for the clean surfaces of well-ordered inorganic single crystals (Chapter 12). Currently, however, considerable interest is centred on developing techniques for STM characterization of organic molecular assemblies with low dimensionality arranged in a predesigned manner on the substrate. From numerous STM observations of organic monolayers, we realized that the close-packed ordered monolayer is one of the most suitable structures for STM visualization of organic molecules. Following the previous results obtained from ordered monolayers,[15–17] STM imagings of epitaxially grown organic monolayers by OMBE technique were carried out, both under UHV conditions and in air.

Figure 11.6(a) is a molecular resolution STM image of an OMBE CuPc layer on highly oriented pyrolytic graphite (HOPG) taken in air. The thickness of the film was around one monolayer, estimated from the intensity change in RHEED patterns during the deposition. This is the first STM image of a MPc lattice in a thin film structure at molecular resolution. Previously, high resolution images had only been obtained of isolated or randomly distributed MPc molecules under UHV.[18,19]

All STM images obtained in OMBE MPc thin films degraded after about ten minutes and subsequently an image of the underlying substrate lattice appeared. Since this change was irreversible, we attribute it to resublimation or destruction of the organic layer, mainly caused by the force between the scanning tip and molecules. From atomic force microscopy (AFM) imagings, it has been realized that nondestructive forces as low as 10^{-9} N for organic monolayers are hardly achieved during STM imagings. Interestingly, however, this permitted the study of the underlying substrate in relation to the deposited film. Figure 11.6(b) shows a graphite substrate image beneath the CuPc layer shown in Fig. 11.6(a). A superposition of Fig. 11.6(a) and (b) revealed that the CuPc lattice is rotated at an angle of 9° with respect to one axis of the graphite lattice. In addition, the sectional

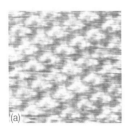

Fig. 11.6   (a) STM image of CuPc on graphite by OMBE. 500 mV and 1.0 nA (tip positive), 9 × 9 nm. (b) STM image of graphite beneath CuPc layer after ten minute scanning, 9 × 9 nm.

view of the STM images allowed more detailed structural analysis including the tilted adsorption to the substrate surface. From such STM images of the phthalocyanine layer and the substrate beneath the organic layer, the orientational correlation between the organic layer and the substrate surface was clearly determined in real space. In Fig. 11.7, a tentative model for heteroepitaxy of CuPc on graphite is shown. This is one of the best fit models to the observed lattice dimensions and angles in STM images.

It is important to consider the possibility of observing artifacts during scanning, especially using graphite as a substrate. As we discussed above, we have an independent check via the RHEED patterns obtained during the actual film growth. It should be noted, however, that the RHEED patterns are recorded under UHV conditions during growth, while the STM image is taken in air; therefore the possibility of some relaxation in the film must be considered. Even so the lattice spacing corresponding to the fine RHEED streak monitored is about 15 Å, close to that measured by the STM image. This would be further evidence that Fig. 11.6(a) is not due to artifacts.

11.5   Prospects for OMBE

Firstly, we will show some experimental results for MBE-grown phthalocyanine films for nonlinear optics. Metallophthalocyanines (MPcs) are macrocyclic π-conjugated systems and show a large macroscopic third order optical nonlinear susceptibility $\chi^{(3)}_{ijkl}(-\omega_4; \omega_1, \omega_2, \omega_3)$ (see Chapter 5). In the electronic spectra of MPcs, two characteristic absorption bands are well known, i.e. the Soret band (300–400 nm) and the Q-band (π–π^* transition in 600–800 nm region) (Fig. 11.8). The latter is sensitive to the orientation and packing of the MPc rings. In general, the central metal has little effect on the electronic state of phthalocyanine but a strong influence on the packing arrangement of the phthalocyanine molecules in the condensed state. Therefore, depending on the central metal, features of the Q-band change remarkably in the condensed state. This band has been widely studied as a probe of the phase transitions induced by thermal and/or solvent vapour treatment. MPcs shown polymorphism, e.g., vanadylphthalocyanine VOPc has three phases. The phase I corresponds to a cofacially stacked form, the phase II to a triclinic crystal structure (slipped-stacked form) (Fig. 11.9), but phase III is not yet determined. Hence it is expected that the structure of the film grown on a solid substrate is easily affected by the interaction with a substrate.

As mentioned before, MBE is a sophisticated vacuum deposition technique in which the lattice matching of deposited molecules to the substrate takes place. Hara et al.[8] reported the heteroepitaxy of CuPc on HOPG or MoS$_2$ substrates and RHEED intensity oscillation during the growth of the film. Tada et al.[20] also studied the initial crystal structure of VOPc thin films grown on KBr substrates, and proposed a square lattice with four-fold symmetry of VOPc molecules and the molecular planes parallel to the cleaved (001) face of KBr. In the condensed state, the

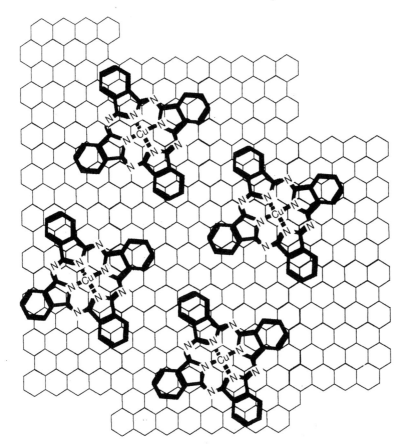

Fig. 11.7 Tentative model for the heteroepitaxial deposition of CuPc on graphite deduced from the superposition of Figures 11.6(a) and 11.6(b).

broadening in the Q-band was observed as a result of exciton splitting due to transition dipole interactions between adjacent VOPc molecules in aggregates.

The exciton decay dynamics in VOPc thin films were investigated using femtosecond pump-probe spectroscopy. The optical layout and the detailed description of the detection of transmittance changes are described elsewhere.[21] The differential transmission spectra (DTS) of the phase II films of VOPc(t-bu)$_{1.1}$ doped in polystyrene (10%) and of MBE-grown VOPc film on KBr are shown in Fig. 11.10. Here, VOPc(t-bu)$_{1.1}$ is a tertiary-butyl substituted vanadylphthalocyanine (1.1 corresponds to the average number of substituted t-bu group). The key feature of DTS is that significant bleaching appears at the lowest energy absorption peak (centred around 810 nm), while the higher energy absorption peaks bleach only weakly. This indicates either that excitons created by the femtosecond

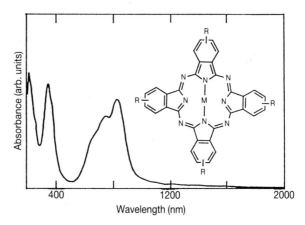

Fig. 11.8 Structure of metallophthalocyanines (MPc) and typical electronic spectrum for a VOPc thin film fabricated by PVD.

pulses at 620 nm undergo very rapid internal conversion to the lowest excited state, or that the induced excited-state absorption in this energy region cancels the bleaching signals. In the spectral range of 500–600 nm

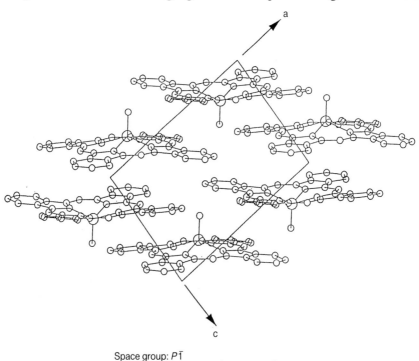

Space group: $P\bar{1}$
a = 12.027 Å, b = 12.571 Å, c = 8.690 Å
α = 96.04°, β = 94.80°, γ = 68.20°

Fig. 11.9 Projection of molecular packing of phase II of VOPc along the b axis.

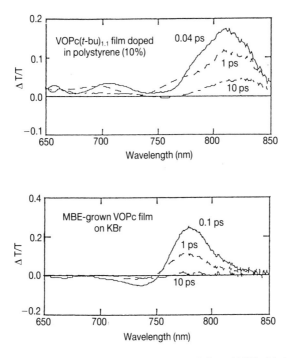

Fig. 11.10 The differential transmission spectra of phase II film of VOPc(*t*-bu)$_{1.1}$ doped in polystyrene (10%) and of MBE-grown VOPc film on KBr.

(corresponding to the optical window of the ground state absorption), the excited state induced absorption was observed, which was much weaker than the bleaching signal. The decay curve of the transmittance change consists of three parts: a fast (hundreds of femtoseconds), a slower (tens of picoseconds), and a long-lived (longer than hundreds of picoseconds) component. The fast component is interpreted as a bimolecular process, i.e. exciton–exciton annihilation, because of its excitation intensity dependence. On the other hand, the slower decay is rather independent of excitation intensity and can be attributed to a unimolecular decay via exciton–phonon coupling. The long-lived component is presumably due to triplet-state formation, widely observed in MPcs. To fit the initial decay curve, we used the exciton–exciton annihilation model with a time-dependent rate of $t^{-1/2}$ based on the long-range dipole–dipole interaction between excitons, or the motion-limited exciton diffusion. In this model, the exciton density (n) was described by the following equation for a delta-function excitation pulse:

$$dn/dt = \gamma t^{-1/2} \kappa n, \tag{11.1}$$

where γ and κ are, respectively, a bimolecular and a unimolecular decay constant. Solving equation (11.1), the following time dependence of the exciton density is obtained:

$$n(t)/n_0 = \exp(-\kappa t)/\{1 + (2n_0\gamma/\kappa^{1/2}) \, erf[(\kappa t)^{1/2}]\} \qquad (11.2)$$

where $erf(x)$ is the error function. We have added a constant term which phenomenologically accounts for the long-lived component to equation (11.2). This function can fit the data over a time range up to 100 ps, and yields fitting parameters of $n_0\gamma = 9.1 \times 10^5 \, s^{-1/2}$, $\kappa = 2 \times 10^{10} \, s^{-1}$ and a constant term which is about 10% of the maximum value of the data. The value of γ for a pure VOPc(t-bu)$_{1.1}$ spin coated film is estimated as $1.6 \times 10^{-14} \, cm^3 \, s^{-1/2}$. MBE-grown VOPc films, on the other hand, showed a rather fast decay of bleaching, presumably due to the different packing arrangements. It is therefore suggested that the long-range order in the molecular packing of MBE-grown VOPc film affects the relaxation of excitons. In this way we can influence the ultrafast electronic excitation available for the enhancement of third order optical nonlinearity.

MBE grown NiPc films (approximately 80 nm thick) on glass substrates, show a molecular packing of α-crystalline phase as shown in Fig. 11.11, and form extremely flat surfaces.[3] A MIM (metal/insulator/metal) cell composed of Al/NiPc(300 nm)/ITO can be prepared. This configuration is a Schottky barrier type cell, and hence shows rectifying J-V characteristics (Fig. 11.12). Under a high applied voltage (about 10 V, corresponding to 0.3 MV/cm), the Al electrode might melt due to Joule heating. However, reconstitution of MIM cell after removing the Al by acid and redepositing Al gives the same J-V characteristics as before. This results from the well packed molecular film without pin-holes.

11.6   Conclusions

The two examples above are only a small aspect of MBE grown organic ultrathin films. Further application of OMBE techniques is quite promising for the development of molecular electronic devices. In addition, when a

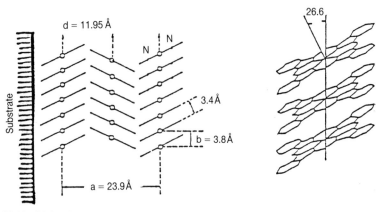

Fig. 11.11   Molecular packing structure of MBE-grown NiPc film on the glass substrate.

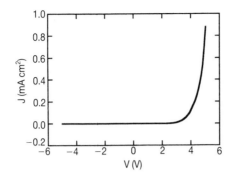

Fig. 11.12   Plot of current density J versus applied voltage V for Al/NiPc (300 nm)/ITO cell.

super-lattice structure of highly polarizable organic molecules and metal is formed, the 2-dimensional high T_c superconductors based on the exciton mechanism[22] might be realized. More experimental effort should be focused on these problems.

Acknowledgements   The authors would like to express their thanks to Prof. A. Koma and Dr H. Tada of University of Tokyo for their kind supply of MBE-growth VOPc sample, and to Drs T. Wada and A. Terasaki of Frontier Research Program, RIKEN, for their ultrafast optical measurements.

References

1. F. L. Carter (Ed.), *Molecular Electronic Devices*, (Marcel Dekker Inc., N.Y., 1982).
2. F. L. Carter (Ed.), *Molecular Electronic Devices II*, (Marcel Dekker Inc., N.Y., 1987).
3. F. L. Carter, R. E. Siatkowski and H. Wholtjen (Eds), *Molecular Electronic Devices*, (North Holland, Amsterdam, 1988).
4. A. Aviram (Ed.), *Molecular Electronics—Science and Technology*, (Engineering Foundation, N.Y., 1989).
5. C. Nicolini (Ed.), *Towards the Biochip*, (World Scientific, Singapore, 1990).
6. H. Sasabe, in *Hybrid Materials—Concepts and Case Studies*, Ed. H. Nogawa, (ASM International, Ohio, 1988), Chap. II-1.
7. K. Misoh, S. Tasaka, S. Miyata and H. Sasabe, *Bull. Chem. Soc. Jpn.*, **763** (in Japanese) (1983).
8. M. Hara, H. Sasabe, A. Yamada and A. F. Garito, *Mat. Res. Soc. Symp. Proc.*, **159**, 57 (1990).
9. M. Hara, P. E. Burrows, A. F. Garito and H. Sasabe, *Nonlinear Optics*, **1**, 75 (1992).
10. A. Koma and K. Yoshimura, *Surf. Sci.*, **174**, 556 (1986).
11. M. Hara, H. Sasabe, A. Yamada and A. F. Garito, *Jpn. J. Appl. Phys.*, **28**, L306 (1989).
12. J. A. Wilson and A. D. Yoffe, *Adv. Phys.*, **18**, 193 (1969).

13. M. Ashida, *Bull. Chem. Soc. Jpn.*, **39**, 2632 (in Japanese) (1966).
14. G. Binnig and H. Rohrer, *Helv. Phys. Acta*, **55**, 726 (1982).
15. M. Hara, Y. Iwakabe, K. Tochigi, H. Sasabe, A. F. Garito and A. Yamada, *Nature*, **344**, 228 (1990).
16. Y. Iwakabe, M. Hara, K. Kondo *et al.*, *Jpn. J. Appl. Phys.*, **30**, 2542 (1991).
17. M. Hara, T. Umemoto, H. Takezoe, A. F. Garito and H. Sasabe, *Jpn. J. Appl. Phys.*, **30**, L2052 (1991).
18. P. H. Lippel, R. J. Wilson, M. D. Miller, Ch. Wöll and S. Chiang, *Phys. Rev. Lett.*, **62**, 171 (1989).
19. J. K. Gimzewski, E. Stoll and R. R. Schlittler, *Surf. Sci.*, **181**, 267 (1987).
20. H. Tada, K. Sakai and A. Koma, *Jpn. J. Appl. Phys.*, **30**, L306 (1991).
21. A. Terasaki, M. Hosoda, T. Wada *et al.*, *Nonlinear Optics*, in press (1992).
22. V. L. Ginzburg, *Contemp. Phys.*, **9**, 355 (1968).

12

Scanning Tunnelling Microscopy

J P Rabe

12.1 Introduction

A challenging goal of molecular electronics, at least in the long run, is to develop devices, which are based on individual molecules or small molecular aggregates, rather than large molecular ensembles. Nature provides the proof of existence of such *molecular devices*. For instance, in a photosynthetic reaction centre, which is a small molecular aggregate, the energy of a photon is used very efficiently to transfer electrons.[1] Synthetic molecular devices remain a challenge so far. However, considerable progress has been achieved for *molecular materials*, which, for instance, can be efficient electron conductors.[2] Moreover, various methods have been developed to assemble functional molecules in ordered ultrathin molecular layers on solid substrates.[3] However, since conventional characterization methods average over some macroscopical number of molecules, it is often difficult to determine whether a certain property is a condensed matter phenomenon, or whether a single molecule would exhibit the same effect. It is desirable, therefore, to address individual molecules, and to determine their properties *in situ* and non-destructively.

The STM[4–8] is a tool which allows a variety of very local investigations on a preselected area at a surface or interface. Its essential feature is a sharp metallic tip, which is brought into close proximity to an electronically conducting solid, and whose position normal as well as parallel to the surface is controlled very accurately on an atomic length scale. Originally STM has been applied to inorganic metal and semiconductor surfaces under ultrahigh vacuum conditions, but the method itself does not require vacuum, and atomically resolved images can also be obtained at interfaces with fluids or soft solids. This renders STM particularly attractive for organic materials,[5,6] which often are neither prepared nor employed under vacuum conditions. The basic requirements for a system to be investigated by STM are twofold: sufficient electrical conductivity and a limited mobility at its surface. Not required, on the other hand, is any long range order. As far as molecular materials are concerned, two classes of systems lend themselves to STM studies: organic conductors and monomolecular layers (thickness $d \leqslant 1$ nm) of organic insulators on conducting substrates. The fact that STM is nondestructive and yields truly local information on the subnanometer scale makes it a unique method to study molecular defects as well as structure and dynamics of only partially ordered phases.

Following the invention of the STM, a number of related scanning microscopies,[7,8] including atomic force microscopy (AFM)[9] and scanning near-field optical microscopy (SNOM)[10] have been introduced. Their common feature is a local probe, which exhibits a strongly distance dependent interaction with a surface. While the probe is scanned across the surface this interaction is measured. Scanning force microscopy has become particularly widespread; compared to STM it has the advantage that it does not require an electronic conductivity of the system under investigation. On the other hand, provided the conductivity is sufficient, STM has the advantage that the probe can be very specific and well controlled, e.g., as far as the energy of the tunnelling electrons or the current density is concerned. This is of particular interest for molecular electronics. Therefore, this chapter deals mainly with STM, even though some of the considerations also apply to other scanning probe microscopies.

In the following, the principles of the method, and its application to molecular materials will be discussed. Moreover, it will be shown that the STM may be employed to address individual molecules at interfaces and to investigate their electronic properties. Finally, some prospects for the controlled manipulation of molecules by means of the STM will be discussed.

12.2   Basic Concepts

The basis of STM is the quantum mechanical phenomenon of electron tunnelling. If the electronic orbitals of the outermost atoms of two solids (a tip and a sample) overlap spatially, a tunnelling current will flow, when an electric potential difference is applied between the two. In a simple picture, the current may be approximated by the solution of the 1-dimensional tunnelling problem

$$I \propto \exp\{-2\kappa s\}, \tag{12.1}$$

where s is the barrier thickness, and κ is the average decay length of the wavefunctions of tip and sample in the barrier. For large distances ($s > 1$ nm), κ is determined by the average work functions (typically about 5 eV), with typical values for κ of the order of 1 Å^{-1}. This means that the tunnelling current decreases by about one order of magnitude for every

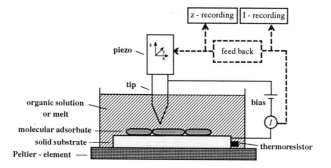

Fig. 12.1   Schematic of an experimental STM set-up at solid-fluid interface.[24]

Ångstrom of increase in gap spacing. For smaller distances s, κ decreases with decreasing s.

Due to its high sensitivity to the gap width, the tunnelling current is a suitable signal to control the gap by means of a feedback loop. If the tip is now scanned across the surface while the current is sensed, a topograph of the surface can be obtained in two ways (see also Fig. 12.1). In the *constant current mode* the tip height is adjusted with the feedback loop maintaining a constant current.[4] The image is obtained as a map of the tip height $z(x, y)$ versus the lateral coordinates x and y. Alternatively, in the *constant height mode* the tip is scanned at constant height and the current $I(x, y)$ is recorded as a function of x and y.[11] In the latter case the limiting frequency of the feedback loop must be smaller than the frequency of current variations due to atomic structure, which means that the feedback loop serves only to maintain the average current constant.

A fairly simple and widely used theory of the STM[12] is a perturbation approach for weak coupling between tip and sample, which is based on the transfer Hamiltonian of Bardeen.[13] Assuming an ideal tunnelling geometry (tip on flat), the tunnelling current is approximated to be proportional to the local density of states (LDOS) of the surface at the centre of curvature of the tip and at the Fermi energy. In this model, an STM image which has been recorded in the constant current mode will simply represent a contour of constant LDOS. In the usual case of an extended Fermi surface of the sample this is equal to the contour of constant total charge density. If an adsorbate modifies the LDOS at the Fermi energy, it becomes visible in the STM image.[14] A remarkable quantitative agreement between experiment and theory has been reported for Xe on Ni(110).[15] Alternatively, it has been suggested to ascribe the STM contrast of different molecular segments to different barrier heights, which may be due to the polarizability of the adsorbate[16] or due to different energy gaps between their front orbitals (HOMO and LUMO).[17–19] It must not be forgotten, however, that most STM experiments are carried out under conditions in which tip-sample interactions cannot be ignored.[20] Since the perturbation approach will break down in this case, more elaborate theories for strong coupling have been formulated.[21,22]

Common to basically all the published instrument designs are piezoelectric elements as x-, y- and z-translators, which allow the control of the tip-position with an accuracy better than 0.1 Å.[23] While external vibration isolation was essential in the earlier instruments, newer designs are often so small and rigid that it became less important or even obsolete. Variable temperature experiments can be carried out easily between room temperature and 120°C,[24] and special instruments have been built to operate at very low (4.2 K[25]) or very high (1200 K[26]) temperatures, respectively.

Perhaps the most crucial component of the STM is the tip. It may be mechanically ground or electrochemically etched from different tip-materials, including Pt/Ir and W. For use in electrolytes, it can be coated with glass or polymer, leaving only the last few microns bare. For rough surfaces, i.e. those with a corrugation of the order of the tip-to-sample distance or more, the tip shape plays an important role, since the part of the tip which interacts most strongly with the sample will not remain the same during the scan. It is necessary, therefore, to deconvolute tip and sample structure, which means that one has to determine the tip shape from a test sample with known structure.[6]

For atomically flat surfaces it is not the general tip-shape but the exact structure of the metallic cluster at the end of the tip, which matters. Unfortunately, this structure is difficult to characterize *in situ*. Moreover, due to thermally activated atomic motion, tip sample interactions, or the applied electric field, it is not perfectly stable during scanning. Therefore, the condition of the tip should be judged by the image contrast it produces on a well known surface. For materials which have not been imaged reliably before, it is necessary to reproduce the image with several different tips. If resolution, contrast or stability is not as expected, a short bias pulse (e.g. 4 V, 100 ns) can be applied to the tip during scanning. This usually alters the image contrast, and can be attributed to a change of the tip apex. It may be associated with field evaporation from the tip.[27] Since it also modifies the sample at this position one has to move to another area of the sample for further investigations. With such a tip-conditioning, no significant differences are observed between tips, which had initially been mechanically ground or electrochemically etched.

The instrument which has been used to obtain the results discussed below has been described elsewhere.[28] It was designed with an emphasis on high scanning speeds (1 ms per line) and low currents (down to 1 pA). The high speed has practical advantages (fast search across the sample as well as tolerance against thermal drift) and, moreover, it allows some molecular dynamics to be followed. The low current is advantageous for non-destructive imaging of relatively poor conductors.

12.3 Solid Surfaces

Originally, the STM was operated under ultrahigh vacuum conditions. However, since tunnelling does not require vacuum, the STM may equally be employed at the interface with a fluid or a soft solid, provided the sample is chemically sufficiently stable. Since the ambient material is displaced during scanning, its viscosity or hardness must not be too large, in order to avoid image distortions.

Organic Conductors One class of molecular materials which can be readily imaged by STM are organic conductors. A simple case are the low index faces of charge transfer crystals. For instance, tetrathiaful-valene-tetracyanoquinodimethane (TTF-TCNQ) (see Chapter 8) has been imaged at atomic scale resolution in air.[29] The image contrast can at least qualitatively be understood, if one assumes that it reflects the highest occupied molecular orbital (HOMO) of the donor (TTF) and the lowest unoccupied molecular orbital (LUMO) of the acceptor (TCNQ), respectively. Similarly, other crystalline organic conductors have been imaged, including BEDT-TTF-salts,[30–32] which incidentally become superconducting at low temperatures, and other organic charge transfer complexes.[33,34] A limitation of the method is due to the stability of the materials. It is quite typical that surface layers get removed during prolonged scanning, and that after some time remarkably extended atomically flat areas appear.[29,30] This observation indicates that molecular steps and defects on the surface promote the removal of material and that it is very difficult to non-destructively image pristine surfaces, which probably exhibit more roughness.

Also polymer surfaces can be imaged by STM, provided the polymer has been coated with a thin metallic overlayer, or it is electronically conductive. The achieveable resolution is in the range of a few nanometres (see also Butt *et al.*[6]). As an example, images of a number of differently prepared polyacetylene films (see Chapter 7)[30,35] are discussed. The polyacetylene films consist of a network of fibrils with a typical diameter of several nanometres. The native material is electronically insulating. For STM imaging it may be coated with a 5 nm thin layer of Pt/C. Figure 12.2(a) displays a STM image, which reveals the fibrils of the material as prepared. Upon stretching the film a pronounced orientation of the fibrils becomes visible (Fig. 12.2(b)). Significantly, characteristically ordered areas can be found (Fig. 12.2(c)), indicative of oligomers on the polymer fibrils, decorated by the Pt/C coating. The resolution in this case is of the order of about 2 nm, which is about the grain size of the conductive overcoat. While in favourable cases this resolution may also be obtained by transmission electron microscopy, it should be pointed out that for STM no thin film preparation is necessary, but instead the metallized samples can be arbitrarily thick.

Since the grain size of the coating may limit the resolution, it is desirable to image the polymer directly without any overcoat. This can be achieved if the polyacetylene is oxidized, in order to render it electronically conductive.[30] Figure 12.2(d) reveals the fibrillar structure of uncoated, iodine doped polyacetylene. The sample had not been stretched, but it had been pressed prior to imaging, since the native network was not rigid enough to be imaged directly by STM. The relatively high degree of order must be partly attributed to this processing. The resolution of the order of

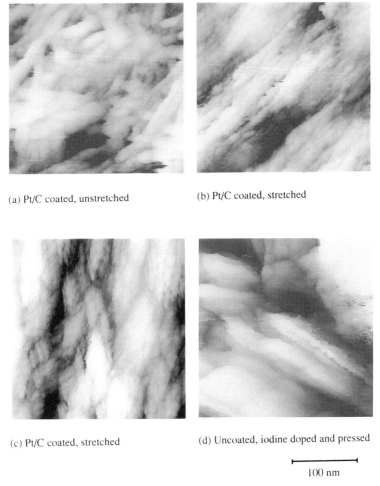

(a) Pt/C coated, unstretched   (b) Pt/C coated, stretched

(c) Pt/C coated, stretched   (d) Uncoated, iodine doped and pressed

⊢———————————⊣
100 nm

Fig. 12.2 STM images of (a–c) Pt/C coated pristine, and (d) uncoated iodine doped polyacetylene.[35]

a few nanometres indicates that it should be possible to use the STM to interface external electronic circuits via the metallic STM-tip to an individual fibril. This should allow low energy electron-spectroscopic experiments to be performed under well defined contact conditions, as opposed to the usual situation, where due to the typically complex morphology of polymers the contacts are rather ill-defined.

Noble Metals Noble metals play an important role as electrode materials, since they do not readily form oxides under ambient conditions. For instance, platinum (alloyed for mechanical reasons with some 10% of iridium) is widely used as a tip material. Moreover, gold and silver are

(a)

(b)

100 nm

Fig. 12.3   STM images of 100 nm thick silver films, evaporated at (a) 50°C and (b) 275°C on to mica.[36] Image sizes: 250 × 250 nm².

suitable solid substrates for molecular monolayers, provided they are sufficiently flat. They may be evaporated onto freshly cleaved mica, which exhibits atomically flat terraces over macroscopical distances.

Figure 12.3(a) displays the STM image of a thin silver film, evaporated onto mica near room temperature.[36] The image reveals the polycrystalline structure of the film, with the largest crystallites about 20 nm in diameter, and a peak to valley roughness of about 5 nm across an area of

250×250 nm^2. One way to obtain larger crystallites is to heat the mica during silver evaporation to 275°C.[36] Figure 12.3(b) displays the STM image of such a silver film. It exhibits Ag(111) terraces, which are flat within a few 0.1 nm and up to 100 nm in diameter. Similarly, the crystallite size in Au(111) films on mica is of the order of 20 nm near room temperature and becomes a few 100 nm at 400°C.[36] It should be pointed out, however, that under ambient conditions the surfaces are not atomically clean, but instead there will be some adsorbates present at the surfaces, which are difficult to characterize.

Layered Materials   An alternative way to obtain atomically flat surfaces of a conductive solid is to cleave a layered conductor. A particularly interesting case is highly oriented pyrolytic graphite (HOPG), because its basal plane is both chemically as well as mechanically very inert. Moreover, highly perfect material is commercially available. HOPG is a semimetal, which has been imaged at atomic resolution under ultrahigh vacuum and at various solid–fluid interfaces. It exhibits atomically flat and perfect crystallites of the order of a few micrometres in diameter, and its characteristic defects have been described, including domain boundaries,[37] $\sqrt{3} \times \sqrt{3}$ R30° superstructures around point defects,[28,38] and large hexagonal superstructures, which arise if the top graphite layer is rotated relatively to the second layer around the normal axis.[39] This careful characterization of the substrate is important to distinguish unambiguously between graphite defects and molecular adsorbates.[28,40,41]

A number of rather stable layered semiconductors belong to the class of transition metal dichalcogenides. Materials which have been investigated by STM include MoS_2,[42] $MoSe_2$,[43,44] and WSe_2.[45] In Fig. 12.4 a typical image of a $MoSe_2$ basal plane is displayed. It indicates that the surface exhibits quite perfect terraces as far as atomical flatness is concerned.

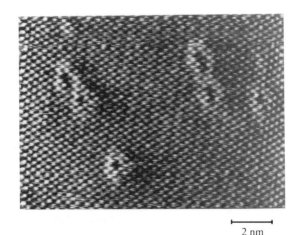

2 nm

Fig. 12.4   STM image of the basal plane of $MoSe_2$.[43,44]

However, there are characteristic single atom defects, which are possibly due to other transition metal atoms substituting a molybdenum atom.[44]

12.4   Molecular Adsorbates under Ultra-High Vacuum

Necessary for an adsorbate to be imaged with the STM is that the molecular mobility is not too high. The advantage of ultrahigh vacuum is that reactive metal surfaces may be prepared and reacted *in situ*. The binding of chemisorbed molecules can be strong enough to guarantee immobilization of individual molecules. Provided the molecular thickness does not exceed the usual tunnelling gap width of the order of 0.5 nm, the tunnelling mechanism is sufficient for electron transport through the molecular layer. The specific electronic structure of adsorbate and substrate determines whether a sizeable contrast due to the molecules can be expected.

Highly resolved images of organic adsorbates have been obtained on an ordered (3×3) superlattice of co-adsorbed benzene and carbon monoxide on a rhodium (111) surface.[46] While images on the close packed monolayer are very stable, the diffusion increases for submonolayer coverage, and it is observed that an individual molecule changes places during the STM scan of one image.

A similar situation is encountered for phthalocyanines on Cu (100). Regular molecular arrays as well as single phthalocyanine molecules could be observed and under some circumstances atomic scale features were resolved, which may be attributed to the molecular front orbitals.[17] Attempts to image multilayers gave neither regular structures nor very high resolution, indicating that in this case the conductivity of the layer may be too small to allow nondestructive observation.

12.5   Molecules at Solid-Fluid Interfaces

An interesting feature of STM is that it allows high resolution imaging not only under ultrahigh vacuum but also at the internal interface between two condensed media, one being a conducting solid and the other being a gas, a liquid or a soft solid. Under these conditions various molecules physisorbed to inert substrates can be investigated. It should be emphasized, however, that the interaction energy between a small molecule and a chemically inert substrate is usually too small to immobilize a single isolated molecule sufficiently at room temperature. One way out is to immobilize a molecule within a monomolecular layer on a layered solid substrate.

Molecular Structure   A model system for atomic scale resolution studies are monolayers of alkanes and alkyl-derivatives, self-assembled at the interface between the basal plane of HOPG and a melt or an organic

lamella

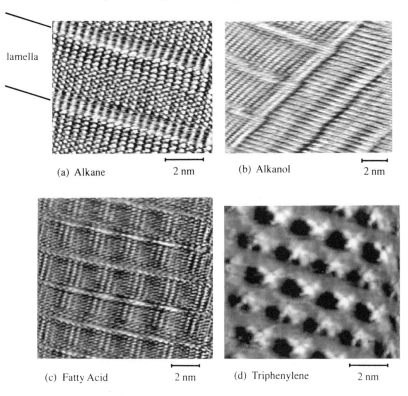

(a) Alkane   2 nm   (b) Alkanol   2 nm

(c) Fatty Acid   2 nm   (d) Triphenylene   2 nm

Fig. 12.5   STM images of (a) an alkane ($C_{27}H_{56}$);[50,55] (b) an alkanol ($C_{18}H_{37}OH$);[50,52] (c) a fatty acid ($C_{17}H_{35}COOH$);[50] and (d) hexakis-heptyloxytriphenylene[35] at the internal interface between HOPG and solutions in phenyloctane.

solution.[18,24,35,47–56] Figure 12.5(a) displays an STM image of a monolayer of an *n*-alkane, obtained *in situ* at the interface between HOPG and a phenyloctane solution. The image reveals a high degree of order, with the molecules in all-*trans* conformation, and packed in lamellae, in which the long molecular axes are oriented parallel to the substrate and perpendicular to the lamella boundaries. In addition, one can conclude that substrate and adsorbate lattices are commensurate in both directions within the plane, since on the one hand, the tunnelling current is determined by both substrate and adsorbate, and on the other hand, each molecule within the monolayer exhibits the same contrast. Moreover, one can conclude that the linear carbon lattices within the HOPG surface and along the alkane chain are incommensurate, since the contrast along an individual molecule varies over several repeat units.[50,53] Finally, one can observe molecularly well defined domain boundaries,[51] as well as molecular mixtures with different chain lengths.[54]

Similarly, many other long chain alkyl-derivatives form highly ordered monolayers at the interface between HOPG and phenyloctane solutions.[47,49,50,52,55] Therefore, a generally applicable method to render a small molecule accessible for an STM investigation is to alkylate the molecule and then to assemble it in a monolayer on HOPG (or any other suitable substrate). Depending on the particular molecule and substrate, different 2-dimensional molecular patterns can be obtained. For instance, long chain alkanols form monolayers on HOPG, in which the long molecular axes are tilted 30° relatively to the lamella normal.[50,52] Figure 12.5(b) shows a domain boundary. Images of fatty acids (Fig. 12.5(c)) exhibit a superstructure along the lamellae, which can be attributed to a beating pattern due to a slight mismatch between adsorbate and substrate lattice.[50] Since the lattice constants of the substrate are well known, the superstructure allows a very accurate determination of the molecular distances in the adsorbate layer. In the case of the fatty acids the interchain distance is about 10% larger than in the case of the alkanes, since the alkyl carbon zig-zag plane is oriented parallel or perpendicular to the graphite plane, respectively.[50,53] Also topologically more complex molecules may form regular patterns. For instance, hexakisheptyloxytriphenylenes pack in double rows with poorly ordered sidechains (Fig. 12.5(d)),[35] while the analogue with hexadecyl-sidechains exhibits sidechain crystallization.[55] Further examples for alkyl-derivatives on HOPG can be found in the literature, including cyanobiphenyl,[18,47] triiodobenzoate,[51] phenyl-benzoate,[35] or azobenzene.[43]

Interestingly, the same molecules may also order on other atomically flat substrates, which are completely different as far as surface chemistry and lattice constants are concerned. Examples include cyanobiphenyls on MoS_2,[56] and alkanes as well as didodecylbenzene on MoS_2 and $MoSe_2$.[43,44] In these cases a very similar contrast has been observed as on HOPG.

Molecular Dynamics   So far the images have been interpreted as static, assuming that the time scales of molecular dynamics and image recording are sufficiently different. Since the time resolution of STM is limited by the digitization speed to about 10 μs, one usually averages over most intra-molecular motions. However, there are also molecular dynamic processes which are slow enough to be resolved. For instance, different packings within a monolayer can be sufficiently long lived that transitions between them may be imaged directly. An example is the motion of domain walls in didodecylbenzene monolayers, which is associated with a screw motion of the individual molecules.[49] Other examples include the cooperative tilt flip within alkanol[50] and alkane lamellae.[53] In the latter case the corresponding type of molecular motion has also been found in molecular dynamics (MD) simulations.[53] However, since the total simulation time, which may be achieved in MD simulations is on the order of 1 ns, as compared to 100 ms for the minimum time required to record a STM image of 100×100 pixels, the simulation has been carried out for a shorter chain length than the STM experiment.

The molecular dynamics within a monolayer may be increased with temperature. For instance, in crystalline alkane lamellae at the interface between HOPG and a phenyloctane solution a reversible increase in

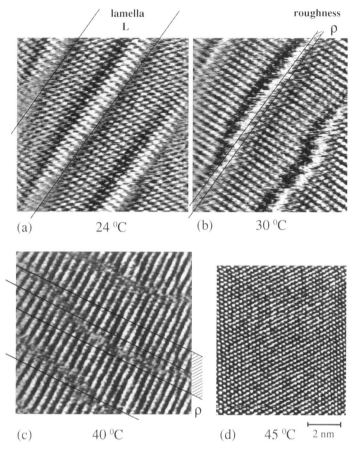

Fig. 12.6 STM images of a $C_{24}H_{50}$ monolayer at the interface between a phenyloctane solution and HOPG, imaged at various temperatures.[24] The average roughness of the lamellae increases continuously until it becomes comparable to the lamella width and the lamella order disappears.

longitudinal molecular motion with increasing temperature is observed, until the amplitude becomes comparable to the lamella width and the order disappears (Fig. 12.6).[24] It should be noted, however, that the individual STM images do not show a snap shot on a molecular time scale, since the correlation times for the molecular motions are shorter than the scan time per image. Since the bottom of an image has been recorded about 1 s later than the top, the lamella width seems to vary substantially within a particular image. Nevertheless it remains constant on average.

A different situation is observed at the interface between HOPG and a melt of a neat alkane at a few degrees above the bulk melting point. Figure 12.7 displays an STM image of a tetracosane ($C_{24}H_{50}$) monolayer adsorbed to HOPG from its melt and images *in situ* at 56°C, which is 6°C

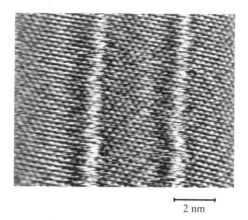

2 nm

Fig. 12.7   STM image of a $C_{24}H_{50}$ monolayer at the interface between its melt and HOPG, imaged at 56°C. The regular hexagonal pattern is due to the graphite lattice, while the vertical lines represent boundaries between alkane lamellae.[24]

above its bulk melting point of 50°C.[24] Individual molecules cannot be recognized any more. Instead the image exhibits the graphite lattice in the background and superimposed straight lines with a separation of the width of the crystalline alkane lamellae, suggesting a smectic phase in two dimensions.[54] These data show that one may obtain molecular scale information on structure and dynamics even for only partially ordered (liquid crystalline) monolayers. This allows the characterization of mixtures of alkanes with different chain lengths, which over a large concentration range exhibit a columnar phase.[54]

Electronic Properties   The interpretation of the images in terms of structure and dynamics has been based to a large extent on symmetry arguments, not really considering the physics of contrast generation. One of the reasons is that the theoretical description of the contrast is a very difficult many-particle problem under fairly extreme conditions and with little symmetry in the system. So far we are only at the very beginning of a good understanding. Different concepts have been put forward, attributing the effect of the molecular adsorbate on the tunnelling current either to a modification in the density of states within substrate or tip,[15] or to a change of the tunnelling barrier.[16-19]

On the other hand, since the structure of the ordered monolayers can indeed be interpreted well without assuming particularly detailed contrast models, experimentation may lead the way. In fact, STM allows a most direct experiment as far as molecular electronics is concerned, namely the measurement of the current-voltage (I–V) characteristics through individual molecular segments. In the case of a number of alkyl-derivatives on HOPG, the experimental results may be correlated to the energy gap between the HOMOs of the substrate and the adsorbate.[19] Since the current will depend on the initial and final states as well as on the barrier

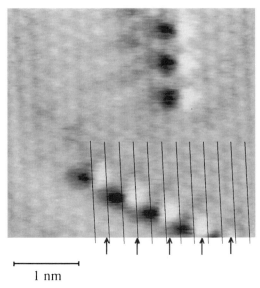

1 nm

Fig. 12.8 STM image of a domain boundary in a didodecylbenzene monolayer on HOPG.[5,55] Under the given tunnelling conditions, the image is dominated by the underlying graphite lattice and the bright benzenes on top (the dark shadow to the left of them is a recording artifact), while the contrast due to the alkyl chains is negligible.

between them, it will contain information on tip, substrate and molecular adsorbate. This is clearly reflected, e.g., in the STM image given in Fig. 12.8, which displays a small area within the didodecylbenzene monolayer on HOPG, recorded under imaging conditions, where the alkyl chains hardly contribute anything to the contrast, while the current through the benzene rings is particularly large.[5,55] This allows the use of the graphite lattice as a very accurate ruler to determine the location of the benzenes at the particular domain boundary. Moreover, images recorded at different biases may be used to calibrate I–V-curves through different molecular segments internally. It is an important advantage of this method that the identical tip is used for the different molecular segments, since it is practically speaking impossible to prepare and maintain a tip of exactly known structure while tunnelling at ambient temperature at a solid fluid interface.

12.6 Prospects for Molecular Manipulation

The STM allows not only the characterization but also the modification at interfaces.[43,47] This may be also of interest for information storage purposes. In this case, several requirements need to be fulfilled: for instance, both writing and reading should be *fast*, the initial as well as the final structures should be sufficiently *stable*, the local change should *not perturb the molecular environment* much, and the change should be *clearly detectable* by STM. The STM itself can also be used to investigate suitable systems. So far, systematic work has been carried out primarily on surfaces of inorganic materials, but the underlying physics does not prevent the application to organic materials in the future. We end this chapter with some interesting cases.

A fundamentally important process is field evaporation, which may be used to create holes and deposits on gold and silver.[27,58] Holes on Ag(111) have been fabricated, for instance, using bias pulses of about 5 V and as short as 10 nanoseconds at a base line current of 2 pA.[27] In this case, the primary writing process occurs at least on the 10 nanosecond time scale, and also the depressions may be read out relatively fast, since the height of the flying tip does not need to be adjusted.

Another interesting example is the local chemical modification of graphite under ambient conditions or at the interface with undried solvents. Metastable surface modifications of less than 1 nm diameter can be produced at a moderate bias of about 2 V,[59] while long-term stable holes of a few nanometres in diameter could be etched with bias pulses of the order of 1 ms and 5 V.[59-61] The processes have been explained in terms of an anodic oxidation of graphite.[61] Similarly, other layered materials have been locally etched, using an STM.[62]

Interestingly, there are a number of modification mechanisms which are particular to molecular materials. An example may be the reorientation of molecules within a molecular pattern. In two-dimensional patterns of long chain molecules several of the prerequisites for information storage schemes described above an be fulfilled. Figure 12.9 shows the image of a monolayer of long chain alkanes with a single square of molecules, which are rotated by 90° with respect to the other molecules within this domain.[43]

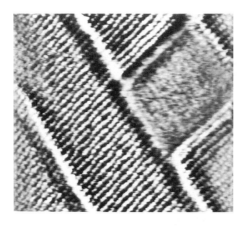

2 nm

Fig. 12.9   STM image of a metastable square of alkanes ($C_{44}H_{90}$), rotated by 90° with respect to the other molecules within this domain.[43]

Obviously, the situation is stable enough to be imaged, the rest of the molecular pattern is not perturbed, and the square is easily detectable. While it is not obvious how to rotate the molecules in this square on purpose, one may imagine related molecules. which should allow this process.

Another molecular property which may be modified is the conformation of a molecule. Azobenzenes, for instance, exhibit a stable *trans*- and a metastable *cis*-conformation. In bulk materials their conformation can be switched back and forth via the appropriate visible light.[63] Monolayers may be formed from an azobenzene derivative in its *trans*-conformation, as evidenced by STM.[43] Upon illuminating these patterns with light, which switches the molecules into their *cis*-conformation, the pattern can be destroyed. A proper molecular design may allow that the *cis*-forms can also order well.

Finally, chemical reactions, including complexations, dimerizations or polymerizations may be employed. While potentially reactive molecules (e.g., diacetylenes[55]) have been organized within molecular patterns, in which they may react, the demonstration of a controlled intermolecular reaction remains a challenge for the future.

References

1. J. Deisenhofer and H. Michel, *Angew. Chem. Int. Ed. Engl.*, **28**, 829 (1989); *Angew. Chem.*, **101**, 872 (1989); R. Huber, *Angew. Chem. Int. Ed. Engl.*, **28**, 848 (1989); *Angew. Chem.*, **101**, 849 (1989).
2. *Conjugated Polymers: The Novel Science and Technology of Highly Conducting and Nonlinear Optically Active Materials*, Eds J. L. Brédas and R. Silbey, (Kluwer, Dordrecht, 1991).
3. A. Ulman, *An Introduction to Ultrathin Organic Films—From Langmuir–Blodgett to Self-Assembly*, (Academic Press, San Diego, 1991).
4. G. Binnig and H. Rohrer, *Rev. Mod. Phys.*, **59**, 615 (1987); *Angew. Chem.*, **99**, 622 (1987); *Angew. Chem. Int. Ed. Engl.*, **26**, 606 (1987).
5. J. P. Rabe, *Ultramicroscopy*, **42–44**, 41 (1992).
6. H.-J. Butt, R. Guckenberger and J. P. Rabe, *Ultramicroscopy*, **46**, 375 (1992).
7. *Scanning Tunneling Microscopy and Related Methods*, Eds R. J. Behm, N. Garcia and H. Rohrer, NATO ASI Series E, (Applied Science 184, Dordrecht, 1990).
8. *Scanning Tunneling Microscopy I & II*, Eds H.-J. Güntherodt and R. Wiesendanger, (Springer Verlag, 1992).
9. G. Binnig, C. F. Quate and Ch. Gerber, *Phys. Rev. Lett.*, **56**, 930 (1986).
10. E. Betzig and J. K. Trautman, *Science*, **257**, 189 (1992).
11. A. Bryant, D. P. E. Smith and C. F. Quate, *Appl. Phys. Lett.*, **48**, 832 (1986).
12. J. Tersoff and D. R. Hamann, *Phys. Rev. Lett.*, **50**, 1998 (1983); *Phys. Rev. B*, **31**, 805 (1985).
13. J. Bardeen, *Phys. Rev. Lett.*, **6**, 57 (1961).
14. N. D. Lang, *Phys. Rev. Lett.*, **56**, 1164 (1986); *Phys. Rev. Lett.*, **58**, 45 (1987).
15. D. M. Eigler, P. S. Weiss, E. K. Schweizer and N. D. Lang, *Phys. Rev. Lett.*, **66**, 1189 (1991).

16. J. K. Spong, H. A. Mizes, L. J. LaComb Jr., M. M. Dovek, J. E. Frommer and J. S. Foster, *Nature*, **338**, 137 (1989).
17. P. H. Lippel, R. J. Wilson, M. D. Miller, Ch. Wöll and S. Chiang, *Phys. Rev. Lett.*, **62**, 171 (1989).
18. D. P. E. Smith, J. K. H. Hörber, G. Binnig and H. Nejoh, *Nature*, **344**, 641 (1990).
19. G. Lambin, M. H. Delvaux, A. Calderone *et al.*, *Mol. Cryst. Liq. Cryst.*, **235**, 75 (1993).
20. S. Ciraci, *Ultramicroscopy*, **42–44**, 16 (1992).
21. G. Doyen, E. Koetter, J. P. Vigneron and M. Scheffler, *Appl. Phys. A*, **51**, 281 (1990).
22. W. Sacks and C.Noguera, *Phys. Rev. B*, **43**, 11612 (1991); *J. Vac. Sci. Technol. B*, **9**, 488 (1991).
23. Y. Kuk and P. J. Silverman, *Rev. Sci. Instrum.*, **60**, 165 (1989).
24. L. Askadskaya and J. P. Rabe, *Phys. Rev. Lett.*, **69**, 1395 (1992).
25. S. A. Elrod, A. L. de Lozanne and C. F. Quate, *Appl. Phys. Lett.*, **45**, 1240 (1984).
26. S. Kitamura, T. Sato and M. Iwatsuki, *Nature*, **351**, 215 (1991).
27. J. P. Rabe and S. Buchholz, *Appl. Phys. Lett.*, **58**, 702 (1991).
28. J. P. Rabe, M. Sano, D. Batchelder and A. A. Kalatchev, *J. Microscopy (Oxford)*, **152**, 573 (1988).
29. T. Sleator and R. Tycko, *Phys. Rev. Lett.*, **60**, 1418 (1988).
30. J. P. Rabe and S. Buchholz, in *Conjugated Polymeric Materials: Opportunities in Electronics, Optoelectronics, and Molecular Electronics*, Eds J. L. Brédas and R. R. Chance, NATO-ARW Series E, (Kluwer, Dordrecht, 1990), Vol. 182 p. 483.
31. H. Bando, S. Kashiwaya, H. Tokumoto, H. Anzai, K. Kinoshita and K. Kajimura, *J. Vac. Sci. Technol. A*, **8**, 479 (1990).
32. S. N. Magonov, G. Bar, E. Keller, E. B. Yaguskii, E. E. Laukhina and H.-J. Cantow, *Ultramicroscopy*, **42–44**, 1009 (1992).
33. R. Fainstein and J. C. Murphy, *J. Vac. Sci. Technol. B*, **9**, 1013 (1991).
34. S. N. Magonov, J. Schuchhardt, S. Kempf, E. Keller and H.-J. Cantow, *Synth. Met.*, **40**, 59 (1991); S. N. Magonov, S. Kempf, H. Rotter and H.-J. Cantow, *Synth. Met.*, **40**, 73 (1991).
35. J. P. Rabe, in *Nanostructures Based on Molecular Materials*, Eds W. Göpel and C. Ziegler, (VCH Weinheim, 1992) p. 313.
36. S. Buchholz, H. Fuchs and J. P. Rabe, *J. Vac. Sci. & Techn. B*, **9**, 857 (1991).
37. D. Tomanek, S. G. Louie, H. J. Mamin *et al.*, *Phys. Rev. B*, **37**, 7790 (1987); T. R. Albrecht, H. A. Mizes, J. Nogami, S.-I. Park and C. F. Quate, *Appl. Phys. Lett.*, **52**, 362 (1988).
38. H. A. Mizes and J. S. Foster, *Science*, **244**, 559 (1989).
39. M. Kuwabara, D. R. Clarke and D. A. Smith, *Appl. Phys. Lett.*, **56**, 2396 (1990); C. Liu, H. Chang and A. J. Bard, *Langmuir*, **7**, 1138 (1991).
40. H. Chang and A. J. Bard, *Langmuir*, **7**, 1143 (1991).
41. C. R. Clemmer and T. P. Beebe Jr., *Science*, **251**, 640 (1991).
42. M. G. Youngquist and J. D. Baldeschwieler, *J. Vac. Sci. Technol. B*, **9**, 1083 (1991).
43. J. P. Rabe, in *Atomic and Nanoscale Modification of Materials: Fundamentals and Applications*, Ed. P. Avouris, NATO ASI Series, (Kluwer Academic Publishers, Dordrecht, 1993) p. 263.
44. S. Cincotti and J. P. Rabe, *Appl. Phys. Lett.*, **62**, 3531 (1993).
45. S. Akari, M. Stachel, H. Birk, E. Schreck, M. Lux and K. Dransfeld, *J.*

Microscopy (Oxford), **152**, 521 (1988).

46. H. Ohtani, R. J. Wilson, S. Chiang and C. M. Mate, *Phys. Rev. Lett.*, **60**, 2398 (1988).
47. J. S. Foster and J. E. Frommer, *Nature*, **333**, 542 (1988).
48. G. C. McGonigal, R. H. Bernhardt and D. J. Thomson, *Appl. Phys. Lett.*, **57**, 28 (1990).
49. J. P. Rabe and S. Buchholz, *Phys. Rev. Lett.*, **66**, 2096 (1991).
50. J. P. Rabe and S. Buchholz, *Science*, **253**, 424 (1991).
51. J. P. Rabe and S. Buchholz, *Makromol. Chem.—Macromol. Symp.*, **50**, 261 (1991).
52. S. Buchholz and J. P. Rabe, *Angew. Chem.*, **104**, 188 (1992); *Angew. Chem. Int. Ed. Engl.*, **31**, 189 (1992).
53. R. Hentschke, B. L. Schürmann and J. P. Rabe, *J. Chem. Phys.*, **96**, 6213 (1992).
54. R. Hentschke, L. Askadskaya and J. P. Rabe, *J. Chem. Phys.*, **97**, 6901 (1992).
55. J. P. Rabe, S. Buchholz and L. Askadskaya, *Synth. Metals*, **54**, 339 (1993).
56. M. Hara, Y. Iwakabe, K. Tochigi, H. Sasabe, A. F. Garito and A. Yamada, *Nature*, **344**, 228 (1990).
57. C. F. Quate, in *Highlights of the Eighties and Future Prospects in Condensed Matter Physics*, NATO Science Forum '90, (Plenum Press, 1991).
58. H. J. Mamin, P. H. Guethner and D. Rugar, *Phys. Rev. Lett.*, **65**, 2418 (1990).
59. J. P. Rabe, S. Buchholz and A. M. Ritcey, *J. Vac. Sci. & Technol. A*, **8**, 679 (1990).
60. T. R. Albrecht, M. M. Dovek, M. D. Kirk, C. A. Lang, C. F. Quate and D. P. E. Smith, *Appl. Phys. Lett.*, **55**, 23 (1989).
61. S. Buchholz, H. Fuchs and J. P. Rabe, *Adv. Mater.*, **3**, 51 (1991).
62. B. Parkinson, *J. Am. Chem. Soc.*, **122**, 7498 (1990).
63. K. Anderle, R. Birenheide, M. Eich and J. H. Wendorff, *Makromol. Chem. Rapid Commun.*, **10**, 477 (1989).

13

The Biological Membrane

J B C Findlay

13.1 Introduction

The evolution and development of the biological membrane was one of the major steps in cellular evolution. The events by which this occurred can only be guessed at but one scenario envisages the entrapment of nucleic acid, protein or a complex of the two in a hydrophobic droplet. This compartment offered a degree of protection from the external environment leading to the conditions under which nucleic acid replication could occur. Over time, the hydrophobic barrier became increasingly refined to the structure known today as the lipid bilayer. In the process, the development of cellular activity and regulation occurred.

The biological cell requires efficient processes through which it can communicate with and respond to its external environment. Being hydrophobic, the cell membrane constitutes an effective barrier to both polar and bulky substances. Thus, specialized mechanisms are needed to effect the transport across this barrier of both the material and information required for metabolism and regulated activity. This chapter will outline briefly the basic structure of the biological membrane and the mechanisms which have evolved to serve its biological function. Many of these highly refined processes can be reproduced, at least crudely, *in vitro* and manipulated in various ways for both practical and didactic purposes.

13.2 Structure

Lipid Bilayers The core structure of the biological membrane is the bilayer composed of a complex mixture of lipids. The essential feature of these lipids is that they consist of polar and nonpolar domains and thus, for energetic and thermodynamic stability in aqueous environments, they spontaneously adopt an organization which maximizes polar interactions with water, whilst minimizing nonpolar exposure. The most common organization is a bilayer but there are various micellar forms that can also be generated. Whether any of these non-bilayer forms are allowed, if only transiently, in native bilayers is an interesting issue on which there are few definitive data. Although the chemical composition of natural bilayers is complex, the lipid constituents can be grouped into a few classes which are illustrated in Fig. 13.1. The greatest degree of variability is due to esterified fatty acids which constitute the hydrophobic core of the bilayer. They can vary both in length (C5 up to C24) and in the degree of unsaturation (0 to 6 double bonds). In addition, there can be some rare modifications.

In the biological situation, this gives rise to heterogeneity within a single membrane, between membranes in a single cell and between cells in a single organism. The precise reason for this heterogeneity is unclear, since bacteria can exist successfully with only a single type of fatty acid in their membrane; but it is generally thought that it provides a flexible yet precise environment for the many diverse processes that the membrane carries out. It may also be subtly but effectively engineered to generate a tight barrier. Addition of cholesterol, for example, to artificial lipid mixtures can significantly increase bilayer impermeability.

The biological bilayer can be symmetric in terms of the polar head groups but more commonly, particularly for the membrane facing the external environment, it exhibits asymmetry. Thus for the red blood cell, sphingomyelin and phosphatidylcholine are disproportionally located in the outward facing monolayer, while phosphatidylserine and phosphatidylethanolamine are mainly in the monolayer facing the cellular cytosol. This pattern is not uniform in all mammalian cells, however, and asymmetry is much less obvious in intracellular membranes. How this asymmetry is created is again fairly obscure. There are some indications that a protein-based machinery may participate in the process, but a more likely mechanism is that it arises spontaneously as a result of the complex interplay of lipid synthesis and composition, protein activity and the nature of the cellular environment.

The exact relationship between lipid composition, organization and function in biological systems has not yet been firmly established. Very few membrane proteins, for example, have an absolute requirement for a specific lipid but their activity can vary considerably in different lipid mixtures. One important feature seems to be that the membranes should have a degree of 'fluidity'. Biological membranes in their native state appear to permit rapid translational and rotational motion within the plane of the bilayer, but movement from one monolayer to the other (called 'flip-flop') is much more restricted. In terms of phase diagrams, therefore, it is assumed that the lipids in cellular membranes under normal 'biological conditions' are generally in a liquid-crystalline rather than a solid-crystalline state, but it should be appreciated that in such complex

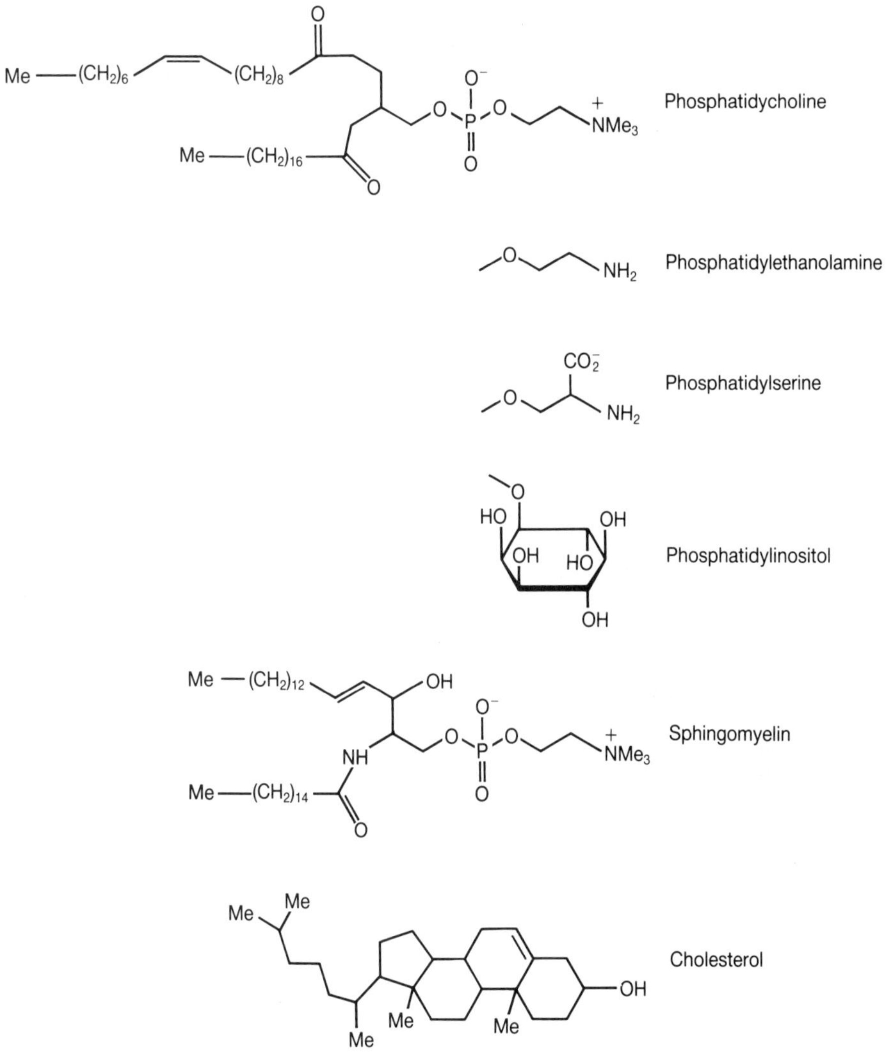

Fig. 13.1 The different types of lipid structure are indicated, showing the polar head groups and fatty acyl moieties. These may vary both in chain length and degree of saturation.

mixtures, there are likely to be transient microscopic domains with a range of physical characteristics.

The most influential factor concerning fluidity is the nature of the esterified fatty acid. Shorter chain lengths and higher unsaturation leads to lower transition temperatures (solid to liquid) whilst longer and more saturated fatty acids have higher transition points. The neutral lipid cholesterol has a modulating effect; it reduces the lipid order in solid crystalline systems while decreasing disorder in the more fluid situations.

This feature is thought to be significant in maintaining the integrity (e.g. very limited leakiness) of the biological membrane whilst maintaining its dynamic properties. Interestingly, cholesterol is an essential component (almost up to its solubility limit) in the plasma membranes of eukaryotic cells; it is much less abundant in intracellular membranes and absent altogether in the prokaryotic membrane.

It is possible to manipulate the lipid composition and physical state, particularly of bacterial membranes, and this does affect the behaviour and activity of membrane-associated processes. An everyday example occurs with hydrophobic anaesthetics whose *in vitro* effects seem to be exerted from within the bilayer and which can markedly affect the transport properties of proteins within the bilayer. The importance of the dynamic balance of the lipid bilayer is also illustrated by the bacterial response to alterations in the temperature of their environment. Raising this temperature by only a few degrees elicits the synthesis and incorporation of longer and more saturated fatty acids, while lowering it enhances the production of shorter and less saturated species. Presumably the exquisite activity of the membrane is preserved by manipulating the properties of the lipid fraction in this way. In general, it also appears that the composition of the eukaryotic membrane is regulated carefully.

In essence then the lipid bilayer is a heterogeneous Langmuir–Blodgett film with features in common with liquid crystals. It forms spontaneously under aqueous conditions, occurring as films, multilamellar sheets, liposomes and vesicles. It can be manipulated both *in vitro* and *in vivo*, and supports the reconstitution of a multitude of biological activities. Despite much study there is still considerable ignorance about behaviour and interactions at the molecular level and how composition is related specifically to activity.

Proteins   The second major component in biological membranes are the proteins illustrated in Fig. 13.2. They fall into two classes, *peripheral* or *extrinsic* species, which adhere to the surfaces of the bilayer and which resemble classical water-soluble proteins, and *integral* or *intrinsic* membrane proteins which penetrate through the hydrophobic phase.[1] Many cartoons of a membrane also depict proteins partly or completely embedded in the hydrophobic core, but there are theoretical reasons why such types are unlikely to exist, at least in monomeric form, for other than very transient periods—and certainly none have so far been clearly demonstrated.

The distribution of these protein species is asymmetric and this asymmetry is absolute, for there are very strong energetic restraints which prohibit diffusion or rotation of proteins through the bilayer (certain small peptides might behave rather differently). Translation along the surface of the bilayer is readily demonstrated, on the other hand, and this lateral diffusion is limited only by the viscosity of the lipid components or by any specific protein–protein interactions which may occur.[2] Indeed, it is clear that the cell has the ability when required to direct its proteins away from or towards particular regions of the membranes which may be participating

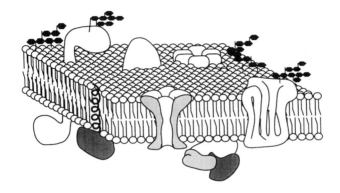

Fig. 13.2 The cartoon illustrates the lipid bilayer containing phospholipid, embedded or associated with which are various types of membrane protein. These may include integral proteins which traverse the membrane or which form channels permitting movement of small molecules across the bilayer. Other extrinsic proteins are associated with the surface of the bilayer. Carbohydrate residues are shown covalently bound to either lipid or protein.

in some activity. This ability largely rests with an underlying network of cytoskeletal or structural proteins which can undergo lever-like redistribution, a phenomenon which is not too different from the concept of molecular machines. Finally, it is possible for integral membrane proteins in the bilayer to form regular 2-D arrays/crystals in the bilayer, under both natural and artificial conditions.

There are considerable structural restraints on polypeptide segments which are in the hydrophobic phase of the bilayer. Energetically, they have to present a hydrophobic external surface to the fatty acid milieu. For stability the hydrogen-bonding potential of polar side-chains or the moieties of the peptide bond should be largely fulfilled. In the absence of sufficient water molecules, these respective demands are best met by hydrophobic side-chains and by the adoption of α-helical or β-sheet conformations where regular internal H-bonding occurs. Crystal structures of integral membrane proteins are very few—the best so far, the photosynthetic reaction centre[3] and bacterial porin,[4] beautifully illustrate these structural features (Fig. 13.3). In both cases, the intramembranous regions are largely regular α-helices and β-sheets respectively, but there is the suggestion in other systems that absolute regularity is not always essential. Where rare distortions do occur, the resultant unsatisfied H-bonding potential may be an important element in the functional properties of the protein. These two secondary structure types will be discussed later in the context of function.

Carbohydrate   The third major membrane component, carbohydrate moieties, are always found covalently associated with either the lipid or protein fractions of the bilayer.[5] Due to the biological synthetic processes involved, their sugar residues are almost always associated with the external surface of the cell or on inner compartment surfaces that are its

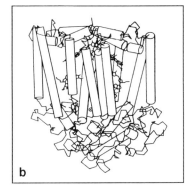

Fig. 13.3 Three-dimensional representation of the polypeptide backbone of (a) bacterial porin (from Cowan, S. W. (1993) *Current Opinion in Structural Biology* **3**, 501) and (b) the bacterial photosynthetic reaction centre (from Rees, D. C., Komaya, H., Yeates, T. O., Allen, J. P. and Feher, G. (1989) *Annual Review of Biochemistry* **58**, 607), in which α-helices are approximated to straight cylinders.

equivalent. There are a few reports of special glycosylation reactions occurring in the cytosolic compartment; cytoskeletal proteins may be one of the targets for such modifications. The size and variety of these carbohydrate side-chains can vary enormously and, for that reason, they can contain highly specific information. As a result, they represent very good targets for signalling processes. Surprisingly, however, there are only a relatively small number of cases, compared to the high amount of glycosylation that takes place, where it can be shown that oligosaccharides or sugars act as messenger systems. Clearly, in view of the energetic expense in their synthesis, there will be other functions. Amongst these are roles in the folding and orientation of polypeptide-chains in the bilayer, the sorting and targeting of vesicles and various recognition phenomena. These sugar moieties are very flexible and there is little knowledge of their 3-D structures.

13.3 Transmembrane Signalling

Most of the communication messages between cells in a multicellular organism, or the tactic signals in unicellular systems, are conveyed by hydrophilic materials. Being readily soluble in water and in some cases being rather large, these signal molecules do not traverse the bilayer. The cell must be able to respond to such signals, however, and as a result, complex, multifactor processes have evolved. The transfer of information across the cell membrane appears to occur via proteins which span the bilayer. The transduction of the external signal into a cellular response involves sequential interactions with other proteins at the other (inner) surface of the bilayer and in the soluble phase of the sealed compartment. The points to appreciate are that there are various stages at which specificity is ensured and that the process possesses an in-built amplification cascade. This means that a small quantity of an external stimulus (e.g. an odour, a hormone etc.) can elicit a defined response in a defined cell.

There are a number of different types of signal transduction system which are used over and over again in biology. Most can crudely be classified into one of three basic processes. The first involves receptors which are linked to guanine nucleotide-binding proteins (G-proteins) and which show specificity in terms of both the external ligand with which they become associated and of the G-protein they subsequently activate.[6] The G-proteins in turn influence specific effector proteins in the cell. An example of this type, visual transduction, is illustrated in Fig. 13.4. The second, quite different structural type becomes chemically modified (usually phosphorylation or methylation) as a result of the conformational changes which follow ligand-receptor interaction. Tyrosine kinases such as the growth factor receptors are an example of this class.[7] The modified receptor is then recognized by other proteins which are the first intracellular components in the amplification cascade (Fig. 13.5). The third type also responds to interaction with a ligand by going through a series of conformational changes, but in this case these changes are directed towards the opening of channels within the same protein which allow the flow of ions down their concentration gradient from one side of the membrane to the other.[8] These so-called 'ligand-gated channels', an example of which is the acetylcholine receptor, are a particular speciality of neural systems although they occur in many other systems.

As with all the processes involved in signal transduction, the receptor is acting as a switch which is activated by the binding of a specific signal molecule. Activation of the switch then initiates specific pathways of information flow. The whole process is rapidly reversible. The details of the mechanisms may be complex but the overall design principles are relatively simple.

13.4   Material Transport[9]

In the above cases of information transfer, the crucial primary event is the binding or alteration of a ligand which subsequently provokes conformational changes in the protein. From work on enzymes, it is not difficult to appreciate the kinds of structural alteration which may occur and that these will be sufficient to activate the 'switch'. Such changes lie also at the heart of the processes by which transport of material may also be effected, but here we have no real precedents to guide the speculation. The mechanisms responsible for ion-channelling in particular present substantial difficulties. This transport of material, as in the ligand-gated systems, is the basis of signal transduction rather than the initial alteration in conformation.

Nonchannel Systems   In this type of process, the transport protein possesses a specific binding site which recognizes the substance to be translocated. Non-covalent association of the substance triggers structural

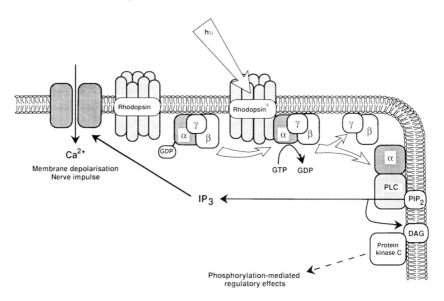

Fig. 13.4 The diagram illustrates the basic features involved in signal recognition and transduction in the invertebrate visual system. The receptor protein—rhodopsin—occurs as a complex with a derivative of vitamin A, 11-*cis*-retinal, and is situated in specialized membranes in the photoreceptor cells of the retina. A photon of the incident light is absorbed by 11-*cis*-retinal, which promptly isomerizes to the all-*trans* form—biochemically it is estimated we can 'see' a single photon! This change in configuration also involves conformational changes in the protein, initially at the ligand binding site but almost instantaneously radiating to other regions of the molecule. These conformational changes generate a complementary high-affinity binding site for a specific guanine nucleotide binding protein (*G*-protein) on the inner surface of the membrane. The *G*-protein, which consists of three subunits (α, β and γ), acts as a transducing link between the cell receptor and the response molecule within the cell. Thus, when the *G*-protein interacts with the activated receptor, the α-subunit dissociates by a process involving the exchange of bound GDP (guanosine diphosphate) for GTP (the triphosphate). The $G\alpha$ then directly activates an enzyme, in this case phospholipase C, which catalyses the removal of the head group of a specialized membrane lipid phosphatidylinositol bisphosphate (PIP_2). The inositol phosphate thus generated appears, itself or after further metabolism to a closely related compound, to bind to a protein channel in the plasma membrane inducing a flow of calcium ions into the cell. This has the effect of changing the membrane potential which in turn causes the final act in the activation cascade, a change in chemical transmitter release from the terminus of the cell. The next stages involve nerve impulse conduction which reflect a regulated flow of K^+ and Na^+ ions across the membrane. The diacylglycerol (DAG) which is also generated from PIP_2 may activate an enzyme (protein kinase C) which can in turn exert additional regulatory effects within the cell.

It is important to stress that these systems can respond in millisecond time scales, to minute quantities of the signal (pico or nanomoles), and with tightly regulated specificity.

As well as an activation cascade, this process must have a deactivation mechanism, but much less is known about it. The $G\alpha$-subunit possesses an inherent ability to hydrolyse GTP to GDP, and thus to regenerate the quiescent state. The receptor itself is also thought to be shut off, perhaps by phosphorylation. Certainly the primed condition can be restored very rapidly.

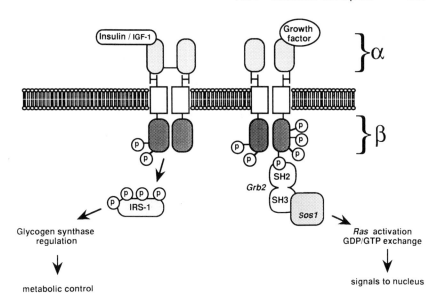

Fig. 13.5 Signal transduction by the tyrosine kinase type of receptor system. Such systems include the insulin receptor (left), and receptors to a variety of protein factors which influence cell growth and division (right). These related receptor proteins comprise α and β domains, connected by a transmembrane section. The interaction of a ligand with the α domain on the outside of the cell is translated into a conformational change, allowing receptor molecules to associate. The β domains of stimulated receptors have an enzymatic activity which can attach phosphate groups to tyrosine amino acid residues of both substrate proteins within the cell and to themselves (autophosphorylation). In the case of the insulin receptor, the β domain can phosphorylate a large number of molecules of a regulatory protein, Insulin Receptor Substrate-1 (IRS-1), which in turn regulates enzymes involved in glucose metabolism (glycogen synthase). The phosphorylated growth factor receptor is able to interact with a particular structural motif (SH2) on a regulatory protein *Grb2*. This protein in turn can associate with another regulator, *Sos1*, giving a chain of interactions which ultimately results in activation of proteins which influence the growth cycle of the cell (for example, *Ras*). In each case the receptor protein can 'activate' a large number of substrate proteins, producing a large amplification of signal whilst maintaining specificity.

alterations by which it is exposed to the other surface of the membrane and can then dissociate from the protein (Fig. 13.6). The degree of movement of the transported entity on the protein surface may be quite small—a few Ångströms—and it is probably not correct to envisage a permanent pore or hole through the protein. Where the flow is down a concentration gradient, no energy input is required, but when the flow is up the gradient some form of energy is needed in order to effect the desired conformational state or change in the protein.

The ATP/ADP shuttle across mitochondrial and membranes[10] is one case where a substantial structural change accompanying transport can be demonstrated, but the closest we have yet come to understanding the process occurs with bacteriorhodopsin, a proton-translocator from the membrane of *Halobacterium halobium*. Like visual rhodopsin, the protein

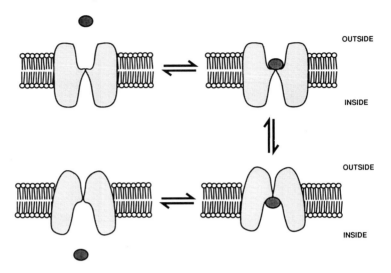

Fig. 13.6 Putative mechanism by which material transport across the bilayer might be effected. The non-covalent association of a substrate molecule with the transporter protein triggers structural changes which result in exposure of the substrate to the other surface of the membrane. Note that only a small structural change may be required to facilitate transport across the bilayer.

absorbs light energy via bound *cis*-retinal, but unlike the visual pigment, the energy is used to promote proton transfer across the lipid bilayer, through conformational changes in the protein molecule. Based on the structure of the protein and the effects of changing particular amino acid residues, a mechanism has been proposed whereby the proton is transferred from amino acid side-chain to side-chain through the protein.[11] The 3-D structure of the protein is constructed such that this transfer occurs readily. A similar process may occur for other H^+-transporting systems, but for larger substrates such as glucose, this shuttle-type procedure is less likely. Instead the binding residues may stay relatively constant but the walls of this site change in a process more akin to a 'car wash'.

Channel Systems[9]   These appear to represent a somewhat different method of membrane transport. Here translocation, usually of ions, is extremely rapid, only an order of magnitude or so slower than free diffusion in water. This feature alone suggests that a simple conformational change model is inappropriate or inadequate. Yet there is clearly a very powerful mechanism for ion selectivity because many of these ion-channels show pronounced specificity, even between ions of similar character, e.g. Na over K etc.[12–14]

The second distinctive feature of these channels is their probable oligomeric nature—they all appear to exist as tetrameric, pentameric or greater assemblies of similar or identical subunits or, as with sodium and calcium channels, as an internally duplicated concatenate.

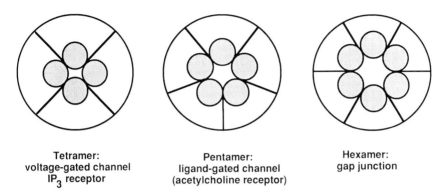

Tetramer:
voltage-gated channel
IP_3 receptor

Pentamer:
ligand-gated channel
(acetylcholine receptor)

Hexamer:
gap junction

Fig. 13.7 The oligomeric nature of channel-forming proteins. Variations in packing of channel-lining helices (shaded) in oligomeric assemblies of identical membrane proteins can produce variability in the dimensions of the transmembrane channel.

These features have given rise to a rather different concept of transport, namely that the transport route is provided by essentially the same structural element—for example, the surface of the same transmembrane helix or β-sheet—provided by each of the subunits (Fig. 13.7). This system would impose definite packing constraints to conform to the demands of symmetry. Thus, with a tetrameric arrangement the packing of channel-lining helices would be quite intimate, but this would become more relaxed for a pentameric structure; while a hexamer could provide a rather large water-filled pore.

The big debates currently concern the molecular basis for ion selectivity, the precise nature of the transport route, and the transport mechanism. The best—but still very imprecise—guess is that a protein-based site or surface confers ion specificity, and that polar amino acid side-chains facilitate this extraordinarily fast movement. For some channels, the latest evidence has implicated a small section of structure joining two putative helical transmembrane segments as crucially important in the selection and transport process.[12,13] This section may have β-like structure and form the contact region between the associating monomers. The ionic species transported is likely to be at best only partly hydrated, but any role for water within a tetra- or pentameric pore has not so far been determined. Unlike nonchannel systems, the assumption is that several ions can be found in the transport 'pore' at any one time.

These biological channels can also be tightly controlled either by membrane potential (voltage) or by specific ligands. There are, for example, both voltage-activated and voltage-inactivated systems for a wide variety of ions (e.g. K, Na, Ca),[9,12–14] while a number of chemical messengers, such as acetylcholine, inositol phosphates, various amino acids or ATP, can activate or inhibit specific channels.[8,9,15] In all cases, the particular proteins contain regions that are capable first of detecting a particular stimulus and then responding to its presence by conformational changes which influence transport activity. Yet again, we see the intimate connection between noncovalent effects and conformation change that is the hallmark of biological communication.

Electron Transport Systems[16]   Apart from the movement of elemental ions, charged species such as phosphate and sulphate ions and larger molecules such as sugars, amino acids and drugs, there are biological mechanisms which facilitate the efficient migration of electrons. Very often this occurs in conjunction with, but separate from, the transport of protons. In general, the detailed path taken by the electron is imprecisely known, but movement principally occurs by way of specific prosthetic groups held at different redox potentials. These prosthetic groups are often relatively small molecules and the mechanism which has developed relies on the precise juxtaposition of these prosthetic groups such that electron transport occurs more readily. Specific proteins have evolved to provide the framework to hold these prosthetic groups in exactly the right orientation, and perhaps to contribute themselves to the route taken by the electron.

Most of the electron transport systems in biology are involved in some way with bioenergetics, the best characterized structurally being the photosynthetic system (Fig. 13.8). The fundamental event here is light-induced primary charge separation. There are special complexes consisting of light harvesting protein and tightly associated porphyrins, chlorophyll, which absorb light energy and transfer it to a special pair of chlorophyll *a* molecules located in the Photosynthetic Reaction Centre. In these an electron is promoted into an excited state and can then pass to an acceptor of lower redox potential called pheophytin and hence to a secondary acceptor which is a quinone. These acceptors are located in the Photosynthetic Reaction Centre (Fig. 13.9), bound to the proteins so that electrons can flow down through the system and be coupled through a tyrosine residue and manganese atoms to the splitting of water.

An analogous system occurs in the respiratory chain, except that most of the prosthetic groups are Fe porphyrins, i.e. haem or Fe/S clusters, both again protein bound.[16] The number of such complexes is quite large and the pathway is quite involved, utilizing diffusable electron carriers of various kinds shuttling between principal respiratory chain complexes. One such shuttle is cytochrome C which contains a covalently bound haem group and on which a considerable amount of effort has been expended to trace the electron path. It is clear that the uptake of an electron by cytochrome C, from the cytochrome bc_1 complex, and its delivery to cytochrome oxidase, implies that the electron is able to travel specifically across considerable distances through moieties with the protein. The identity of these moieties has proved elusive, although chemical modification of lysine side-chains close to the exposed edge of the bound haem does inhibit the process. What is clear is that the biological system retains control over the electrons whose energy is not inefficiently dissipated.

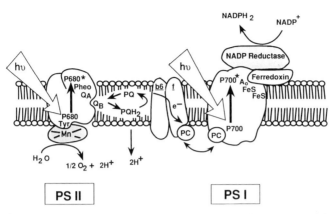

Fig. 13.8 Schematic arrangement of membrane proteins in the photosynthetic electron transport chain of green plants. The two principal membrane protein species, photosystem II (PS II) and photosystem I (PS I), contain specialized Reaction Centre chlorophylls (P680 and P700 respectively) in which an electron is promoted to an excited state by absorption of light. This generates a species with higher redox potential which can then pass electrons thermodynamically 'down' through a series of redox couples. In the Reaction Centre of PSII (which is structurally and functionally similar to the bacterial Reaction Centre (Fig. 13.3b)), this electron moves through pheophytin and quinone molecules. A mobile quinone donates electrons to a membrane-cytochrome complex (b6/f). The Reaction Centre is rereduced by electrons extracted from water via a manganese-containing protein. In photosystem I, the excited special pair electron passes through quinones and iron–sulphur centres, ultimately being used in the reduction of a soluble carrier nicotinamide dinucleotide phosphate (NADP). The PS I Reaction Centre is rereduced by electrons from b6/f complex, donated via an extrinsic carrier protein plastocyanin (PC). Two quanta are therefore used to move each electron from water to NADP across a total redox span of 1140 mV, with concomitant movement of protons across the membrane.

This highly organized system of redox centres, associated with both integral and extrinsic proteins, ensures rapid stabilization of the primary charge separation at the Reaction Centres and efficient transport of electrons from water to NADP in discrete, tightly-controlled steps.

13.5   Conclusions

Biological membranes clearly contain systems and concepts that could be of value in the development of molecular electronics. The basic components are small, specific, switchable and reversible. The systems used have built-in amplification, can be constructed in three dimensions (and conceptually extended to more), and can produce graded responses that are related to the strength of the signal. Potentially, biological membranes are also capable of possessing 'memory'. In addition, they possess the important ability to undergo self-assembly spontaneously, which should facilitate construction on a molecular scale.

But there are also disadvantages, some of them severe, which suggest that biological systems in general will not be incorporated easily into robust working devices. The two most obvious ones are the complexity and the instability of the components and the systems into which they are integrated. In some cases, the problems may not be unsurmountable. There

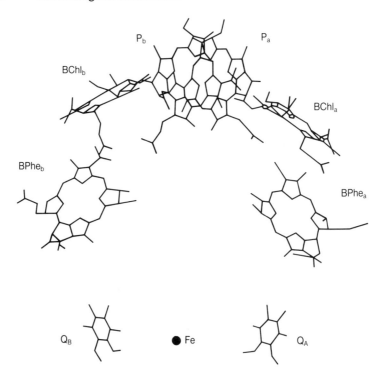

Fig. 13.9 Organization of cofactors within the Reaction Centre of photosynthetic bacteria. $P_{A,B}$; Special pair chlorophylls. $BChl_{A,B}$; Bacteriochlorophyll. $BPh_{A,B}$; Bacteriopheophytin. Q_A and Q_B; Primary and secondary acceptor quinones, respectively. The phytyl and isoprenoid chains have been omitted for clarity. Each cofactor is coordinated to specific amino acid residues of the transmembrane helices of Reaction Centre proteins (see Fig. 13.3b) and closely aligned for efficient energy transfer. (From Diner, B., Nixon, P. J. and Farchaus, J. W. (1991) *Current Opinion in Structural Biology* **1**, 546.)

are enzymes, for example, which are unaffected by wide temperature fluctuations—some can withstand boiling. It is not beyond the realms of possibility, therefore, that techniques in protein engineering will permit us to produce quite cheaply large amounts of synthetic proteins with stable nonbiological functions. Unusual properties, such as the ability to be specifically emplaced either as a surface or in an assembly, or to be resistant to poisoning or modification, could be incorporated into their structures.

The lipid bilayer, too, can be modified to introduce a wide range of fluidity and stability characteristics. The challenge is to produce stable, self-assembling protein-lipid systems with particular functional properties. It is possible, for example, to form organized Langmuir–Blodgett films containing 3-dimensional arrays of proteins, which adhere to specifically modified lipids or which can be precisely positioned. Or again, we could incorporate proteins into LB type films whereby the conformational changes undergone when protein binds ligand can be sensed and communicated by the supporting matrix. These proteins could possess a wide variety of functional and structural characteristics to suit different requirements.

In one example, it is becoming possible to build information processing and ion conducting systems into artificial bilayers. It has been known for some time that small unusual peptides produced by microorganisms are capable of partitioning into bilayers and catalyzing specific ion translocation. These peptides contain nonstandard amino acid residues and have distinctive structures. Valinomycin, a K^+-ionophore, has a circular, dough-ring type of structure, the central cavity of which has a high affinity for K^+ over Na^+. The external surface of the molecule is very hydrophobic and it diffuses readily in a lipid bilayer, carrying the K^+ ion from compartments of high to those of low K^+ concentration. Similar movement is catalyzed by gramicidin but here the process is rather different. Gramicidin adopts a beta-helix type structure which spans a single monolayer. Two molecules can then aggregate tail to tail to provide a continuous route across the bilayer. K^+ ions are thought to move down through the core of the gramicidin dimer[18] but the details of the mechanism and the basis of selectivity for K^+ over Na^+ is not yet clear. There are a range of such specific ionophores in nature, some not really peptidic, which adopt a number of different conformations and aggregation states. There are even examples of voltage-dependent association and dissociation. Together they provide a rich source of information for synthetic mimics, some of which already exist.[19] The biological machinery for their synthesis is complex and cannot yet be reproduced readily in the laboratory. This currently restricts the scope for experimental analysis and structural manipulation, but nevertheless they represent a potential for future exploitation.

There are also other more classical peptides which are produced by organisms. These too have the ability to self-associate to provide trans-membrane channels, but in many cases specificity is lost and these agents act more as lethal cell toxins by mediating more general leakage. However, drawing on these examples, a number of research groups have shown that the putative channel-forming segments from several biological channels can confer specific transport functions on an otherwise impermeable bilayer, even in the absence of the rest of the polypeptide chain.[19] Clearly, there is some inbuilt self-assembling property which allows such peptides to form active complexes. What we do not yet fully appreciate are the key elements in self-assembly, the basis for ion selectivity, the structural nature and mechanism of action of the transporting system, and the influence of the hydrophobic environment. By understanding these fundamental features, we will be in a much better position to move further away from biology towards completely artificial conducting systems. Ultimately, it should be possible to build means of regulation by voltage, light, ligands etc. into these systems.

Finally, such membrane-based structures can extend over considerable distances and can be designed to incorporate functional specificity at particular positions. They are thus much more realistic systems for the development of communicating networks than whole cells.

Concerns as to the relevance of biology for molecular electronics include the need for an aqueous environment, the complexity of the molecules and the relatively slow speed of systems using ions (even protons). Currently, these aspects raise legitimate doubts—but how crippling are these constraints? It is already clear that a minimalist approach towards more stable components is feasible. It is conceivable also that only vanishingly small amounts of water may be required for the stability and function of engineered products and that different, more malleable, substances may be substituted. Finally while ions do move more slowly, they also possess the advantages of selectivity, concentration-dependent responses and amplification. The important issue to consider at this stage in this new science is that the behaviour and potential of such systems should be strenuously and rigorously explored to unfetter the imagination for other novel developments.

References

1. R. B. Gennis, *Biomembranes: Molecular Structure and Function*, (Springer-Verlag, New York, 1989), Chapters 1 and 2.
2. F. Zhang, G. M. Lee and K. Jacobson, *Bioessays*, **15**, 579–88 (1993).
3. J. Diesenhofer, O. Epp, K. Miki, R. Huber and H. Michel, *Nature*, **318**, 618–21 (1985).
4. M. S. Wiess, T. Wacker, U. Nestel, D. Woitzik, J. Weckesser, W. Kreutz, W. Welte and G. E. Schulz, *FEBS Lett.*, **256**, 143–46 (1989).
5. A. Varki, *Glycobiology*, **3**, 97–130 (1993).
6. Y. Kaziro, H. Itoh, T. Kozasa, M. Nakafuku and T. Satoh, *Annu. Rev. Biochem.*, **60**, 349–400 (1991).
7. J. Schlessinger and A. Ullrich, *Neuron*, **9**, 383–91 (1992).
8. D. Bertrand, J-L. Galzi, A. Devillers-Thièry, S. Bertrand and J-P. Changeux, *Curr. Opin. Cell Biol.*, **5**, 688–93 (1993).
9. W. D. Stein, *Channels, Carriers and Pumps. An Introduction to Membrane Transport*, (Academic Press, San Diego, 1989).
10. M. Klingenberg, *Arch. Biochem. Biophys.*, **270**, 1–14 (1989).
11. R. Henderson, J. M. Baldwin, T. A. Ceska, F. Zemlin, E. Beckmann and K. H. Downing, *J. Mol. Biol.*, **213**, 899–929 (1990).
12. C. Miller, *Science*, **252**, 1092–96 (1991).
13. L. Y. Jan and Y. N. Jan, *Annu. Rev. Physiol.*, **54**, 537–55 (1992).
14. M. Stephan and W. S. Agnew, *Curr. Opin. Cell Biol.*, **3**, 676–84 (1991).
15. E. A Barnard, *Trends Biochem. Sci.*, **17**, 368–74 (1972).
16. D. G. Nicholls and S. J. Ferguson, *Bioenergetics*, **2** (Academic Press, London, 1992).
17. A. W. Rutherford, *Trends Biochem. Sci.*, **14**, 227–32 (1989).
18. G. A. Woolley and B. A. Wallace, *J. Membrane Biol.*, **129**, 109–36 (1992).
19. M. Montal, *FASEB J.*, **4**, 2623–35 (1990).

14

Biosensors

M Aizawa

14.1 Introduction

An increasing interest has arisen in biosensing technology which may fall in three major fields: (1) biosensors composed of biological materials, (2) biosensory mimicking systems, and (3) sensing technology for biological systems, which has grown rapidly over the past two decades. Although these three fields of biosensing technologies have evolved from quite different roots, they are based on the understanding that biological systems should be endowed with superior information transducing and processing capabilities. The purpose of this chapter is not to overview the whole scope of biosensing technology, but to concentrate on science and technology of biosensors composed of biological materials.

A biosensor is defined as a sensor that makes use of biological or living material for its sensing function. Biological materials are used due to improve selectivity that is hardly attained by any other materials such as semiconductor and metals. A variety of proteins capable of molecular recognition have been implemented successfully in sensing devices. A biosensor should be the first realization of biomolecular-electronic devices. Since mid 1980s extensive research has been made on development of biosensors, which has brought the study of biosensors right up to date.[1-5]

This chapter describes the design principle of biosensors and the characteristics of electrochemical and optical biosensors with their current status of research and development.

14.2   Molecular Information Transduction in Biological Systems

The design principle of biosensors has been derived from the molecular communication mechanism in biological systems, where instead of electrons and photons such molecules as hormones and neurotransmitters serve as information carriers. The molecular information system should be provided with molecular information release, transport and transduction functions.

Molecular communication is the characteristic information system in biological systems. The endocrine system may represent the features of molecular communication as illustrated schematically in Fig. 14.1.

The gland is a collection of specialized cells that synthesize, store and release hormones. The hormone is released into the extracellular fluid and transported via the blood to two types of cell: target cells where the hormone acts, and other cells that degrade the hormone. In some systems, the target cell and the degradation site are in the same organ or even the same cell. Both activities may even be located on the same plasma membrane. The receptor for the hormone is located on the surface of the plasma membrane, and is associated with such proteins as GTP-binding protein (G-protein) and phosphokinase to generate a second messenger in an amplified manner at the interior of a cell (Fig. 14.2).

Another example of molecular communication is found in a neural synapse, which is a communicating junction between two neurons. The presynaptic membrane releases the neurotransmitter that is bound by the receptor of the postsynaptic membrane. The neurotransmitter causes the receptor to open the ion channel, which is followed by an influx of sodium ions to generate an impulse across the membrane (Fig. 14.3).

The molecular sensing mechanisms of these receptor proteins may provide us with the principles with which to design highly selective and sensitive biosensors.

14.3   Design Principles of Biosensors

Molecular Recognition and Signal Transduction   The molecular sensing mechanism of a living cell indicates that a receptor protein is essential to recognize a specific molecule selectively, which provides us with the design principle of biosensors.

A biosensor has been designed in such a manner as to have two functional parts for molecular recognition and signal transduction (Fig. 14.4). Either a biocatalyst or a bioaffinity substance is used as the major material for molecular recognition to attain extremely high selectivity. The signal transducing part typically involves an electrochemical, an optical, a thermal and a piezoacoustic device. To design a highly sensitive biosensor it is important to link efficiently the molecular recognition capabilities of a biosubstance with a signal transducing device.

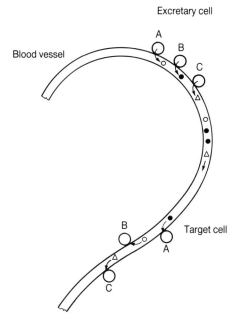

Excretary cell

Blood vessel

Target cell

○ Information molecule from cell A
● Information molecule from cell B
△ Information molecule from cell C

Fig. 14.1   Molecular communication in biological systems.

Two Types of Biosensor   A biocatalyst recognizes the corresponding substrate (analyte) and immediately generates products by a specific reaction. The complex of the catalyst and the analyte remain stable in a transition state. A change in either analyte or product is detected in the signal transducing device of a biosensor. It is so difficult to quantify the transition state of the complex that parameters such as oxygen, hydrogen

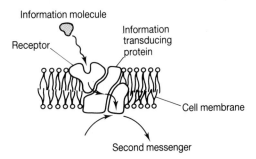

Information molecule

Receptor

Information transducing protein

Cell membrane

Second messenger

Fig. 14.2   Information transduction by a receptor embedded in a cellular membrane.

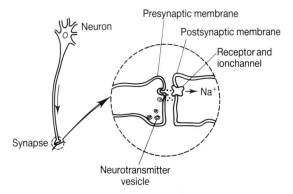

Fig. 14.3 Information transduction at a presynaptic membrane.

peroxide, heat and photons are commonly used in such detection in biocatalytic biosensors.

A bioaffinity sensor, on the other hand, involves an antibody, binding protein or receptor protein which forms a stable complex with a corresponding ligand. The bioaffinity protein–ligand complex is stable enough to result in signal transduction. A change in receptor–ligand complex formation is detected without any label or with the help of a label such as an enzyme, fluorophore or electroactive substance in a bioaffinity biosensor (Fig. 14.5).[6]

Biocatalytic Sensors   Redox enzymes are recognized as the major material in constructing both biocatalytic and bioaffinity sensors. Biocatalytic sensors for glucose, lactate and alcohol utilize glucose oxidase, lactate dehydrogenase (lactate oxidase) and alcohol dehydrogenase (alcohol oxidase) as molecular recognizable material. Since these redox enzymes are mostly associated with the generation of electrochemically active substances, many electrochemical enzyme sensors have been developed by linking redox enzymes for molecular recognition with electrochemical devices for signal transduction. These enzymes, however, have been linked in an indirect manner with electrochemical devices, which may cause a loss of sensitivity (Fig. 14.6).

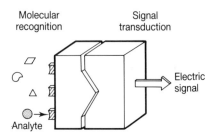

Fig. 14.4 Molecular recognition and signal transduction parts of a biosensor.

(a) Biocatalytic sensor
   Enzyme sensor, microbial sensor

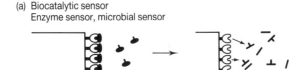

(b) Bioaffinity sensor
   Immunosensor, DNA sensor

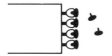

Fig. 14.5   Biocatalytic sensors and bioaffinity sensors.

Extensive research has been devoted to the direct linking of redox enzymes for molecular recognition with electrochemical devices for signal transducing devices. Despite efficient electron transfer from redox enzymes with corresponding electron carrier molecules, few redox enzymes can transfer electrons directly to metal or semiconductor electrodes even if the energy correlation is satisfied. The authors have proposed a molecular interface between a redox enzyme and an electrode to promote electron transfer. Several molecular interfaces have been developed, which include electron mediators and conductive polymers. These findings have offered a new design principle for realizing an electron transfer type of enzyme sensor (Fig. 14.7).

Bioaffinity Sensors   Immunosensors take advantage of the high selectivity provided by the molecular recognition of antibodies. Because of significant differences in affinity constants, antibodies may confer an

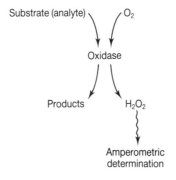

Fig. 14.6   Principle of amperometric enzyme sensors.

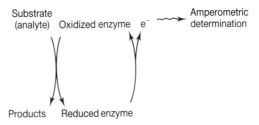

Fig. 14.7 Electron transfer type of amperometric enzyme sensors.

extremely high sensitivity to immunosensors in comparison to enzyme sensors. Furthermore, antibodies may be obtained in principle against an unlimited number of determinants. Immunosensors are thus characterized by high selectivity, sensitivity and versatility.

Immunosensors can be divided in principle into two categories: non-labelled and labelled immunosensors.[6,7] Nonlabelled immunosensors are designed so that the immunocomplex, that is, the antigen–antibody complex, is directly determined by measuring the physical changes induced by the formation of the complex. In contrast, in a labelled immunosensor a sensitively detectable label is incorporated. The immunocomplex is thus determined through measurement of the label.

Nonlabelled immunosensors are based on several principles (Fig. 14.8). Either the antibody or the antigen is immobilized on the solid matrix to

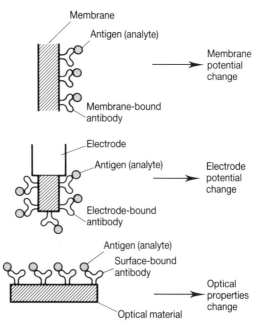

Fig. 14.8 Principles of non-labelled immunosensors.

form a sensing device. The solid matrix should be sensitive enough, in its surface characteristics, to detect the immunocomplex formation. Electrode, membrane, piezoelectric material, or optically active material surfaces may be used to construct nonlabelled immunosensors. The antigen or the antibody to be determined is dissolved in a solution and reacts with the complementary matrix-bound antibody or antigen to form an immunocomplex. This formation thus alters the physical properties of the surface, such as the electrode potential, the transmembrane potential, the intrinsic piezofrequency, or the optical properties.

A sufficiently high selectivity may be obtained with nonlabelled immunosensors, although such problems as nonspecific adsorption onto the matrix-bound antibody surface remain unsolved. To meet the requirements of immunosensors, intensive research has been conducted to enhance sensitivity as well as selectivity. Various labels have been incorporated into immunosensors to achieve high sensitivity, but radioisotopic labels have essentially been avoided (Fig. 14.9). Several nonisotopic labels for immunosensors are listed in Table 14.1. These involve enzymes, catalysts, fluorophores, electrochemically active molecules and liposomes. Among these labels enzymes, catalysts and liposomes provide chemical amplification. Highly sensitive immunosensors may be designed by incorporating these labels.

Labelled immunosensors are basically designed so that immunochemical complexation takes place on the surface of the sensor matrix. There are several variations of the procedure to form an immunocomplex on this matrix. In the final step, however, the label should be incorporated into the immunocomplex, which can thus be determined.

When an enzyme is used as a label it is covalently bound to either the antibody or the antigen, and catalase is one of the labels that has been successfully used for such a purpose. An oxygen electrode can be coupled with an antibody binding membrane to form an immunosensor. In the final

Table 14.1   *Nonisotopic labels for immunosensors*

Label		Determination
Enzyme	Catalase	Amperometric determination of O_2
	Glucose oxidase	
	Alkaline phosphatase	Amperometric determination of phenol
	Peroxidase	Luminescent determination
	Luciferase	
	Urease	Potentiometric determination of phenol
Catalyst	Hemin	Luminescent determination
Fluorophore	FITC	Fluorescent determination
Electroactive substances	Ferrocene	Amperometric determination
	Pyrene	Electrochemical luminescent determination
	Luminol	
Liposome	Ionic marker	Potentiometric determination of ionic marker

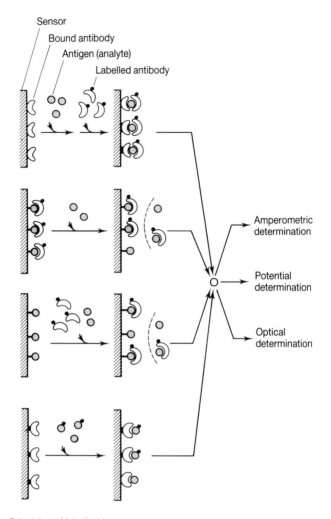

Fig. 14.9 Principles of labelled immunosensors.

step of the analytical procedure, both nonlabelled and labelled immuno-complexes are adsorbed on the surface of the antibody binding membrane.

14.4 Electrochemical Biosensors

Electrochemical Sensing Principle Electrochemical sensors make use of electrochemical reactions for signal transduction. As a parameter, the potential or conductance between two electrodes or the current through a polarized electrode is measured. A great variety of chemical substances can be measured in this way.

The important principles of electrochemical analytical chemistry are the following.

- *Potentiometry*
  Potentiometric sensors are based on the measurement of the potential at an electrode in a solution. This potential is measured in an equilibrium situation, i.e. no current is allowed to flow during the measurement. According to the Nernst equation, the potential is proportional to the logarithm of the concentration of the electroactive species.

  Ion-selective electrodes and ion-selective field effect transistors (ISFET) are typical potentiometric sensors that are coupled with biological materials to form biosensors.

- *Voltammetry*
  Voltammetric sensors are based on the measurement of the current–voltage relationship in an electrochemical cell consisting of electrodes in a solution. A potential is applied to the sensor and a current proportional to the concentration of the electroactive species of interest is measured. Amperometry is a special case of voltammetry where the potential is kept constant as a function of time.

  Oxygen and hydrogen peroxide are electrochemically active species, and these are most commonly associated with enzyme reactions. A Clark-type oxygen electrode and hydrogen peroxide electrode thus provide biosensors with effective signal transducing devices.

  An oxygen electrode consists of a gold or platinum cathode, separated from a silver anode, which is housed in a plastic or glass casing. This comes in contact with a test solution only through an oxygen-permeable polymer membrane. When oxygen diffuses through the membrane it is electrochemically reduced at the cathode by an applied potential of -0.6 V vs. Ag/AgCl. The electrochemical reduction of oxygen causes a current to flow between the anode and cathode which is proportional to the partial pressure of oxygen in a solution.

- *Conductometry*
  Conductometric sensors are based on the measurement of the conductance betwen two electrodes in a solution. The conductance is measured by applying an AC potential with a small amplitude to the electrodes in order to prevent polarization. The presence of ionic elements are detected as an increase in the conductance.

- *Ion-sensitive field effect transistors*
  Ion-sensitive field effect transistors (ISFET) are a special type of metal oxide field effect transistors (MOSFET), where the gate oxide of the device is in direct contact with an electrolyte. Usually the gate oxide layer of an ISFET is covered with a Si_3N_4 or Al_2O_3 layer to improve the characteristics such as stability and Nernstian behaviour.

Amperometric Glucose Sensors   A glucose sensor is the best known example of a biosensor. The realization of a glucose sensor is the gateway to more complex and more sensitive biosensors. The possibilities are outstanding.

The basic structure of a glucose sensor consists of either an oxygen or hydrogen peroxide electrode which is covered with a glucose oxidase membrane. Glucose oxidase catalyzes the following reaction:

$$\beta\text{-D-Glucose} + O_2 + H_2O \rightarrow \beta\text{-D-Gluconate} + H_2O_2 \qquad (14.1)$$

Glucose in a sample solution is oxidized with a resulting consumption of oxygen when contacted with the membrane-bound glucose oxidase. The decrease of dissolved oxygen is sensitively detected with the oxygen electrode. The output change of the sensor reflects the concentration of glucose in the solution.

Hydrogen peroxide generates in the glucose oxidase-catalyzed reaction. Detection of hydrogen peroxide can also be performed using a platinum anode polarized at about 0.7 V vs. Ag/AgCl. The enzyme membrane is positioned on the sensor tip confining a buffered and chlorinated solution. The sensor responds linearly to the hydrogen peroxide generated by the enzyme reaction, the output current being correlated with the analyte concentration.

Amperometric glucose sensors have long been installed in various table-top glucose analyzers which are widely used for clinical analysis, food analysis, and so on. These have been evaluated in such practical applications.

In contrast with the continuous use of glucose sensors, a disposable type of glucose sensors have recently been introduced on the market. These are mostly used for personal diagnosis of diabetes. Disposable glucose sensors have been realized by innovative technology including screen-printing fabrication of electrodes and enzyme layers, which has brought economically feasible mass-production.

Molecular Interfacing for Electron Transfer of Enzymes The pioneering work of Hill and Eddows have opened the way to realize fast and efficient electron transfer of enzymes at the electrode surface.[8] They modified a gold electrode with 4,4'-bipyridyl, an electron promotor, not a mediator since it does not take part in electron transfer in the potential region of interest. This accomplishes rapid electron transfer of cytochrome c. Their work has triggered intensive investigation of electron transfer of enzymes using modified electrodes.[9]

Apart from electron promoters a large number of electron mediators have long been investigated to make redox enzymes electrochemically active on the electrode surface. In such research, electron mediators such as ferrocene and its derivatives have been successfully incorporated into an enzyme sensor for glucose. The mediator was easily accessible to both glucose oxidase and an electrode to transfer electron in an enzyme sensor. Heller and Degani have chemically modified glucose oxidase with an electron mediator.[10] They presumed that an electron tunnelling pathway could be formed within the enzyme molecule. The present author and co-workers[11] and Lowe and Foulds[12] used a conducting polymer as a molecular wire to connect a redox enzyme molecule to the electrode surface.

These progresses in electron transfer of enzymes have led us to conclude that a molecular level assembly should be designed to facilitate electron transfer at the interface between an enzyme molecule and an electrode. Such a molecular level of assembly at the interface may be termed 'molecular interface'.[13]

There are several molecular interfaces for redox enzymes to promote electron transfer at the electrode surface (Fig. 14.10).

- *Electron mediator*
  Either electrode or enzyme is modified with an electron mediator in various manners.
- *Molecular wire*
  The redox centre of an enzyme molecule is connected to an electrode with such a molecular wire as a conducting polymer chain.

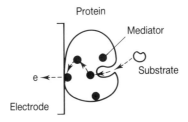

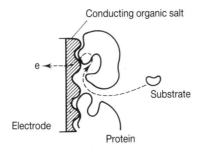

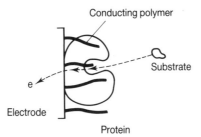

Fig. 14.10   Electron transfer of electrode-bound redox enzymes through the molecular interface.

- *Organic salt electrode and conducting polymer electrode*
  The surface of an organic electrode may provide enzymes with smooth electron transfer.

The first application of the molecular interfacing should be fabrication of disposable glucose sensors if glucose oxidase is used. Some other enzyme sensors have also been fabricated using the molecular interface. However, these have been limited to oxidases.

Such dehydrogenases as alcohol dehydrogenase (ADH) are associated by nicotinamide adenine dinucleotide (NAD) as cofactor. For fabrication of an enzyme sensor coupled with NAD, NAD should be immobilized within a matrix in which the corresponding dehydrogenase is immobilized, and should be regenerated for continuous enzyme reaction. A novel method has been developed in the author's laboratory to fulfill these requirements.[14]

The method is characterized by electrochemical fabrication of conductive membrane-embedded enzyme on the electrode surface. The conductive polymer used as membrane matrix works to promote the electron transfer from the enzyme to the electrode surface through NAD. The electrons passed can be quantified by amperometry, which results in quantifying the corresponding enzyme substrate (i.e., the analyte). In addition, embedded NAD may be electrochemically regenerated.

On the basis of the characterization a biosensor for gaseous ethanol was assembled. The platinum electrode covered with the ADH/meldola blue/NAD/polypyrrole membrane, a platinum counter electrode and an Ag/AgCl reference electrode were installed in a plastic sheath, so as to locate the membrane-covered electrode at the terminal. Furthermore the terminal was sealed by a gas-permeable polymer membrane. An internal electrolytic solution was filled in the sheath.

The sensor output current was measured under a constant potential application. Ethanol either in solution or in gas phase was selectively determined with this dehydrogenase sensor.

The principle of the dehydrogenase sensor for ethanol may be applicable when assembling any other dehydrogenase sensors.

Potentiometric Enzyme Sensors   As many enzyme reactions produce pH changes, a vast amount of effort has been devoted to the construction of pH electrode-based enzyme sensors.

In addition to pH-glass electrodes various metal–metal oxide electrodes have been advantageously employed in combination with enzymes. The pH variation that occurs within the biological layer due to the enzymatic reaction is supposed to affect the structure of the oxide layer with concomitant modification of the electrode potential. Antimony oxide, palladium oxide, and titanium electrodes coated with oxides (irridium oxide and ruthenium oxide) have been proposed,[15] and have been shown to offer interesting advantages over the classic pH-glass electrode in terms of response time, miniaturization, and robustness.

An all-solid-state electrode for detecting ammonium ions has been proposed and combined with urease for the assay of urea.[16] The ammonium ion electrode consists of a conductive resin (epoxy + graphite) covered with a nonacting-polyvinyl chloride matrix.

Enzyme FETs Enzyme FETs have been developed by immobilizing an enzyme layer on the surface of the gate of a pH-sensitive ISFET.[17–20] These sensors are based on a change in pH in an enzyme layer by molecular recognition. An increase in the glucose concentration, for instance, will result in a decrease of the pH in the glucose oxidase layer due to generation of gluconic acid.

Potentiometric Immunosensors Three types of potential immunosensor have been proposed. The first is based on the determination of transmembrane potential across an antibody (or antigen) membrane that specifically binds the corresponding antigen (or antibody) in solution. Concentrations of either the target antigen or antibody can be determined, by measuring a change in the transmembrane potential occurring when the immunocomplex forms on the membrane surface (Fig. 14.11).

The second type is based on determination of the electrode potential. The surface of an electrode is modified by an antibody or antigen to bind specifically the corresponding antigen or antibody. Immunocomplex formation causes the electrode potential to vary, primarily as a result of a

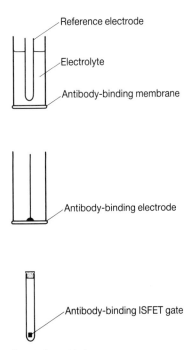

Fig. 14.11 Three types of potentiometric immunosensors.

change in surface charge resulting from the concentration of the analyte in solution.

The third type is based on the determination of surface potential of the gate of an FET covered by a thin antibody binding membrane. The surface potential of the FET gate may vary with the concentration of the corresponding antigen in solution.

Results for potentiometric immunosensors for syphilis and blood typing have been reported.[21] They are based on the determination of the transmembrane potential across an immunoresponsive membrane. For blood typing, for instance, the immunoresponsive membrane incorporates blood group substances and a pair of reference electrodes for measuring the transmembrane potential.[22] A, B and O blood samples have been typed using this potentiometric immunosensor.

Yamamoto et al. proposed a potentiometric immunosensor with an antibody against human chorionic gonadotropin (HCG) hormone, covalently bound to the surface of an electrode.[23] The electrode is reported to respond to HCG in solution.

Amperometric Immunosensors A variety of labelled immunosensors have been proposed. Enzyme labels can enhance the sensitivity of immunosensors owing to chemical amplification. Since an enzyme rapidly converts a substrate into a product, the product that accumulates can be detected with various electronic and optoelectronic devices.[24–27]

Some enzyme labels are electrochemically detected with a high sensitivity. These include catalase, glucose oxidase, and peroxidase. Both catalase and glucose oxidase may be associated with an oxygen electrode.

When catalase, which catalyzes the decomposition of hydrogen peroxide into oxygen, is the labelling enzyme for an antigen, an enzyme immunosensor can be constructed by assembling an antibody-bound membrane and an oxygen-sensing electrode. In heterogeneous enzyme immunoassays the labelling enzyme activity is measured by amperometry with the oxygen-sensing device. Because only a short time is required for measuring the labelling enzyme, a rapid and highly sensitive enzyme immunoassay can be achieved.

An enzyme immunosensor was prepared by attaching the antibody-bound membrane to a Clark type oxygen electrode involving an oxygen-permeable synthetic (e.g., Teflon) membrane on the cathode surface (Fig. 14.12).

An antibody membrane with anti-α-fetoprotein (AFP) was attached to an oxygen electrode to form an enzyme immunosensor for AFP. The sensor is immersed in a test solution to which a known amount of catalase-labelled AFP was added. After a thorough washing the sensor was placed in phosphate buffer. After a background current due to dissolved oxygen was obtained, hydrogen peroxide was injected. The sensor responded very rapidly to the generated oxygen, and a steady-state current was obtained within 30 s.

A calibration curve for competitive enzyme immunoassay for AFP with this sensor shows that AFP can be determined in the range

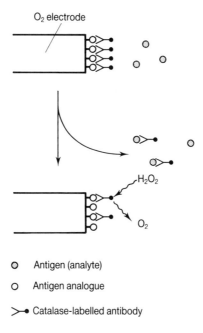

O₂ electrode

⊙ Antigen (analyte)

○ Antigen analogue

>• Catalase-labelled antibody

Fig. 14.12 Principle of an enzyme immunosensor.

5×10^{-11}–5×10^{-8} g/ml. The standard deviation for 25 assays of 10^{-9} g AFP is 15%.

Ochratoxin A (OTA), a secondary metabolite of *Aspergillus ochraceus*, *Penicillium viridicatum*, and strains of some other species of both genera, causes nephropathy in animals and probably also in humans. Recently carcinogenicity was also demonstrated with feeding experiments, and a simplified and rapid analysis for OTA is urgently needed.

An enzyme immunosensor for OTA was recently developed. The competitive immunoassay of OTA is carried out by using the toxin sensor, which consists of an amperometric oxygen electrode and an OTA-bound membrane. OTA is determined under a constant concentration of catalase-labelled antibody. The standard curve shifts to the lower concentration range with a decrease in antibody concentration. The detection limit is of the order of 10^{-10} g/ml when the concentration of labelled antibody is fixed at 20 μg/ml.

Enzyme labels are efficiently detected with optoelectronic devices in a manner similar to that for electrochemical devices. Peroxidase and luciferase, for instance, catalyze luminescent reactions of luminol and luciferin, respectively, to generate photons. These enzymes may be incorporated as labels to form optical enzyme immunosensors.

14.5 Optical Biosensors

Optical Sensing Principles Optical sensors make use of the effect of chemical reactions on optical phenomena. Photodiodes, phototransistors and interferometers in combination with optical fibres and optical guides are typical detector structures. Optical fibre sensors are prominent in these structures.

The design of the optical biosensor and the choice of the proper sensing approach depend on both the analyte to be determined and the matrix. An important issue is the proper choice of the materials including the solid support, like an optical fibre which may function as a light guide and as a support on which to immobilize biological molecules.[28,29]

Optical biosensors have been designed in such a manner that various optical transducers are incorporated for signal transduction. These include the pH, ammonia and oxygen optical fibre sensors.

One of the first optical fibre oxygen sensors was described by Lubbers *et al.* in 1975.[30] Oxygen partial pressure was measured by the dynamic quenching of an indicator, pyrenebytyric acid, fluorescence by molecular oxygen. Extensive improvement has been made by many researchers, which has made it possible to use the oxygen optical fibre sensor in practical applications.

pH optical fibre sensors are based on changes in the optical properties of an acid/base indicator. Phenol red, 1-hydroxypyrene-3,6,8-trisulphonate (HPTS), and eosin/phenol red are typical indicator dyes that are used in pH optical fibre sensors.

Optical Enzyme Sensors Oxygen optical fibre sensors can be used as transducers for enzyme sensors if the enzyme reaction follows the general scheme of consuming oxygen. The consumption of oxygen is measured with the oxygen optical fibre sensor. It is necessary, however, that the enzyme layer on top of the oxygen sensor be permeable to the analyte. Second, it should be kept in mind that the sensor responds to the total oxygen concentration. Hence the response of the sensor to the substrate concentration is strongly affected by the oxygen supply of the analyte solution.

Optical enzyme sensors for glucose, lactate, alcohol, cholesterol and others have been assembled using the corresponding oxidase in conjunction with a fibre optic oxygen sensor.

An optical fibre pH sensor has also been applied to construct enzyme sensors for various analytes, which are selectively converted by the corresponding enzyme to change pH. These include optical fibre sensors for penicillin, urea and others.

Optical Immunosensors If the solid surface is sensitive enough to allow changes in its optical properties with immunocomplex formation, optical immunosensors without label may be constructed. Surface plasmon resonance (SPR) is so sensitive that an immunocomplex may be detected on the surface of a silver-coated solid (Fig. 14.13). It is noted that an

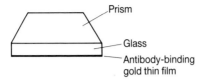

Prism

Glass

Antibody-binding
gold thin film

Fig. 14.13   An optical immunosensor based on surface plasmon resonance (SPR).

immunosensing system based on SPR has been on the market primarily for basic research.[31,32]

Acoustic immunosensors have attracted many researchers because of their simplicity of operation. The surface of a piezoelectric material is coated with antibodies or antigens. The intrinsic frequency of the piezoelectric material is expected to shift when responding to the corresponding antigen or antibody in solution. Although several papers have been published on the subject, further confirmation is required because of the uncertainty due to numerous artifacts.

Since homogeneous immunoassays require to bound/free species separation process, they may be advantageous over heterogeneous immunoassays. Although intensive efforts have been concentrated on the development of homogeneous immunosensors, a very limited number of papers has been published on the subject.

An optical fibre structure has been proposed for designing a homogeneous immunosensor. An antibody is immobilized on the surface of an optical fibre core, and a fluorescence compound is used as label (Fig. 14.14). Both labelled antigen and free antigen to be determined react competitively with bound antibody to form an immunocomplex on the core surface.

The surface-bound label can be excited by an evanescent wave that passes through the optical fibre core. On the other hand, fluorescence labels in the bulk solution cannot be excited, even if the excitation beam comes through the optical fibre core. Labels attached to the surface-bound immunocomplex are thus discriminated from labels in solution. Although

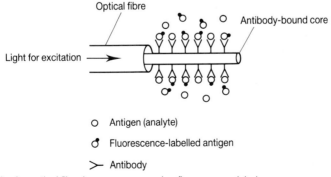

Optical fibre

Light for excitation

Antibody-bound core

○   Antigen (analyte)

♂   Fluorescence-labelled antigen

⊱   Antibody

Fig. 14.14   An optical fibre immunosensor using fluorescence label.

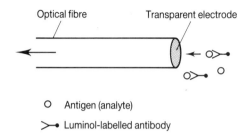

Optical fibre

Transparent electrode

○   Antigen (analyte)

>•   Luminol-labelled antibody

Fig. 14.15   An optical immunosensor based on electrochemiluminescence label.

the details have not been reported, the principle is elegant and homogeneous immunoassay may be possible.

Some electrochemically active substances able to generate photons on an electrode surface can be used as labels for homogeneous immunoassays. A labelled antigen exhibits an electrochemical reactivity and generates luminescence, but when it is immunochemically complexed the labelled antigen loses its electro-chemiluminescence property (Fig. 14.15).

An optical immunosensor for homogeneous immunoassay was assembled by sputtering platinum on the end surface of an optical fibre.[34,35] Sputtered platinum maintains optical transparency and works as an electrode. An optical fibre electrode efficiently collects photons generated on the surface of the transparent electrode.

Luminol as label offers excellent characteristics for designing a homogeneous immunosensor, and generates luminescence by two different types of electrochemical excitation. One is based on a two-step electrochemical excitation, cathodic excitation followed by anodic excitation. The cathodic excitation produces hydrogen peroxide that causes anodically generated luminol radical to emit photons. The other type is a single-step electrochemical excitation in the presence of hydrogen peroxide. Luminol is simply oxidized by anodic excitation to generate radicals, followed by emission of photons. The single-step electrochemical excitation provides a very high sensitivity with a limit of detection equal to 10^{-13} mol/L of luminol.

Homogeneous immunoassay with immunoglobulin G (IgG) as a model antigen, labelled with luminol, was thus performed by the single-step electrochemical excitation using an optical fibre electrode. As with free luminol, labelled IgG generates electrochemical luminescence in the presence of hydrogen peroxide by anodic oxidation. Electrochemical luminescence sharply decreases by immunocomplexation with anti-IgG antibody. The addition of 10^{-13} g/L of antibody results in an appreciable suppression of luminescence. The lower limit of detection may then fall in the range of 10^{-13} g/L of antibody.

14.6   Conclusions

Biosensor technology has made prominent progress, primarily in focusing on enzyme sensors in the last two decades. One of the results should be realization of disposable glucose sensors in mass production, which has encouraged many researchers to continue their research and development of biosensors for practical applications.

Current progress of immunosensing technology has been crystallized as an immunosensing system based on SPR. The technology has shown that solid surface-bound molecules can be detected selectively and sensitively by a combination of molecular immunochemistry and physical measurement. Since that first report on nonlabelled immunosensors there have been a huge number of papers on immunosensing technology. Now we should focus our efforts on the development of basic technology which makes it possible to quantify the states of specific molecules on the solid surface. This should be possible by harnessing molecular biotechnology and nanotechnology.

References

1. S. J. Updike and G. P. Hicks, *Nature*, **214**, 986 (1967).
2. A. P. F. Turner, I. Karube and G. Wilson, Eds, *Biosensors—Fundamentals and Applications*, (Oxford University Press, Oxford, 1987).
3. R. P. Buck, W. E. Hatfield, M. Umana and E. F. Bowden, *Biosensor Technology*, (Marcel Dekker, New York, 1990).
4. A. P. F. Turner, Ed., *Advances in Biosensors, Vol. 1*, (JAI Press, London, 1991).
5. L. J. Blum and P. R. Coulet, Eds, *Biosensor—Principles and Applications*, (Marcel Dekker, New York, 1991).
6. M. Aizawa, *Anal. Chem. Acta*, **250**, 249 (1991).
7. M. Aizawa, *Philos. Trans. Royal Soc. London, Ser. B*, **316**, 121 (1987).
8. M. J. Eddows and H. A. O. Hill, *J. Chem. Soc., Chem. Commun.*, **771** (1977); *J. Am. Chem. Soc.*, **101**, 4461 (1979).
9. A. E. G. Cass, G. Davis, G. D. Francis *et al.*, *Anal. Chem.*, **1880** (1984).
10. Y. Degani and A. J. Heller, *J. Phys. Chem.*, **91**, 1285 (1987).
11. M. Aizawa, S. Yabuki and H. Shinohara, *Proc. 1st Int. Symp. Electro-organic Synthesis*, Ed. S. Torii, (Elsevier, Amsterdam, 1987), pp. 353–60; *J. Chem. Soc., Chem. Commun.*, 945 (1989).
12. N. C. Foulds and C. R. Lowe, *J. Chem. Soc., Faraday Trans.*, 1, **82**, 1259 (1986).
13. M. Aizawa, S. Yabuki, H. Shinohara and Y. Ikariyama, in *Molecular Electronics: Science and Technology*, Ed. A. Aviram, (Eng. Foundation, New York, 1989); *Molecular Electronics—Biosensors and Biocomputers*, Ed. F. T. Hong, (Plenum Press, New York, 1988), pp. 269–76.
14. T. Ishizuka, E. Kobatake, Y. Ikariyama and M. Aizawa, *Technical Digest 10th Sensor Symp.* (1991), 73.
15. M. T. Flanagan and N. J. Carrel, *Biotechnol. Bioeng.*, **26**, 642 (1984).
16. S. Alegret and E. Martinez-Fabregas, *Biosensors*, **4**, 287 (1989).
17. S. Cares and J. Janata, *Anal. Chem.*, **52**, 1935 (1980).

18. Y. Miyahara, F. Matsu, T. Moriizumi, H. Matsuoka, I. Karbe and S. Suzuki, *Proc. 1st Int. Meeting Chem. Sensors*, Kodansha, Tokyo 19 (1983).
19. Y. Hanazato and S. Shiono, *Proc. 1st Int. Meeting Chem. Sensors*, Kodansha, Tokyo 22 (1983).
20. J. Kimura, T. Kuriyama and Y. Kawana, *Sensors Actuators*, **9**, 373 (1986).
21. M. Aizawa, S. Suzuki, Y. Nakagawa, R. Shiohara and I. Ishiguro, *Chem. Lett.*, 779 (1977).
22. M. Aizawa, S. Kato and S. Suzuki, *J. Membrane Sci.*, **2**, 125 (1977).
23. N. Yamamoto, Y. Nagasawa, M. Sawai, T. Suda and H. Tsubomura, *J. Immunol. Methods*, **22**, 309 (1978).
24. M. Aizawa, A. Morioka, H. Matsuoka *et al.*, *J. Solid-Phase Biochem.*, **1**, 319 (1976).
25. M. Aizawa, S. Suzuki and Y. Nagamura, *Anal. Biochem.*, **94**, 22 (1979).
26. M. Aizawa, A. Morioka and S. Suzuki, *Anal. Chim. Acta*, **115**, 61 (1980).
27. M. Aizawa, S. Suzuki, T. Kato, T. Fujiwara and Y. Fujita, *J. Appl. Biochem.*, **2**, 190 (1980).
28. W. R. Seitz, *Anal. Chem.*, **56**, 16A (1984).
29. B. P. H. Schaffar and Q. S. Wolfbeis, in *Biosensors—Principles and Applications*, Eds L. J. Blum and P. R. Coulet, (Marcel Dekker, New York, 1991), pp. 163–94.
30. D. W. Lubbers, *Z. Naturforsche*, **20C**, 532 (1975).
31. B. Liedberg, C. Nylander and I. Lundstrom, *Sensors Actuators*, **4**, 299 (1983).
32. S. Lofas and B. Johnson, *J. Chem. Soc., Chem. Commun.*, **1526** (1990).
33. A. Shons, F. Dorman and J. Najarian, *J. Biomed. Mater. Res.*, **6**, 565 (1972).
34. Y. Ikariyama, H. Kunoh and M. Aizawa, *Biochem. Biophys. Res. Commun.*, **128**, 987 (1987).
35. M. Aizawa, M. Tanaka, Y. Ikariyama and H. Shinohara, *J. Bioluminesc. Chemiluminesc.*, **4**, 535 (1989).

15

Biomolecular Optoelectronics

R R Birge and R B Gross

15.1 Introduction

This chapter overviews the use of the biological photochrome bacteriorhodopsin in optoelectronic devices. The principal goal is to provide the reader with a perspective on the use of this protein in selected applications that include holography, spatial light modulators, neural network optical computing, nonlinear optical devices and optical memories. There are significant advantages inherent in the use of biological molecules, either in their native form, or modified via chemical or mutagenic methods, as active components in optoelectronic devices.[1–8] These advantages derive in large part from the natural selection process, and the fact that nature has solved through trial and error problems of a similar nature to those encountered in harnessing organic molecules to carry out logic, switching or data manipulative functions. Light transducing proteins such as visual rhodopsin,[5] bacteriorhodopsin,[4] chloroplasts,[6,7] and photosynthetic reaction centres[8] are salient examples of biomolecular systems that have been investigated for optoelectronic applications. This review will concentrate primarily on the current and proposed optoelectronics applications of bacteriorhodopsin. The emphasis is justified because this protein has received more attention with respect to optoelectronics than any other biomolecule studied to date. The significance of bacteriorhodopsin stems from its biological function as a photosynthetic proton pump in the bacterium, *Halobacterium halobium*. A combination of serendipity and natural selection has yielded a native protein with characteristics near optimum for many linear and nonlinear optical applications. The development of genetic engineering, combined with chemical modification and chromophore substitution methods, also provides an additional flexibility in the optimization of this material for individual applications. We will explore a number of optoelectronic applications of bacteriorhodopsin as well as the methods of optimizing the protein in this chapter.

15.2   Linear Optoelectronic Devices

Applications of bacteriorhodopsin that operate via one-photon absorption are defined for the purposes of this review as linear. Most of these devices operate by selecting a pair of the ground states that populate the complex photocycle of the protein. In order to understand their operation, it is necessary to understand the nature of the bacteriorhodopsin photocycle. The following sections are devoted to a brief overview of the photochemistry of this protein.

The Function and Photochemistry of Bacteriorhodopsin   Bacterio-rhodopsin (MW $\cong 26000$) is the light harvesting protein in the purple membrane of a micro-organism called *Halobacterium halobium*, or more formally called *Halobacterium salinarium*.[9,10] This bacterium thrives in salt marshes where the concentration of salt is roughly six times higher than sea water. The purple membrane, which consitutes a specific functional site in the plasma membrane of the bacterial cell, houses semi-crystalline protein trimers in a phospholipid matrix (3:1 protein to lipid). The bacterium synthesizes the purple membrane when the concentration of dissolved oxygen in its surroundings becomes too low to sustain ATP production through aerobic respiration. The light absorbing chromophore of bacter-iorhodopsin is all-*trans* retinal (Vitamin A aldehyde) (Fig. 15.1). It is bound to the protein through a protonated Schiff base linkage to a lysine residue attached to one of the seven α-helices which make up the protein's secondary structure. The absorption of light energy by the chromophore initiates a complex photochemical cycle, characterized by a series of spectrally distinct thermal intermediates and a total cycle time of approx-imately 10 milliseconds (see Fig. 15.2). As a result of this process, the protein expels a proton from the intracellular to the extracellular side of the membrane. This light-induced proton pumping generates an elec-trochemical gradient which the bacterium uses to synthesize ATP in accordance with Mitchell's chemiosmotic model for energy transduction. Thus, bacteriorhodopsin provides the necessary molecular machinery for *H. halobium* to convert light energy into a metabolically useful form of energy vital for the cell's survival under anaerobic conditions. Accordingly, *H. halobium* can switch from aerobic respiration to photosynthesis in response to changing environmental conditions. The fact that the protein must function in the harsh environment of a salt marsh requires a robust protein resistant to both thermal and photochemical damage. The cyclicity of the protein (i.e. the number of times the protein can be photochemically cycled between intermediates before denaturing) exceeds 10^6, a value considerably higher than those observed in known synthetic photochromic materials. This high value is due to the protective features of the integral membrane protein which serves to isolate the chromophore from potential-

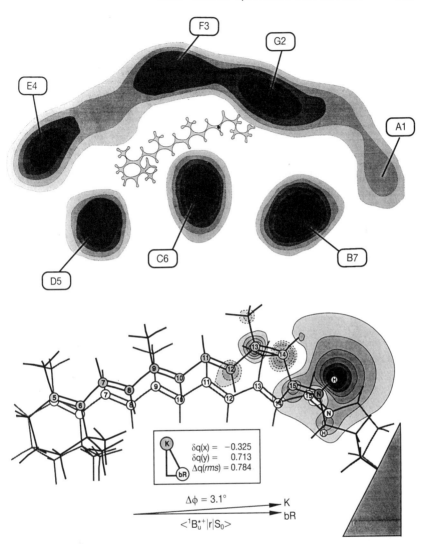

Fig. 15.1 The chromophore binding site and the primary photochemical event in bacterio-rhodopsin. The upper diagram shows electron density profiles (data from Hayward *et al.*[65]) of bacteriorhodopsin viewed from the cytoplasmic side, showing the seven transmembrane spanning segments and the presumed location of the chromophore in relation to the helices based on the available experimental data.[5] FTIR studies indicate that the polyene chain of the chromophore in bacteriorhodopsin lies roughly perpendicular to the membrane plane.[66] The retinyl chromophore is rotated artificially into the membrane plane to show more clearly the polyene chain (the imine nitrogen is indicated with a black dot) and the β-ionylidene ring. The bottom diagram shows a model of the primary photochemical event [**bR** (grey bonds; underneath)→ **K** (black bonds; above)] and the shift in charge that is associated with the motion of the positively charged chromophore following 13-*trans*→ 13-*cis* photoisomerization. It is believed that the initial photoelectric signal is due primarily to the motion of the chromophore. The conformations of the lysine residue in the **bR** and **K** states are tentative.[67]

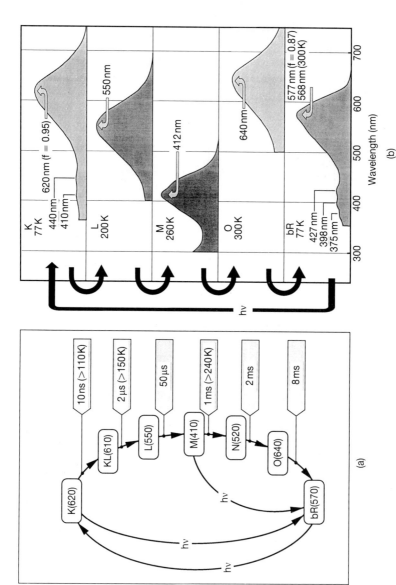

Fig. 15.2 A simplified model of the light adapted bacteriorhodopsin photocycle (a) and the electronic (one-photon) absorption spectra of selected intermediates in the photocycle (b). The height of the symbols in (a) is representative of the relative free energy of the intermediates, and the key photochemical transformations relevant to device applications are shown. Note that not all of the intermediates are shown, and that there are in fact two species of M (M_{fast} and M_{slow}), but only one is shown for convenience. (M_{fast} and M_{slow} have virtually identical absorption spectra.) Band maxima are indicated in nanometres. Oscillator strengths (f) determined by log-normal fits of selected λ... bands are indicated in parentheses.

ly reactive oxygen, singlet oxygen and free radicals. Thus, the common misperception that biological materials are too fragile to be used in technological devices does not apply to bacteriorhodopsin. The optoelectronic characteristics of bacteriorhodopsin have been reviewed in detail,[4,5] and our discussion will be selective. The key optical properties of the protein are summarized in Table 15.1.

Although the bacteriorhodopsin photocycle is comprised of at least five thermal intermediates (Fig. 15.2(a)), only three of the intermediates (**bR**, **K** and **M**) have recognized potential in device applications. The absorption spectra of the key intermediates are shown in Fig. 15.2(b). The unique absorption spectra exhibited by each intermediate are associated with the changing electronic environment of the chromophore binding site during the course of the photocycle. The internally bound chromophore carries a net positive charge and interacts electrostatically with neighbouring charged amino acids in the binding site. These interactions in large part determine the protein's photochemical and spectral properties during the photocycle, and are therefore targeted areas for genetic engineering and/or chemical modification. When the chromophore absorbs a photon of light energy, an instantaneous ($<10^{-15}$ s) shift of electron density occurs with negative charge moving along the polyene chain towards the nitrogen atom. This shift in electron density interacts with nearby negatively charged residues and activates a rotation around the $C_{13}=C_{14}$ double bond thereby generating a 13-*cis* chromophore geometry (see Fig. 15.1). The result of this photoisomerization process, which occurs in less than one picosecond,[11,12] is the formation of the **K** spectral intermediate. The reason for the unusually high isomerization speed is due to a barrierless excited state potential surface.[5] In this regard, bacteriorhodopsin is the biological analogue of high electron mobility transistor (HEMT) devices. The isomerization of the protonated chromophore induces a shift in positive charge perpendicular to the membrane sheet containing the protein, and generates a measurable and potentially useful electrical signal. The rise time of this signal is less than 5-picoseconds and correlates with the formation time of **K**.[13] Another feature of the chromophore *trans-cis* isomerization process is its photoreversibility. Thus, irradiation of the protein with a wavelength within the absorption band of **K** results in the reformation of the **bR** state. Many of the early proposed optical memories based on bacteriorhodopsin utilized the photochemical switching between **bR** and **K**. These devices suffered, however, from the requirement that liquid nitrogen temperatures were needed to arrest the photocycle at the **K** intermediate (e.g. Birge *et al.*[14]). While these devices were potentially efficient and very fast (the **bR**⇔**K** interconversions take place in a few picoseconds), the use of cryogenic temperatures and the small change in absorption maxima associated with the **bR** to **K** transition mandated expensive operating hardware and precluded general use.

The most significant photochemical intermediate, both from a physiological and optical application point of view, is the blue light-absorbing **M** intermediate. The formation of **M** ensues after a series of protein conformational changes occurring ~50 μs after the absorption of a photon of light

Table 15.1 *Properties and characteristics of bacteriorhodopsin and bacteriorhodopsin based materials**

Property	Description and discussion	Values
Molecular weight and structure	Membrane bound protein with 248 amino acids arranged in 7α-helical segments; purple membrane patch (10 lipid molecules per protein; 40 nm × ~500 nm × ~500 nm) is a two-dimensional crystalline lattice consisting of trimers	MW ≈ 26 000 Dalton
Chromophore	Protonated Schiff base of all-*trans* retinal attached via lysine-216 embedded inside highly polar binding site	MW ≈ 280 Dalton
Biological function	Light-driven proton pump used by bacterium as a source of energy when low oxygen concentration prevents respiration	
Stability	Protein can withstand constant illumination, oxygen, pH 3–10, temperatures below 80°C, most hydrophilic polymers	$3 < \text{pH} < 10$ $T < 80°C$
Instabilities	Uv irradiation ($\lambda < 370$ nm), most organic solvents, pH extremes, temperatures above 80°C	$\lambda > 370$ nm $T > 80°C$ $\Phi > 0.64$
Photochemistry	Absorption of light generates photocycle via *cis-trans* isomerization of the chromophore; high quantum efficiency	
Wavelength range	Range of write (λ_w) and read (λ_R) wavelengths extends throughout visible region of the spectrum (values in nanometers); chromophore analogues can extend range	$400 < \lambda_w < 660$ $410 < \lambda_R < 800$
Resolution	Diffraction limited performance is typical	5000 lines/mm
Diffraction efficiency	Holographic performance for absorption holograms is absorption limited (<3.7%), phase performance is better	$\eta_{abs} \leqslant 3.7\%$ $\eta_{phase} \leqslant 8\%$
Reversibility	The protein is photochemically robust due to protective characteristics of the protein binding site, cyclicity (total number of write, read, erase cycles) is excellent	cyclicity $\geqslant 10^6$
Sensitivity	Sensitivity can be optimized for specific application	1–80 mJ/cm^2
Thin films	Thin films can be prepared with diffraction limited optical properties using hydrophilic gels, polymers and blends	thickness = 20–500 μm
Write time	This parameter is determined primarily by light intensity and pulse versus cw operation (values shown are typical limits)	$\tau_{pulse} \geqslant 4$ ns $\tau_{cw} \geqslant 50$ μs
Relaxation time	Thermal relaxation time of the blue intermediate (M) can be varied from 10 ms to infinity and is affected by pH, chemical environment, site specific mutation and temperature	τ_T variable $\tau_{h\nu} \leqslant 50$ μs
One-photon absorptivity	Values for the molar absorptivity at λ_{max} in units of M^{-1} cm^{-1} are given for **bR** ($\lambda_{max} = 570$ nm) and M ($\lambda_{max} = 410$ nm)	$\varepsilon_{bR} \cong 66 000$ $\varepsilon_M \cong 45 000$
Two-photon absorptivity	Values for the two-photon absorptivity at λ_{max} (1 GM = 10^{-50} cm^4 s molec^{-1} photon^{-1}) are given for **bR** and **M**	$\delta_{bR} \cong 290$ GM $\delta_M \cong 100$ GM
Second polarizability	The values of β_π for **bR** at 1.06μ (1.17 eV) and 1.9μ (0.66 eV) are given in units of 10^{-30} cm^5 esu^{-1}	$\beta_{1.06\mu} \cong 2250$ $\beta_{1.9\mu} \cong 329$
Third polarizability	The values of γ_π for **bR** at 1.9μ (0.66 eV) and at zero frequency are given in units of 10^{-36} cm^7 esu^{-2}	$\gamma_{1.9\mu} \cong 14 900$ $\gamma_0 \approx 2500$

*Data from text and Oesterhelt *et al.*,[4] Birge[10] and Birge *et al.*[53]

by **bR**. In this stage of the photocycle, the Schiff base proton on the chromophore is transferred to an amino acid of the protein. In doing so, the electrostatic nature of the chromophore and the electric potential of the chromophore binding site are dramatically changed, as reflected in this intermediate's highly blue-shifted absorption spectrum. Under normal conditions **M** thermally reverts to the ground state with a time constant of about ten milliseconds. Most importantly, **bR** can also be photochemically regenerated from **M** by the absorption of blue light. This property of a material, where a ground state photo-initiated reaction results in a relatively long-lived thermal intermediate, which can also be photochemically driven back to the ground state, is called photochromism. The photochromic properties of bacteriorhodopsin are summarized below:

$$\textbf{bR}\,(\lambda_{max} \cong 570 \text{ nm})\,(\textit{State 0})\,\underset{\Phi_2 \sim 0.65}{\overset{\Phi_1 \sim 0.65}{\rightleftarrows}}\,\textbf{M}\,(\lambda_{max} \cong 410 \text{ nm})\,(\textit{State 1})\,(\text{scheme 1})$$

where the quantum yields of the forward reaction (**bR** to **M**) and reverse reaction (**M** to **bR**) are indicated by Φ_1 and Φ_2, respectively. The quantum yield is a measure of the probability that a reaction will take place after the absorption of a photon of light. A photochromic material possessing a quantum yield of unity and a comparatively long thermal intermediate lifetime is extremely light sensitive. One inherent advantage of bacteriorhodopsin as an optical recording medium is the high quantum efficiency with which it converts light into a state change. Complementing this property is the relative ease at which the thermal decay of **M** can be prolonged. The $\textbf{M} \rightarrow \textbf{bR}$ thermal transition is highly susceptible to temperature, chemical environment, genetic modification and chromophore substitution. This property is exploited in many optical devices based on bacteriorhodopsin.

Holographic Optical Recording   Thin films of bacteriorhodopsin fabricated by incorporating the protein into optically transparent polymers and polymer blends have shown good holographic performance and the capability of real time optical processing.[4,5,15–20] The mechanism of diffracting light from a volume hologram produced in a photochromic material is important to understand as it is fundamental to many optical applications. We therefore provide a brief overview of this process. Readers interested in a more detailed discussion of the fundamentals of holography in photochromic materials are referred to the work by Tomlinson[21] or the excellent book by Collier et al.[22]

The optical protocol used to record a plane wave hologram in a thin film of bacteriorhodopsin is schematically shown in Fig. 15.3. Two laser beams derived from the same laser and of a wavelength (λ_w) absorbed by **bR** are overlapped at the plane of the film. Both beams are polarized perpendicular to the plane of incidence and make an angle ϕ_W with respect to the film normal. Due to the coherence properties of laser light, a three-dimensional interference pattern is imposed on the film. The resulting periodic spatial

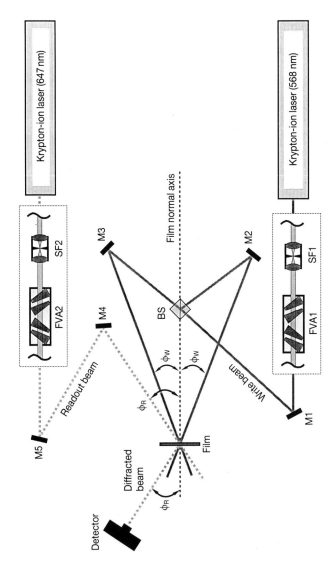

Fig. 15.3 A schematic representation of the experimental apparatus used in writing and reading a hologram. A bacteriorhodopsin hologram is written by overlapping two 568 nm beams derived from a krypton-ion laser. The hologram is nondestructively read at the Bragg angle by using a probe beam not strongly absorbed by the protein (see text).

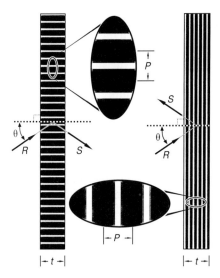

Fig. 15.4 Volume transmission (left) and reflection (right) gratings and the key variables that define the properties of the gratings. The dark regions represent regions of high refractive index (or increased absorptivity) and the light regions represent regions of low refractive index (or decreased absorptivity). The magnitude of the difference in refractive index (or absorptivity) determines in part the diffraction efficiency of the phase (or absorption) hologram. The incident light vector is indicated by **R** and the scattered light vector is indicated by **S**.

light intensity distribution is schematically shown in Fig. 15.4 and can be mathematically described in one dimension by using Equation (15.1):

$$I(x) = (I_1 + I_2)[1 + V \cos(2\pi x/P)] \tag{15.1}$$

where I_1 and I_2 represent the intensities of the individual beams, V is the contrast ratio of the interference pattern $V = 2(I_1 I_2)^{1/2}/(I_1 + I_2)$ and P is the fringe spacing of the grating (Fig. 15.4) given by $\lambda_w/(2 \sin \phi_W)$. The film records both the amplitude and phase information contained in the two incident beams as a periodic spatial concentration distribution of **bR** and **M** (since neither beam contains an object the hologram is 'structure-less' as compared to most eye-striking holograms). Thus, in places of constructive interference **bR** is driven to **M** and in regions of destructive interference no photochemistry is initiated. The spatial concentration distribution of **bR** and **M** can be more conveniently viewed as a spatial modulation of the material's absorption coefficient and index of refraction. These material properties, which are fundamental to the diffraction or reconstruction process in holography, are not independent in most photochromics and are related through the Kramers–Kronig transform:[23,24]

$$\Delta n(\lambda) = \frac{2.3026}{2\pi^2 t} \text{P.V.} \int_0^\infty \frac{A_M(\lambda') - A_{bR}(\lambda') d\lambda'}{1 - \frac{\lambda'^2}{\lambda^2}} \tag{15.2}$$

where P.V. represents the principal value of the Cauchy integral, Δn is the readout wavelength dependent change in refractive index, A_M and A_{bR} represent the absorbances of the ground state and thermal intermediate, respectively, t is the thickness of the hologram and λ is the wavelength of the readout beam. The absorbance, $A(\lambda')$, is related to the absorption coefficient of a material, $\alpha(\lambda')$ (units of reciprocal length), by noting that:

$$\alpha(\lambda') = 2.3026 \frac{A(\lambda')}{t} \tag{15.3}$$

In general, for a photochromic material to exhibit useful holographic properties it should possess a high quantum efficiency and a large shift in absorption maxima between photochromic states. The latter property generally results in a large photochemically induced change in refractive index. As we have seen in the previous section, the large blue shift in absorption maxima, generated by deprotonation of the chromophore during the **bR** to **M** phototransformation, makes bacteriorhodopsin an ideal optical recording material. Figure 15.5 shows a simulation of the refractive and diffractive properties of a thin film of bacteriorhodopsin based on application of the Kramers–Kronig transform and coupled wave theory (see below). It should be clear from the figure that the largest change in the refractive index is expected when the hologram is produced with a write wavelength that efficiently drives the **bR** to **M** photoconversion, and when readout wavelengths are used that yield nondestructive readout (not strongly absorbed by **bR** or **M**). The photodiffractive process can be analysed using the coupled wave theory developed by Kogelnik.[25] The spatial modulations of the absorption coefficient and the index of refraction are described by the following truncated Fourier expansions:

$$\alpha(x) = \alpha_{avg} + \alpha_1 \cos(2\pi x/P) \tag{15.4}$$
$$n(x) = n_{avg} + n_1 \cos(2\pi x/P) \tag{15.5}$$

where P has been defined previously as the fringe spacing of the grating (Fig. 15.4), $a(x)$ and $n(x)$ are the spatial dependent values of the absorption constant and index of refraction, respectively, α_{avg} and n_{avg} are the average values of the absorption coefficient and the refractive index, respectively, and α_1 and n_1 represent the modulation amplitudes of the absorption coefficient (amplitude) and index of refraction. The latter parameters contribute to the total diffraction, the absorptive part through absorptive modulation of the light electric field amplitude, and the refractive component through phase or optical path modulation of the light electric field amplitude. These parameters can be estimated through the use of Equation (15.2) and taking into account the electric field description of the absorption coefficient described in Equation (15.4).

The diffraction efficiency of a hologram is defined as the ratio of the diffracted light intensity I_D to the intensity of the reading beam I_0. As before, the diffraction process can have both an absorption and a phase component, and in the case of bacteriorhodopsin, both contribute:[26]

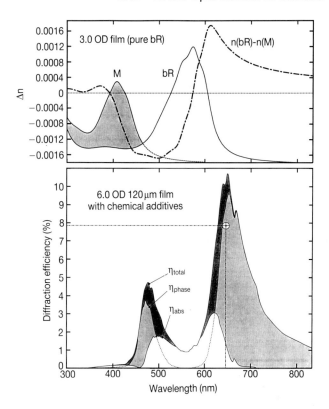

Fig. 15.5   The change in refractive index associated with the **bR**→**M** photoisomerization for a 30 μm film of bacteriorhodopsin with an optical density (OD) of ~3 is shown as a function of wavelength in the upper panel. The refractive index change is expressed as the value for pure **bR** minus the value for pure **M** and is calculated by using the Kramers–Kronig transformation. The absorption spectra of **bR** and **M** are shown for reference. The diffraction efficiency associated with a 6 OD film for **bR** (100%)→**bR** (50%) + **M** (50%) photoconversion is shown in the lower panel, and is calculated based on the observed absorption spectra by using the Kramers–Kronig relationship and Kogelnik approximation.[25] The dot at ~640 nm and ~8% diffraction efficiency represents a recent experimental result from our laboratory using the holographic spatial light modulator described in Fig. 15.7.

$$\eta_{total} = \frac{I_D}{I_0} = \eta_{abs} + \eta_{phase} \tag{15.6}$$

$$\eta_{abs} = \sinh^2 \left\{ \frac{\alpha_1(\lambda_R)t}{2\cos\theta_R} \right\} D \tag{15.7}$$

$$\eta_{phase} = \sin^2 \left\{ \frac{\pi n_1(\lambda_R)t}{\lambda_R 2\cos\theta_R} \right\} D \tag{15.8}$$

$$D = \exp \left\{ \frac{-\alpha_{ave}(\lambda_R)t}{\cos\theta_R} \right\} \tag{15.9}$$

where η_{total} is the total diffraction efficiency $(1 = 100\%)$, η_{abs} is the diffraction efficiency due to absorption, η_{phase} is the diffraction efficiency due to refraction, t is the thickness of the hologram (Fig. 15.4), λ_R is the wavelength of the read laser, θ_R is the angle of incidence of the read laser (Fig. 15.3), $\alpha_1(\lambda_R)$ is the modulation amplitude of the absorption coefficient at the read wavelength, $n_1(\lambda_R)$ is the modulation amplitude of the refractive index at the read wavelength, and $\alpha_{ave}(\lambda_R)$ is the average absorption coefficient of the hologram. Although the read angle θ_R is an experimentally adjustable variable, maximum efficiency is achieved by satisfying the Bragg condition:

$$\theta_R = \sin^{-1}\left\{\frac{\lambda_R \sin(\phi_W)}{\lambda_W}\right\} \tag{15.10}$$

where ϕ_W is the angle of the write beam relative to the hologram film normal (Fig. 15.3) and λ_W is the wavelength of the write beam. The D term defined in Equation (15.9) and which appears in Equations (15.7) and (15.8) places a constraint on the maximum value of the absorptive component of the diffraction efficiency because the absorption modulation change, $\alpha_1(\lambda_R)$, that is required to generate diffraction, also contributes to the average absorption, $\alpha_{ave}(\lambda_R)$. The contribution of D limits the η_{abs} value to 3.7% or less. In contrast, η_{phase} is determined entirely by the change in refractive index, and values approaching 100% are possible. Thus, for applications requiring diffraction efficiencies exceeding 3.5%, phase holograms or mixed absorptive and phase holograms are preferred. This situation is found in the more traditional type of irreversible recording materials such as silver halide photographic films and dichromated gelatin. Figure 15.5 shows the results of the theoretically predicted diffraction efficiency of a 6 OD film of chemically enhanced bacteriorhodopsin as a function of varying readout wavelength. An experimental measurement is also shown indicating that the excellent diffraction efficiency that is predicted can, in fact, be experimentally realized. Holograms can be recorded in pure phase, pure absorption or mixed modes with recording wavelengths from 400–700 nm and readout from 400–850 nm. The recording sensitivity at ambient temperature is in the range 1–80 mJ/cm^2. An additional advantage of using this protein as an optical recording medium is its small size ($\sim$50 nm diameter) relative to the wavelength of light. This results in diffraction limited performance ($>$5000 lines/mm for thin films). Diffraction efficiencies can also be improved by using genetically modified proteins,[15,16] chromophore analogues and chemical enhancement of the native protein (for reviews see Oesterhelt et al.[4] and Birge[5]).

It should be emphasized that maximizing holographic efficiency requires careful adjustment of the laser write and read beam intensities. In contrast to the ideal plane wave situation described above, two interfering Gaussian laser beams do not produce the simple sinusoisal concentration grating depicted in Fig. 15.4, but one which has an overall profile that is determined by the Gaussian intensity distribution within the interacting beams as depicted in Fig. 15.6. If the intensity is too low, the extent of

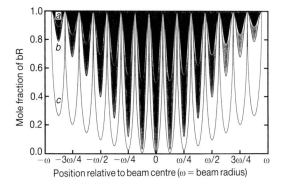

Fig. 15.6 A schematic representation of the grating pattern induced in a thin film of bacteriorhodopsin by using two interfering green laser beams. In regions of constructive interference, the forward (**bR**→**M**) photochemical reaction is driven decreasing the mole fraction of **bR** relative to **M** (vertical axis). The horizontal axis represents the position relative to the centre of the overlapping (interfering) Gaussian laser beams where ω represents the beam radius. Three situations are depicted. The top graph (a) shows a situation where the intensity of the laser light is sufficiently low to generate only partial photochemical conversion even in regions of maximal constructive interference. The middle graph (b) shows a grating which is generated with a laser intensity that is at the highest value possible without 'overdriving' the photochemistry. This situation generates the optimal diffraction efficiency without introducing any nonlinearity. The lowest graph (c) shows a situation where the intensity of the laser light is sufficiently high to 'overdrive' the photochemistry and generate nonlinear performance. In general, a small amount of nonlinearity will improve diffraction efficiency without diminishing the quality of the hologram. However, the situation depicted in graph (c) represents a serious level of nonlinearity which will both diminish diffraction efficiency and holographic image quality.

conversion may be adequate at the beam centre, but inadequate off-centre (graph (a) in Fig. 15.6). If the intensity is too high, photochemistry is overdriven in the higher intensity regions (graph (c) in Fig. 15.6). The optimal linear photochemical transformation is shown in graph (b) of Fig. 15.6. In some cases, however, overdriving the photochemistry to generate a nonlinear grating as shown in graph (c) of Fig. 15.6 can produce enhanced diffraction efficiencies. Unfortunately, this is often accompanied by a loss of resolution that can have a deleterious impact on the image quality of data density when the higher laser intensities are used in optical memory applications or in pattern recognition systems. Simply stated, optimizing the laser intensity to generate a maximum diffraction efficiency is not always optimal for optoelectronic applications.

There are a number of polymer matrices which can be used to solubilize bacteriorhodopsin including polyvinyl alcohol, bovine skin gelatin, methylcellulose and polyacrylamide. Polymeric films containing bacteriorhodopsin are usually sealed from the outside environment to prevent humidity and pH changes which dramatically influence the photochromic properties exhibited by the film. The design shown in Fig. 15.7 yields bacteriorhodopsin films with excellent long-term stability and high diffraction efficiencies (see Fig. 15.5).

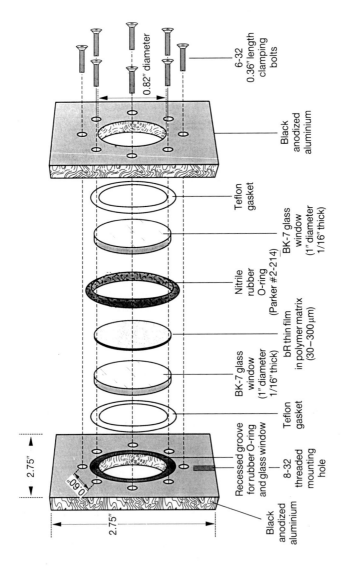

Fig. 15.7 Schematic design of a reversible holographic spatial light modulator based on a thin film of bacteriorhodopsin. A key feature of this design is the use of a compressed nitrile rubber O-ring to seal the protein thin film in order to prevent dehydration of the polymer matrix. The long term optical and shelf stability of the holographic media is excellent.

Spatial Light Modulators   Research in optical engineering during the past decade has demonstrated the unique capability of two-dimensional optical processing systems to perform complex mathematical processing functions such as pattern recognition, image processing, solution of partial differential and integral equations, linear algebra and nonlinear arithmetic.[27,28] Interest in exploring optical processing architectures is prompted by the inherent speed and unique functionalities derived from the massive parallel-processing and interconnection capabilities of optical systems. Spatial light modulators (SLMs) are integral components in the majority of one-dimensional and two-dimensional optical processing environments. These devices modify the amplitude, intensity, phase and/or polarization of a spatial light distribution as a function of an external electrical signal and/or the intensity of a secondary light distribution. The observation that a thin film of bacteriorhodopsin can act as a photochromic bistable optical device (either $\mathbf{bR} \Leftrightarrow \mathbf{K}$ or $\mathbf{bR} \Leftrightarrow \mathbf{M}$ photoreactions), or as a voltage controlled bistable optical device ($\mathbf{bR} \Leftrightarrow \mathbf{M}$ photoreaction), suggests that it has significant potential as the active medium in SLMs.[4,16,18] Soviet scientists were the first to exploit this potential and deserve much of the credit for bringing bacteriorhodopsin to the attention of researchers working in optical engineering.[17,19,20,29–31] The most successful bacteriorhodopsin based SLM device has been recently demonstrated by German researchers. Their work exploits the $\mathbf{bR} \Leftrightarrow \mathbf{M}$ photoreaction of a mutant protein film in a Fourier optical architectural scheme that implements edge enhancement (spatial frequency filtering) on an input image.[32]

Holographic Associative Memories   Associate memories operate in a fashion quite different from the serial memories that dominate current computer architectures. These memories take an input data block (or image), and independently of the central processor, 'scan' the entire memory for the data block that matches the input. In some implementations, the memory will find the closest match if it cannot find a perfect match. Finally, the memory will return the data block in memory that satisfies the matching criteria. Because the human brain operates in a neural, associative mode, many computer scientists believe that the implementation of large capacity associative memories will be required if we are to achieve artificial intelligence fully. Optical associative memories using Fourier transform holograms have significant potential for applications in optical computer architectures, optically coupled neural network computers, robotic vision hardware and generic pattern recognition systems. The ability to change rapidly the holographic reference patterns via a single optical input, while maintaining both feedback and thresholding, increases the utility of the associative memory, and in conjunction with solid state hardware, opens up new possibilities for high speed pattern recognition architectures.

One application currently under investigation is the use of bacteriorhodopsin thin films as the holographic storage components in a real-time optical associative memory.[5,18] Our current design is shown in Fig. 15.8.[26] The optical design, which employs both feedback and thresholding, is

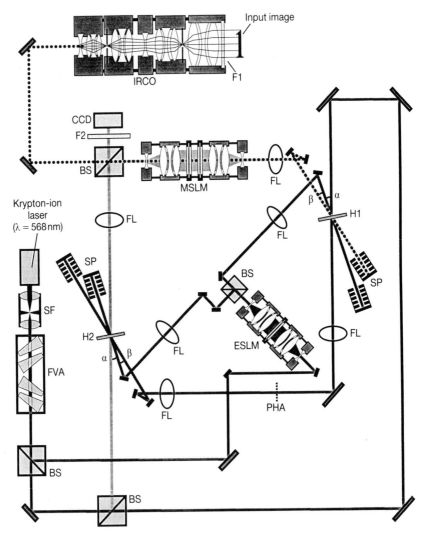

Fig. 15.8 Schematic diagram of a Fourier transform holographic (FTH) associative memory with read/write FTH reference planes using thin polymer films of bacteriorhodopsin to provide real-time storage of the holograms. The optical design is a modification of one proposed by Izgi.[26] The following symbols are used: BS (beam splitter), CCD (charge coupled device two-dimensional array), ESLM (electronically addressable spatial light modulator), FL (Fourier lens), FVA (Fresnel variable attenuator), F1 (broadband filter for image), F2 (interference filter with transmission maximum at laser wavelength; different from λ_{max} of F1), H1 and H2 (holographic spatial light modulator, Fig. 15.7), IRCO (image reduction and collimation optics), MSLM (microchannel plate spatial light modulator), PHA (pin hole array). SF (spatial filter to select TEM_{00}), SP (beam stop).

based on the closed-loop autoassociative design of Paek and Psaltis.[33] During the write operation, reference images stored in an electronically addressable spatial light modulator [ESLM] are optically fed into the loop by plane wave illumination ($\lambda_W = 568$ nm) from a krypton-ion laser. The reference images are stored as Fourier transform holograms on thin polymer films containing bacteriorhodopsin [H1 and H2] (Fig. 15.7). For this real-time application, no chemical additives are used to enhance the **M** state lifetime. Accordingly, the hologram stores the reference image for approximately 10 ms before reverting to the ground state. During the readout operation, the input image (from transparencies or another ESLM) is read into the loop by using the optical imaging system shown in Fig. 15.8. Best results are obtained by illuminating the object by using plane wave illumination from a second krypton-ion laser operating at a wavelength of 676.5 nm. The input image beam is passed through a microchannel plate spatial light modulator [MSLM] operating in threshold-ing mode. Thereafter, the Fourier transformed product of the image-reference is formed and retransformed at the plane of a pinhole array [PHA]. The resulting correlation patterns are sampled by the pinholes (diameter $\sim$500 μm) which are precisely aligned with the optical axis of the reference images. Light from the pinhole plane is retransformed and superimposed with the reference image stored on the second bacterio-rhodopsin hologram [H2]. The resulting cross correlation pattern repre-sents the superposition of all images stored on the multiplexed holograms, and is fed back through the microchannel plate spatial light modulator for another iteration. Thus, each image is weighted by the inner product between the pattern recorded on the MSLM from the previous iteration and itself. The output locks on to that image stored in the holograms which produces the largest correlation flux through its aligned pinhole.

The real-time capability of the associative loop is made possible by using bacteriorhodopsin films as the transient holographic medium. The high speed of phototransformation during write operation ($<$50 μs) coupled with the quick relaxation time of the **M** state ($\sim$10 ms) allow for framing rates up to 33 frames/sec. Slower or faster framing rates can be attained by simply altering the **M**-lifetime with chemical additives, and/or by intermittently erasing the hologram with an external blue light source. The write and read wavelengths of the krypton-ion lasers as well as the respective angle of incidence are chosen to optimize the diffraction efficiency of the bacteriorhodopsin holograms. During the process of sampling the correlation patterns, it is interesting to note that the inclusion of the pinholes destroys the shift invariance of the optical system. If the input pattern is shifted from its nominal position, the correlation peak shifts as well, and the correlation light flux will miss the pinhole. If the pinholes were removed, however, ghost holography would seriously impair image quality. The two apertures within the image reduction and collima-tion optics (IRCO) serve to provide correct registration, but the input image must still be properly centred to generate proper correlation. The problem of shift invariance represents one of the fundamental design issues that must be resolved before optical associative memories will reach their full potential. While there are a number of optical 'tricks' that can be used to counteract poor registration, the most easily implemented approach is to use the controller of the ESLM to scale and translate the reference images to maximize the correlation light flux as measured by the intensity of the image falling on the CCD output detector.

15.3   Nonlinear Optoelectronic Devices

Recent studies have demonstrated the significant potential of organic molecules in applications requiring enhanced second order [34–46] and third order[34,36,42,47–53] polarizabilities. Many of these studies have concluded that substituted polyene chromophores are optimal for nonlinear optical applications due to a combination of electronic, conformational and synthetic advantages. Relatively few studies have been carried out on naturally occurring polyenes, despite the fact that these chromophores have yielded some of the largest second-order[37,39,45] and third-order[47,50,53] polarizabilities. Very few nonlinear applications involving proteins have been reported, but preliminary investigations are encouraging.[37,39,45,53] As we demonstrate below, bacteriorhodopsin has interesting nonlinear optical properties that encourage further study.

Two-photon properties   The two-photon double-resonance spectrum of light adapted bacteriorhodopsin in D_2O at room temperature is shown in Fig. 15.9.[46] This spectrum is unique relative to other two-photon spectra measured for the visual chromophores and pigments[54–60] in that it exhibits two low-lying band maxima. The lowest energy band maximum at 560 nm ($\delta = 290$ GM; 1 GM $= 10^{-50}$ cm^4 s molec^{-1} photon^{-1}) corresponds within experimental error with the one-photon absorption maximum at 568 nm, and is assigned to the $^{'1}B_u^{*+'} \leftarrow S_0$ transition. The higher energy two-photon band at ~488 nm ($\delta = 120$ GM) does not correspond to a resolved one-photon feature and is assigned to the $^{'1}A_g^{*-'} \leftarrow S_0$ transition. Not only is it surprising to find that both the $^{'1}A_g^{*-'}$ and $^{'1}B_u^{*+'}$ states generate discernible two-photon maxima, but the observation that the $^{'1}B_u^{*+'}$ state two-photon absorptivity is more than twice as large as that associated with the $^{'1}A_g^{*-'}$ deserves special notice. This result indicates that the two-photon absorptivities are dominated by initial and final state contributions (see discussion in De Melo and Silbey[59]), and thus the two-photon absorptivity is proportional to the change in dipole moment upon excitation into the final state. A detailed analysis yields $\Delta\mu$ ($^{'1}B_u^{*+'}$) $= 13.5 \pm 0.8$ D and $\Delta\mu$ ($^{'1}A_g^{*-'}$) $= 9.1 \pm 4.8$ D.[46] Thus, the low-lying, strongly allowed $^{'1}B_u^{*+'}$ state is a $\pi^* \leftarrow \pi$ charge transfer state that has long been recognized as a key contributor to the second order molecular polarizability of conjugated molecules.[34–44] Under certain conditions, a low-lying excited state with these properties can enhance the third-order polarizability via Type III enhancements (see, for example Garito et al.[48]). These issues, as well as the potential of using bacteriorhodopsin in two-photon, three-dimensional optical memories, are discussed below.

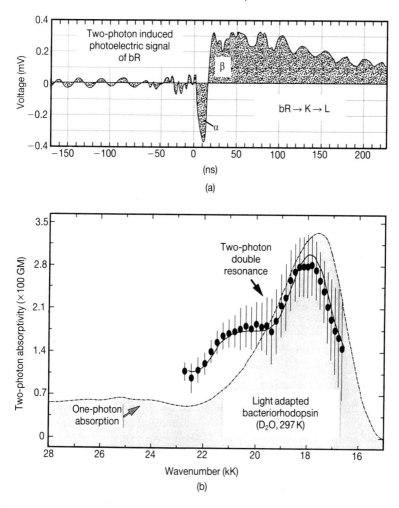

Fig. 15.9 Dispersion of the two-photon absorptivity (b) and the two-photon induced photo-electric signal (a) of bacteriorhodopsin. The photoelectric signal is identical under one-photon and two-photon excitation, and consists of two components, a fast component (α) that follows the laser pulse and an oppositely polarized component (β) that has a temporal profile that is pH and impedance dependent. Comparison of the one-photon absorption and the two-photon double resonance spectra of light-adapted bacteriorhodopsin is shown in (b).

Higher-order Polarizabilities Molecules respond to applied external fields through linear and nonlinear components of the polarizability. The total molecular dipole moment along the ith axis, μ_i^{tot}, is given by the following expansion:

$$\mu_i^{tot} = \mu_i^0 + \alpha_{ij} : E_j + \tfrac{1}{2}\beta_{ijk} : E_j E_k + \tfrac{1}{6}\gamma_{ijkl} : E_j E_k E_l + \ldots \qquad (15.11)$$

where μ_i^0 is the static (zero-field) dipole moment, α_{ij} is the molecular polarizability, β_{ijk} is the second-order polarizability, γ_{ijkl} is the third-order polarizability and E_ζ is the applied electric field in the ζ direction. The terminology and units used to identify the higher polarizabilities are potentially confusing. For example, the second-order polarizability, β_{ijk}, is also called the first hyperpolarizability or the second-order hyperpolarizability. Occasionally, β_{ijk} is referred to as the 'second hyperpolarizability', but this terminology is not correct because α_{ij} is a linear component of the polarizability and therefore not a *hyper*polarizability. Fortunately it is almost always clear from the context what the author means. The units are also a source of confusion. The vast majority of the literature on nonlinear optics has maintained the use of cgs units rather than to adopt the admittedly confusing SI units. It is also customary to simply use the term 'esu' to indicate that the units are cgs(esu) based rather than SI based. This approach is certainly convenient, but causes considerable confusion to students, who are likely to assume that 'esu' stands for an electrostatic unit of charge rather than to electrostatic (cgs) units. The actual (cgs/esu) units for the first three polarizabilities are as follows: α (cm^3); β $(cm^5\,esu^{-1})$; γ $(cm^7\,esu^{-2})$.

It is rarely possible to measure all of the individual components of the hyperpolarizabilities experimentally, and most experimental methods yield the orientationally averaged values:

$$\beta = \beta_{xxx} + (1/3)[\beta_{xyy} + 2\beta_{yyx} + \beta_{xzz} + 2\beta_{zzx}] \tag{15.12}$$

$$\gamma = (1/5)[\gamma_{xxxx} + \gamma_{yyyy} + \gamma_{zzzz} + 2\gamma_{yyzz} + 2\gamma_{zzxx} + 2\gamma_{xxyy}] \tag{15.13}$$

The above equations assume that the x axis is aligned along the dipole moment vector. The reader is referred to Chapter 5 of this book for a more detailed discussion of nonlinear optics.

Second-order Polarizability of Bacteriorhodopsin Values of β for light adapted bacteriorhodopsin have been determined from second harmonic generation[39] and analysis of the absolute two-photon absorptivity data.[46] The two-photon double resonance determination of β for $E_\lambda = 1.17\,eV$ $(\lambda = 1.06\,\mu,$ the Nd:YAG fundamental) yields $\beta = (2250 \pm 240) \times 10^{30}\,cm^5/esu,$[46] a value in good agreement with the oriented thin film measurement of Huang et al.,[39] which gave $\beta = 2500 \times 10^{30}\,cm^5/esu.$ The two-photon based value at $E_\lambda = 0.654\,eV$ is $\beta = (329 \pm 130) \times 10^{30}\,cm^5/esu.$ The dispersion in the second-order polarizability is shown in Fig. 15.10.

The leading term in the second order perturbation analysis of the second-order polarizability defines the magnitude of β in terms of the properties of the low-lying strongly-allowed charge transfer excited state (e), relative to those of the ground state (g):

$$\beta \cong \frac{9e^2h^2}{16\pi^2 m_e} \frac{\Delta\mu_{ge}f_{ge}\Delta E_{ge}}{(\Delta E_{ge}^2 - E_\lambda^2)(\Delta E_{ge}^2 - 4E_\lambda^2)} \tag{15.14}$$

where $\Delta\mu_{ge}$ is the change in dipole moment upon excitation, f_{ge} is the oscillator strength, ΔE_{ge} is the transition energy and E_λ is the photon

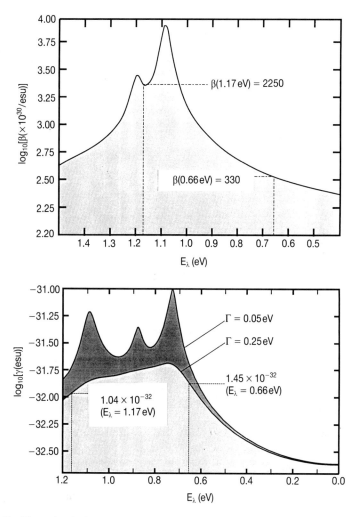

Fig. 15.10 Dispersion in the second-order (top) and third-order (bottom) π-electron polarizability of bacteriorhodopsin as a function of the energy of the incident irradiation in electron volts. The dispersion data for β are from Birge et al.,[45] the data for γ are from Birge et al.[53] The data for γ are plotted as a function of two values of the damping function [$\Gamma = 0.25$ eV, foreground (lower) curve; $\Gamma = 0.05$ eV, background (upper) curve]. The former is more experimentally realistic, but the curve for $\Gamma = 0.05$ eV is included to help identify the resonances.

energy. Bacteriorhodopsin gains its relatively large second-order polarizability from the large change in dipole moment that accompanies excitation into the lowest-lying strongly allowed ''$^1B_u^{*+}$'' state ($\Delta\mu_{ge} = 13.5 \pm 0.8$ D, see above). Potential application of the protein as the active element in second harmonic generators is enhanced by the ease with which the protein can be oriented in optically transparent polymer matrices. This capability follows from the fact that the protein has a very large dipole moment ($\sim$100–500 D depending upon pH) and carries a negative charge of -3 on the cytoplasmic side. Modest electric fields are known to generate near perfect orientation. The fact that the protein absorbs light at wavelengths throughout the visible region precludes general application. Nevertheless, a number of research groups are investigating the use of bacteriorhodopsin and bacteriorhodopsin analogues for frequency doubling applications in the infrared and far infrared.

Third-order Polarizability   The third-order π-electron polarizability, γ_π, of bacteriorhodopsin in the 0.0–1.2 eV optical region has been assigned, based on an analysis of the experimental two-photon properties of the low-lying singlet state manifold.[53] The following selected values of γ_π (units of 10^{-36} cm^7 esu^{-2}) were observed: $\gamma(0; 0, 0, 0) = 2482 \pm 327$; $\gamma(-3\omega; \omega, \omega, \omega) = 2976 \pm 385$ ($\omega = 0.25$ eV), 5867 ± 704 ($\omega = 0.5$ eV), 14863 ± 1614 ($\omega = 0.66$ eV), 15817 ± 2314 ($\omega = 1.0$ eV), 10755 ± 1733 ($\omega = 1.17$ eV). The dispersion of γ as a function of incident photon energy is shown in Fig. 15.10.

In evaluating the third-order properties of bacteriorhodopsin, it is instructive to compare the γ_π values for bacteriorhodopsin with those measured by Hermann and Ducuing for a series of polyenes by using third-harmonic generation.[47] The comparisons suggest that bacteriorhodopsin has a surprisingly large third-order polarizability given the fact that the chromophore has only six double bonds. In fact, it has a γ_π value that is comparable to that measured for dodecapreno β-carotene, a polyene with eleven double bonds.[47]

Garito and coworkers have examined the electronic contributions to the third-order polarizability of polyenes theoretically and note that the principal contributions to γ_π can be divided into three types.[48] Type I is associated with the allowedness of the low-lying *ungerade* (or *ungerade*-like) excited singlet state. If we label this state B, the Type I contribution is proportional to $|\langle B|\mathbf{r}|o\rangle|^4$. Type II is associated with the sequence $S_0 \rightarrow B \rightarrow nA \rightarrow S_0$, and is assigned by Pierce to be the dominant contributor to γ_π in linear polyenes.[50] (The nA state is a high energy *gerade* or *gerade*-like state.)   The   Type II   contribution   is   proportional   to $|\langle B|\mathbf{r}|o\rangle|^2 |\langle B|\mathbf{r}|nA\rangle|^2$. Type III is only relevant to polar molecules and is associated   with   the   product $\langle o|\mathbf{r}|B\rangle\langle B|\Delta\mathbf{r}|B\rangle\langle B|\Delta\mathbf{r}|B\rangle\langle B|\mathbf{r}|o\rangle$, where $\langle B|\Delta\mathbf{r}|B\rangle = \Delta\mu_B/e$ ($\Delta\mu_B$ = the change in dipole moment upon excitation into the B state). Garito and coworkers have noted that this term typically contributes more than one order of magnitude to γ_π in polar long-chain polyenes when $\Delta\mu_B \geqslant 10$ D. The protein-bound chromophore in bacteriorhodopsin exhibits a large change in dipole moment upon excitation into the lowest-lying, strongly-allowed $^1B_u^{*+}$-like state ($\Delta\mu = 13.5$ D).[46] We conclude that Type III enhancement of the third-order polarizability yields at least a 20-fold increase in γ_π.

More subtle is the enhancement associated with the protonation of the chromophore in bacteriorhodopsin. De Melo and Silbey have demonstrated theoretically that polarons and bipolarons have enhanced third-order polarizabilities.[49] While protonated and 'polaronic' chromophores are not identical, the effect of protonation is similar with respect to bond alternation effects and excited state manifold perturbation. Based on simulations, we conclude that protonation yields approximately a five-fold enhancement in the third-order polarizability.

The potential of using bacteriorhodopsin in third-order nonlinear devices has not been explored. Nevertheless, there is one interesting property that might prove useful. We note that the γ_π polarizability of bacteriorhodopsin at zero frequency is predicted to be fairly large $\{\gamma_\pi(0; 0, 0, 0) \cong 2500 \times 10^{-36} \text{ cm}^7 \text{ esu}^{-2}\}$. Preliminary studies indicate that the **M** state has a significantly lower value $\{\gamma_\pi(0; 0, 0, 0) \cong 70 \times 10^{-36} \text{ cm}^7 \text{ esu}^{-2}\}$. Thus, the third-order polarizability of the protein can be optically adjusted over a ~35-fold differential. There may be an application for which this characteristic can be exploited.

Nonlinear Optical Volumetric Memories Based on Bacteriorhodopsin
Although bacteriorhodopsin has competitive second and third order polarizabilities, the principal nonlinear property of the protein with respect to device applications is the large two-photon absorptivity. As noted above, the two-photon absorptivity is anomalously large. This observation suggests the potential of using the protein as the photoactive component in volumetric two-photon memories. We explore this application below.

Two-photon three-dimensional optical addressing architectures offer significant promise for the development of a new generation of ultra-high density random access memories.[61-63] These memories read and write information by using two orthogonal laser beams to address an irradiated volume (1–50 μm^3) within a much larger volume of a nonlinear photochromic material. Because the probability of a two-photon absorption process scales as the square of the intensity, photochemical activation is limited to a first approximation to regions within the irradiated volume. (Methods to correct for photochemistry outside the irradiated volume are described below.) The three-dimensional addressing capability derives from the ability to adjust the location of the irradiated volume in three dimensions. Two-dimensional optical memories have a storage capacity that is limited to ~$1/\lambda^2$, where λ is the wavelength, which yields approximately 10^8 bits/cm^2. In contrast, three-dimensional memories can approach storage densities of $1/\lambda^3$, which yields storages in the range 10^{11} to 10^{13} bits/cm^3. The volumetric memory described below is designed to store 18 GBytes (1 GByte = 10^9 Bytes) within a data storage cuvette with dimensions of 1.6 cm × 1.6 cm × 2 cm. Our current storage capacity is well below the maximum theoretical limit of ~512 GBytes for the same ~5 cm^3 volume.

Bacteriorhodopsin has four characteristics that combine to yield a comparative advantage as a two-photon volumetric medium.[63] First, as noted above, it has a large two-photon absorptivity due to the highly polar environment of the protein binding site and the large change in dipole moment that accompanies excitation.[46] Second, bacteriorhodopsin exhibits large quantum efficiencies in both the forward and reverse direction. Third, the protein gives off a fast electrical signal when light-activated that is diagnostic of its state (see Fig. 15.9). Fourth, the protein can be oriented in optically clear polymer matrices permitting photoelectric state interrogation.[63] The two-photon induced photochromic behaviour is summarized in the scheme below.

$$\textbf{bR} \text{ (State 0) } (\lambda_{max} \cong 1140 \text{ nm}) \underset{h\omega^2;\ \Phi_2 \geq 0.65}{\overset{h\omega^2;\ \Phi_1 \sim 0.65}{\rightleftarrows}}$$

$$\textbf{M} \text{ (State 1) } (\lambda_{max} \cong 820 \text{ nm}) \quad \text{(scheme 2)}$$

We arbitrarily assign **bR** to binary state 0 and **M** to binary state 1. The chromophore in **bR** has an unusually large two-photon absorptivity which permits the use of much lower intensity laser excitation to induce the forward photochemistry. The above wavelengths are correct to only ± 40 nm, because the two-photon absorption maxima shift as a function of temperature and polymer matrix water content.

The optical design of the two-photon three-dimensional optical memory is shown in Fig.15.11.[64] The bacteriorhodopsin is contained in a cuvette and is oriented by using electric fields prior to polymerizing the polyacrylamide gel matrix. This orientation is required in order to observe and use the photoelectric signal (Fig. 15.9) to monitor the state of the protein molecules occupying the irradiated volume. A write operation is carried out by simultaneously firing the two 1140 nm lasers (to write a 1) or the two 820 nm lasers (to write a 0). To eliminate unwanted photochemistry along the laser axes, non-simultaneous firing of the lasers not used in the original write operation is carried out immediately following the write operation. The position of the cube is controlled in three dimensions by using a series of actuators which independently drive the cube in the x, y or z direction. For slower speed maximum density applications, electrostrictive micrometers are used. For higher speed, lower density applications, voice-coil actuators are used. Parallel addressing of large data blocks can also be accomplished by using holographic lenses or other optical architectures.[62]

A key requirement of the two-photon memory is to generate an irradiated volume which is reproducible in terms of xyz location over lengths as large as 2 cm. In the present case, our cubes are typically ~ 1.6 cm in the x and y dimension and ~ 2 cm in the z direction (see Fig.15.11). These dimensions are variable up to 2 cm on all sides, and may be as small as 1 cm on a side depending upon the desired storage capacity of the device. By using a set of fixed lasers and lenses, and moving the cube by using orthogonal translation stages, excellent reproducibility can be achieved (± 1 μm for electrostrictive micropositioners, ± 3 μm by using voice-coil actuators). Refractive inhomogeneities which develop within the protein-polymer cube as a function of write cycles adversely affect the ability to position the irradiated volume with reproducibility. This problem is due to the change in refractive index associated with the photochemical transformation (see Fig. 15.3). The problem is minimized by operating with a relatively large irradiated volume (30 μm^3) and by limiting the photochemical transformation to 60:40 versus 40:60 in terms of relative **bR:M** percentages.

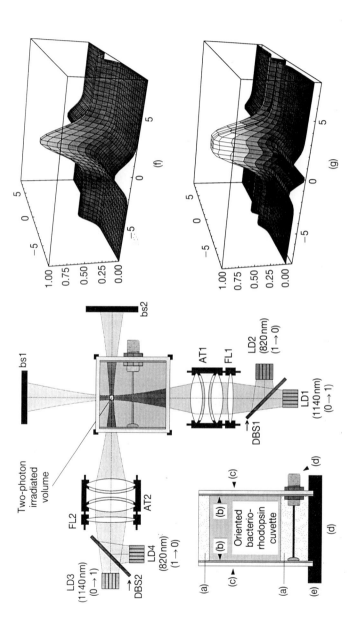

Fig. 15.11 Schematic diagram of the principal optical components of a two-photon three-dimensional optical memory based on bacteriorhodopsin. The write operation involves the simultaneous activation of LD_1 and LD_3 ($0 \to 1$) or LD_2 and LD_4 ($1 \to 0$) to induce two-photon absorption within the irradiated volume and partially convert either **bR** to **M** ($0 \to 1$) or **M** to **bR** ($1 \to 0$). The write operation uses a 10 ns pulse and a pulse simultaneity of 1 ns. The protein is oriented within the cuvette by using an electric field prior to polymerization of the polyacrylamide gel. A polymer sealant is then used to maintain the correct polymer humidity. The *SMA* connector is attached to the indium–tin-oxide conducting surfaces on opposing sides of the cuvette, and is used to transfer the photoelectric signal to the external amplifiers and box-car integrators. Symbols and letter codes are as follows: (a) sealing polymer; (b) indium–tin-oxide conductive coating, (c) *BK7* optical glass; (d) *SMA* or *OS50* connector; (e) Peltier temperature controlled base plate (0–20°C); AT (achromatic focusing triplet); bs (beam stop); DBS (dichroic beam splitter); LD (laser diode); FL (adjustable focusing lens). Computer simulations of the probability of two-photon induced photochemistry (vertical axis) as a function of location relative to the centre of the irradiated volume (Δx_{focus} and Δy_{focus}) in microns are shown in (f) and (g). The upper right contour plot (f) shows the probability after two 1140 nm laser beams have been simultaneously directed along orthogonal axes crossing at the centre of the irradiated volume. The lower right contour plot (g) shows the probability after two 820 nm 'cleaning pulses' have been independently directed along the same axes. The maximum conversion probability at $x = 0$, $y = 0$ is normalized to unity for both contour plots.

Technique of Writing Two-Photon Data The use of two-photon absorption provides for the selection of a data element in three dimensions. It is the ability to control the location of the two-photon irradiated volume in three dimensions that distinguishes the two-photon absorption process from the one-photon absorption process with respect to photochromic data storage. However, it is incorrect to assume that two-photon induced photochromism is limited to the irradiated volume. The actual situation is shown in Fig. 15.11(f), which shows the probability of two-photon induced photochemistry as a function of location relative to the centre of the irradiated volume.

As can be seen from an analysis of Fig. 15.11(f), photochemistry occurs outside of the irradiated volume along the laser beam axes. Nearby memory cells along the laser axes are transformed to a (worst case) ~25% of the irradiated volume (desired memory cell) transformation, and repeated read/write operations may irreparably destroy the contents of unaddressed memory cells after multiple write operations. The optical protocol described below must be used to restore data integrity after each write operation. After the initial two-photon write operation, which involves the use of simultaneous laser excitation at 1140 nm $(0 \rightarrow 1)$ or 820 nm $(1 \rightarrow 0)$, two (non-simultaneous) cleaning pulses are fired without changing the location of the cube. If a $0 \rightarrow 1$ write operation was carried out, the 820 nm lasers are fired. If a $1 \rightarrow 0$ write operation was carried out, the 1140 nm lasers are fired. The intensity and/or duration is adjusted to compensate as accurately as possible for the undesired photochemistry that occurred during the write operation. The end result of the cleaning operation is shown in Fig. 15.11(g). The cleaning operation serves to reduce unwanted photochemistry from ~25% to ~2%, the latter level being more than adequate to maintain data integrity outside of the irradiated volume. The entire write operation, including the subsequent cleaning pulses, is complete in less than 50 ns, and in principle, can take less than 35 ns.

The Read Operation By firing all four lasers simultaneously, the state of the irradiated volume can be probed by monitoring the differential photovoltage $(\alpha - \beta$ in Fig. 15.9(a)). If $\alpha - \beta$ is negative, then the memory cell defined by the irradiated volume is in the 0 state. If $\alpha - \beta$ is positive, then the memory cell defined by the irradiated volume is in the 1 state. (Note that we arbitrarily assign the voltage signs as shown in Fig. 15.9(a).) Careful adjustment of the relative intensities of the four lasers permits minimal disturbance of memory cells outside of the irradiated volume. A standard write operation is then performed to reset the memory cell to the correct state. This procedure serves to enhance data integrity by reducing the risk that multiple read/write cycles along the axis occupied by the interrogated memory cell corrupt the data in that memory cell. The total read process can be completed in ~50 ns, which means that the maximum serial data rate is ~20 Mbit/s. This data rate is decreased by the time necessary to move the cube to the next memory cell. This latency is determined by the extent to which the data to be read are contiguous and by the translation actuators (electrostrictive versus voice-coil).

To maximize reliability, two additional bits are stored after each 8-bit byte to provide single-bit error correction and double-bit error detection. Additional data reliability can be provided by adding checksums at the end of each data block, but the checksum generation and verification process is handled by controller software and not by the two-photon three-dimensional memory control firmware.

15.4 Conclusions

This chapter has overviewed the linear and nonlinear optoelectronic applications of bacteriorhodopsin. Examples discussed include holographic recording, spatial light modulators, three-dimensional random-access memories and holographic associative memories. Natural selection has optimized this protein to function as a high efficiency light-to-energy transducer in the purple membrane of a salt marsh bacterium. The high light intensities and thermal stress associated with a salt marsh environment impose severe requirements on the protein that lead to the excellent cyclicities and thermal stabilities of the devices developed based on the native material. Furthermore, the use of genetic engineering and/or chemical modification can fine tune the optoelectronic properties of the protein for specific applications. These factors have led to the successful application of bacteriorhodopsin in the broad range of optoelectronic devices explored here. The field of bimolecular optoelectronics is in its early stages of development, and the successful use of bacteriorhodopsin in a broad range of device environments will hopefully lead to a recognition that nature has provided a number of other biomaterials that deserve investigation. Other light-transducing proteins including visual rhodopsin,[5] chloroplasts[6,7] and photosynthetic reaction centres[8] have already been investigated, but we should anticipate that many other biomolecules have equal if not greater potential. It should be noted that all of the biomolecules that have been discovered for optoelectronic applications were isolated and characterized because of biological relevance rather than optoelectronic potential. Only during the past few years has biomolecular electronics received direct funding, and the resulting research and development has generated significant progress. The vast potential of this field remains to be appreciated fully.

Acknowledgements The authors thank Deshan Govender and John Izgi for interesting and helpful discussions. The research from the authors' laboratory was sponsored in part by grants from the W. M. Keck Foundation, US Air Force Rome Laboratory, the National Institutes of Health and the Industrial Affiliates Program of the W. M. Keck Center for Molecular Electronics.

References

1. I. Amato, *Science*, September 13, 1213 (1991).
2. D. Kirkpatrick, *Fortune*, December 16, 117 (1991).
3. D. H. Freedman, *Discover*, November, 67–72 (1991).
4. D. Oesterhelt, C. Bräuchle and N. Hampp, *Quart. Rev. Biophys.*, **24**, 425–78 (1991).
5. R. R. Birge, *Annu. Rev. Phys. Chem.*, **41**, 683–733 (1990).
6. E. Greenbaum, *J. Phys. Chem.*, **96**, 514–16 (1992).
7. E. Greenbaum, *J. Phys. Chem.*, **94**, 6151–53 (1990).
8. S. G. Boxer, J. Stocker, S. Franzen and J. Salafsky, in *Molecular Electronics—Science and Technology*, Ed. A. Aviram, (Am. Inst. Phys., New York, 1992), Vol. 262, pp. 226–41.
9. D. Oesterhelt and W. Stoeckenius, *Nature (London), New Biol.*, **233**, 149–52 (1971).
10. R. R. Birge, *Biochim. Biophys. Acta*, **1016**, 293–327 (1990).
11. R. A. Mathies, J. Lugtenburg and C. V. Shank, in *Biomolecular Spectroscopy*, Eds R. R. Birge and H. H. Mantsch, (The International Society for Optical Engineering, Bellingham, Washington, 1989), Vol. 1057, pp. 138–45.
12. R. A. Mathies, C. H. Brito Cruz, W. T. Pollard and C. V. Shank, *Science*, **240**, 777 (1988).
13. R. Simmeth and G. W. Rayfield, *Biophys. J.*, **57**, 1099–101 (1990).
14. R. R. Birge, C. F. Zhang and A. F. Lawrence, in *Molecular Electronics*, Ed. F. Hong, (Plenum, New York, 1989), pp. 369–79.
15. N. Hampp, A. Popp and C. Bräuchle, *J. Phys. Chem.*, **96**, 4679–85 (1992).
16. N. Hampp, C. Bräuchle and D. Oesterhelt, *Biophys. J.*, **58**, 83–93 (1990).
17. N. N. Vsevolodov and V. A. Poltoratskii, *Sov. Phys. Tech. Phys.*, **30**, 1235 (1985).
18. R. R. Birge, P. A. Fleitz, R. B. Gross *et al.*, *Proc. IEEE EMBS*, **12**, 1788–89 (1990).
19. V. Y. Bazhenov, M. S. Soskin and V. B. Taranenko, *Sov. Tech. Phys. Lett.*, **13**, 382–84 (1987).
20. V. Y. Bazhenov, M. S. Soskin, V. B. Taranenko and M. V. Vasnetsov, in *Optical Processing and Computing*, Eds H. H. Arsenault, T. Szoplik and B. Macukow, (Academic Press, New York, 1989), pp. 103–44.
21. W. J. Tomlinson, *Appl. Opt.*, **14**, 2456–67 (1975).
22. R. J. Collier, C. B. Burchhardt and L. H. Lin, *Optical Holography*, (Academic Press, San Diego, 1971).
23. R. Loudon, *The Quantum Theory of Light*, (Clarendon, Oxford, 1973).
24. L. D. Landau and E. M. Lifshitz, *Electrodynamics of Continuous Media*, (Pergamon, New York, 1960).
25. H. Kogelnik, *Bell Syst. Tech. J.*, **48**, 2909–47 (1969).
26. R. B. Gross, K. C. Izgi and R. R. Birge, *Proc. SPIE*, **1662**, 186–96 (1992).
27. A. R. Tanguay Jr., *Optics News*, **Feb.**, 23–26 (1988).
28. A. R. Tanguay Jr., *Opt. Eng.*, **24**, 2–18 (1985).
29. A. B. Druzhko and S. K. Zharmukhamedov, in *Photosensitive Biological Complexes and Optical Recording of Information*, Eds G. R. Ivanitskiy and N. N. Vsevolodov, (USSR Academy of Sciences, Biological Research Center, Institute of Biological Physics, Pushchino, 1985), pp. 119–25.
30. G. R. Ivanitskiy and N. N. Vsevolodov, *Photosensitive Biological Complexes and Optical Recording of Information*, (USSR Academy of Sciences,

Biological Research Center, Institute of Biological Physics, Pushchino, 1985), pp. 1–209.

31. V. V. Savranskiy, N. V. Tkachenko and V. I. Chukharev, in *Photosensitive Biological Complexes and Optical Recording of Information*, Eds G. R. Ivanitskiy and N. N. Vsevolodov, (USSR Academy of Sciences, Biological Research Center, Institute of Biological Physics, Pushchino, 1985), pp. 97–100.

32. R. Thoma, N. Hampp, C. Bräuchle and D. Oesterhelt, *Opt. Lett.*, **16**, 651–53 (1991).

33. E. G. Paek and D. Psaltis, *Opt. Eng.*, **26**, 428–33 (1987).

34. D. S. Chemla and J. Zyss, *Nonlinear Optical Properties of Organic Molecules and Crystals*, (Academic Press, Orlando, Florida, 1987).

35. C. W. Dirk, R. J. Twieg and G. Wagniere, *J. Am. Chem. Soc.*, **108**, 5387–95 (1986).

36. R. A. Hann and D. Bloor, *Organic Materials for Non-linear Optics*, (Royal Society of Chemistry, London, England, 1989).

37. J. Y. Huang, A. Lewis and T. Rasing, *J. Phys. Chem.*, **92**, 1756–59 (1988).

38. J. Y. Huang, A. Lewis and L. Loew, *Biophys. J.*, **53**, 665 (1988).

39. J. Y. Huang, Z. Chen and A. Lewis, *J. Phys. Chem.*, **93**, 3314–20 (1989).

40. H. E. Katz, C. W. Dirk, K. D. Singer and J. E. Sohn, *Mol. Crys. Liq. Cryst. Inc. Nonlin. Opt.*, **157**, 525–33 (1988).

41. J. L. Oudar, *J. Chem. Phys.*, **67**, 446–57 (1977).

42. D. J. Williams, *Nonlinear Optical Properties of Organic and Polymeric Materials*, (American Chemical Society, Washington, D.C., 1985).

43. C. W. Spangler, R. K. McCoy, R. R. Birge, P. A. Fleitz and C. F. Zhang, in *Molecular Electronics—Science and Technology*, Ed. A. Aviram, (Engineering Foundation, New York, 1989), pp. 175–80.

44. K. D. Singer and A. F. Garito, *J. Chem. Phys.*, **75**, 3572–80 (1981).

45. R. R. Birge, P. A. Fleitz, A. F. Lawrence, M. A. Masthay and C. F. Zhang, *Mol. Cryst. Liq. Cryst.*, **180**, 107–22 (1990).

46. R. R. Birge and C. F. Zhang, *J. Chem. Phys.*, **92**, 7178–95 (1990).

47. J. P. Hermann and J. Ducuing, *J. Appl. Phys.*, **45**, 5100–102 (1974).

48. A. F. Garito, J. R. Helfin, K. Y. Wong and O. Zamani-Khamiri, in *Organic Materials for Non-linear Optics*, Eds R. A. Hann and D. Bloor, (Royal Society of Chemistry, London, 1989), pp. 16–27.

49. C. P. De Melo and R. Silbey, *Chem. Phys. Lett.*, **140**, 537–41 (1987).

50. B. M. Pierce, *J. Chem. Phys.*, **91**, 791–811 (1989).

51. P. N. Prasad, P. Chopra, L. Carlacci and H. F. King, *J. Phys. Chem.*, **93**, 7120–30 (1989).

52. J. F. Ward and D. S. Elliott, *J. Chem. Phys.*, **69**, 5438 (1978).

53. R. R. Birge, M. B. Masthay, J. A. Stuart, J. R. Tallent and C. F. Zhang, *Proc. SPIE*, **1432**, 129–40 (1991).

54. R. R. Birge, J. A. Bennett, B. M. Pierce and T. M. Thomas, *J. Am. Chem. Soc.*, **100**, 1533–39 (1978).

55. R. R. Birge, J. A. Bennett, L. M. Hubbard *et al.*, *J. Am. Chem. Soc.*, **104**, 2519–25 (1982).

56. R. R. Birge, L. P. Murray, R. Zidovetzki and H. M. Knapp, *J. Am. Chem. Soc.*, **109**, 2090–101 (1987).

57. L. P. Murray and R. R. Birge, *Canadian J. Chem.*, **63**, 1967–71 (1985).

58. R. R. Birge, L. P. Murray, B. M. Pierce *et al.*, *Proc. Nat. Acad. Sci. USA*, **82**, 4117–21 (1985).

59. R. R. Birge, *Accts. Chem. Research*, **19**, 138–46 (1986).

60. R. R. Birge, *Methods in Enzymology*, **88**, 522–33 (1982).

61. D. A. Parthenopoulos and P. M. Rentzepis, *Science*, **245**, 843–45 (1989).
62. S. Hunter, F. Kiamilev, S. Esener, D. A. Parthenopoulos and P. M. Rentzepis, *Appl. Opt.*, **29**, 2058–66 (1990).
63. R. R. Birge, R. B. Gross, M. B. Masthay, J. A. Stuart, J. R. Tallent and C. F. Zhang, *Mol. Cryst. Liq. Cryst. Sci. Technol. Sec. B. Nonlinear Optics* **3**, 133–47 (1992).
64. R. R. Birge, *IEEE Computer* **25**, 56–67 (1992).
65. B. S. Hayward, D. A. Grano, R. M. Glaeser and K. A. Fisher, *Proc. Natl. Acad. Sci. U.S.A.*, **75**, 4320–24 (1978).
66. T. N. Earnest, P. Roepe, M. S. Braiman, J. Gillespie and K. J. Rothschild, *Biochemistry*, **25**, 7793–98 (1986).
67. R. R. Birge, L. A. Findsen and B. M. Pierce, *J. Am. Chem. Soc.*, **109**, 5041–43 (1987).

16

Molecular Electronic Logic and Architectures

J R Barker

16.1   Introduction

There are now very good prospects for the synthesis, positioning and interfacing of organic molecules with a prescribed switching function.[1–8] Such structures might be the prototypes for a future molecular-scale logic. Charge, spin, conformation, colour, reactivity and lock-and-key recognition are just a few examples of molecular properties, which might be useful for representing and transforming logical information. To be useful, molecular scale logic will have to function close to the information theoretic limit of one bit on one carrier. In this limit the conventional logic operations such as NOT, AND, OR are better constructed from conservative logic modules which more closely represent the physical representation of information. Experimental practicalities suggest that it will be easiest to construct regular molecular arrays, preferably by chemical or physical self-organization. This suggests that the natural logic architectures should be cellular automata: regular arrays of locally connected finite state machines, where the state of each molecule might be represented by colour or by conformation. Schemes such as spectral hole burning already exist for storing and retrieving information in molecular arrays using light. The general problem of interfacing to a molecular system remains problematic. Molecular structures may be the first to take practical advantage of novel logic concepts such as 'emergent computation' and 'floating architecture' in which computation is viewed as a self-organizing process in a fluid-like medium.

In this chapter we describe the principles behind logic devices and systems architectures so far as they impinge on molecular electronics. Practical routes forward are discussed.

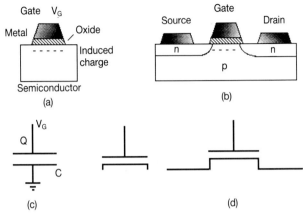

Fig. 16.1 Devices. (a) Memory cell; (b) field effect transistor; (c) capacitor/transistor equivalent; (d) pass transistor notation.

16.2 Inorganic Electronic Systems

The relentless progress of silicon microelectronics technology has been based on the continuing ability to shrink the scale of transistor cells within integrated circuits yet retaining the basic mode of device operation. The transitor cell functions as a memory (for storing charge), as a switch, as an amplifier and by combining transistors it is possible to construct arbitrary families of digital logic devices to build systems. In Fig. 16.1(a) we sketch a simplified picture of a field effect transistor used as a memory cell. Suppose the substrate is *p*-type silicon; by applying a positive voltage between ground and the upper electrode a charge is induced on the electrode, which attracts electrons in the silicon into a channel region at the interface between the silicon and the insulating silicon dioxide. The cell thus functions as a capacitor—it stores charge—it is therefore the basic memory cell for storing sufficient charge to represent 1 bit of information. At the same time we note that applying the gate voltage induces a conducting channel in the silicon. It follows that if we add source and drain electrodes on either side of the gate and contact to an *n*-type conducting region we may switch on a conducting path between source and drain by applying a gate voltage (Fig. 16.1(b)): the basic transistor switch (or pass transistor). A small change in gate voltage will give rise to a large change in source–drain current leading to the amplification property.

In Fig. 16.2(a) we illustrate how to construct a NAND logic gate from two pass transistors plus a resistor (white box in diagram, usually formed from a permanently turned on pass transistor). A and B represent the voltage state at the inputs with a high voltage corresponding to logic '1' (sufficient to switch on a pass transistor) and a low voltage to logic '0'. The truth table in Fig. 16.2(c) shows all the logical outcomes of a NAND (NOT-AND) gate which are reproduced by the circuit. For example if both A and B are 'high'

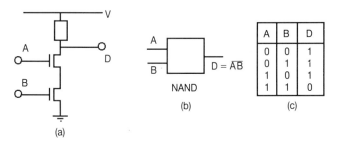

Fig. 16.2 Logic. (a) Layout of NAND gate; (b) NAND gate module; (c) NAND gate truth table.

the two pass transistors turn on, opening a conducting path between the power supply V and ground. The output voltage is therefore brought to zero, 'logic 0'. In any other combination of A and B at least one transistor is switched off and so the voltage seen at D is given by the supply voltage V, 'logic 1'.

By combining a variety of logic gates (AND, OR, NOT, XOR, NOR, NAND, etc.) it is possible to build a combinational logic module which returns at its output a prescribed digital function of the voltages on its inputs. In very large scale integrated (VSLI) circuits there may be as many as a million logic gates on a chip. The timing of the logic operations and the management of the complex function of the chip are often handled by building combinations of clocked data paths and blocks of combinational logic/memory. A data path may be constructed from sequences of pass transistors and inverters clocked by a broadcast clock voltage signal. The clock allows data to flow into a register, and at a later time the data is clocked into a combinational logic module the output from which is clocked onwards via another register. The systems layout of such a path is sketched in Fig. 16.3. An excellent introduction to such systems concepts is given in Mead and Conway.[9]

The biggest problem with VLSI systems is managing the complexity of the design. The trend is towards much more parallelism with regular layouts of simple logic or memory cells. The need for parallelism becomes more urgent at high circuit densities when conventional wiring leads to long dissipative interconnect paths. In parallel arrangements of cells and

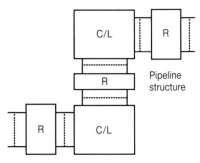

Fig. 16.3 Systems: sections of a pipeline data path.

interconnects (architectures) the interconnects are short and generally set between nearest neighbour cells.

16.3   Molecular Electronic Systems

The earliest approaches to molecular electronics followed the paradigm of inorganic electronics in seeking for a molecular analogue of 'wires', 'transistors' and 'logic gates', based on storing, transmitting and combining information in the form of charge packets. This approach is not necessarily the best way to view a future system of the complexity implied by molecular scales. If instead we look at those properties of molecules which might be useful in a systems context, rather than the bottom-up wires and transistors approach, we find significant possibilities for both systems architecture and for methods of assembly.

The huge range of functionality of molecular materials, particularly the properties of single molecules and interacting molecules, suggest many other ways of storing and transmuting information by representing the information in different physical/chemical forms. In Fig. 16.4 we imagine an array of 'designer molecules or molecular complexes' which have been self-assembled into an array of communicating cells (indicated by the lines). Each molecule (or complex) may be prescribed a well-defined *state* given, for example, by one or more of the following: the local charge, spin, conformation, absorption band (colour), reactivity and so on. There is a well-defined *neighbourhood* of molecules which communicate with any given molecule. Information may flow between adjacent molecules where the flow may be mediated by electrons, protons, ions, moieties, or indirectly via polarization fields, changes in local fields, phonons. *Transitions* will occur between states as the system is driven out of equilibrium by the input of physically coded information. The transition rules might be designed in. Such a system would correspond to a parallel array of finite state machines in inorganic electronics: a system of tremendous complexity, but realized here by a simple molecular matrix.

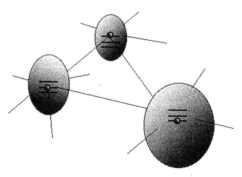

Fig. 16.4   Molecular electronic information processing.

16.4   Guidelines from Biological and Chemical Systems

Inorganic solid state electronics already sets impressive scales of complexity. It is clear from developments in conventional silicon CMOS and bipolar technology that we can expect bit densities as high as 100–1000 Mbit cm^{-2} within the next decade. Optical storage media are currently diffraction-limited theoretically to about 100 Mbit cm^{-2}, and lithographic limits point to ultimate bit densities of about 10 Gbit cm^{-2}. The major limitations here are to do with definable line width (limit $\sim$10 nm) and the restriction to 2D geometry. But truly massive complexity is implied for molecular systems; it is already known that molecular optical storage is, in principle, capable of 1 Tbit cm^{-2} using spectral hole burning (see Section 16.8).

It is conceivable that molecular units could be defined on 1 nm scales leading to surface densities of 1 Tbit cm^{-2}, and with 3D molecular structures the ultimate stable chemical limits may be as high as 10^6 Tbit cm^{-3}. These huge figures are comparable with the 'bit-densities' reached in biological systems and, although comparisons are quite dangerous in general, there are some instructive lessons to be drawn from biology. For example the average adult human brain comprises some 10^{11} neurons, ranging in size from 0.2 μm linear dimension (the olfactory neurons) to about 100 μm, each with an average connectivity of 10^4, giving a crude 'bit-count' of 10^{11} to 10^{15}. An equivalent artificial 'brain' might be built from $10^9 \times 1$ Mbit chips or in the future from $\sim$$10^7 \times 256$ Mbit chips. The power output would be in the range 50 MWatt–5 GWatt; the cost would be in millions to billions of dollars; and the wiring problem would be unamaginable.

In Fig. 16.5 we sketch some of the established features of biological cells which suggest that the neural network is only one, and a macroscopic one at that, of many highly complex 'information processing' systems at work in biology, and from which we might learn the concepts that will push forward molecular electronics.

Coming down in scale, we observe that important switching and recognition (i.e. gating or logic) mechanisms are at work in the transport of ions through ion channels in the cell membrane (Fig. 16.5(b)).

Within the cell itself (Fig. 16.5(a)) are to be found complex networks of microtubules which have been filmed using edge-detection optical microscopy to reveal the transport of chemical moieties along the network. Microtubules (Fig. 16.5(c) and Fig. 16.6) occur as extensive networks or as bundles of long parallel filaments. Microtubules are hollow polymeric cylinders of the order of 25 nm in diameter. The assembly of microtubules is generally reversible and is controlled by the cell. As a result of their underlying polarity microtubules have a preferred end for growth, and one for disassembly. Microtubules also demonstrate the fact that proteins interact to produce complex cellular activities. One function of microtubules is to transport membrane vesicles; this is well-developed in nerve

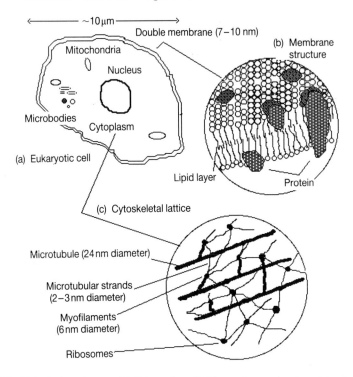

Fig. 16.5 Biological analogues. (a) Eukaryotic cell; (b) membrane structure; (c) cytoskeletal lattice.

cells. The long process of nerve cells, the axon, contains a bundle of microtubules which transport material from the cell body to the terminals, anterograde transport, and from the terminals to the cell body, retrograde transport (see sketch in Fig. 16.6). Two protein complexes have been identified which produce this transport, dynein, for retrograde transport and kinesin for anterograde transport. There have been arguments,[10] that microtubules could act as chemical cellular automata, although other studies suggest otherwise.

At even smaller scales we might refer to the example of repressor molecules which can run along the DNA-helix and 'recognize' base-pair patterns (this remarkable process is described for non-biologists in Ptashne[11]).

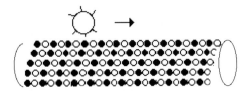

Fig. 16.6 Chemical transport along microtubule.

Fig. 16.7 Lock and key logic.

Once one considers the properties of chemical, biochemical and bio-logical systems there emerges the whole range of self-organizing processes and chemical recognition events, including chemical docking and lock-and-key recognition processes (Fig. 16.7). Since logic involves the ability to detect a state and alter it, we begin to realize that in their most abstract forms logical processes are already ubiquitous within molecular media. There are thus new opportunities for devising radical new information processing systems.

16.5   Complexity and Systems Specification

The massive complexity of a molecular electronic system will require architectures which come to terms with that complexity.[4-8] Some ideas may be gleaned from developments in advanced VLSI circuit design[9] neuroinformatics,[12] regular cellular arrays, neural networks, hierarchical structures. But additional constraints will be required. It will be necessary for the overall circuit function to be insensitive to the precise number of subsystems and to the boundary conditions (for these will be virtually uncontrollable and 'dead zones' may have to be set up). The connection network must be very simple, ideally of a nearest neighbour variety. Figure 16.8 illustrates some of the high-level logic module forms and architectures used in VLSI design. Figure 16.8(a) shows a module which accepts data flows from the north, east and west, and outputs data flows to the south and east and west. Figure 16.8(d) shows a systolic array of such modules in which data is regularly clocked in and out of the modules working in parallel. Figure 16.8(c) shows a pipeline processor comprising a sequence of modules each contributing to a computational flow passing through the array in an assembly-line fashion. Figure 16.8(b) shows a basic cell for combining data in a hexagonally arranged neighbourhood.

A minimum number of molecular complexes must represent the logic modules available to the designer. The necessity for molecular level addressing will put severe constraints on the input/output problem.

16.6   Physical Representations of Data and Logic

The Charge Packet Model   In silicon microelectonics each information bit is encoded as a very large number (not necessarily fixed) of electrons

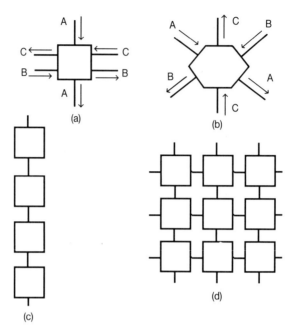

Fig. 16.8 Architectures used in VLSI design.

within a charge pulse. The associated reservoirs and sinks of charge carriers may be tapped and manipulated to provide macroscopic electrical currents, which may be readily amplified or attenuated. Restoring logic is deployed to offset signal leakage and decay. There is a ubiquitous and continuous creation, destruction and restoration of the physical signals. Molecular scale logic cannot afford such luxury: it will require a logic scheme that closely matches the physical representation of information; conservative logic in either its unitary or binary form seems the most profitable route. There is an associated cost in complexity because garbage lines are required to recover many of the usual logic functions. These play an analogous role to power rails and power distribution lines in conventional circuits.

Molecular Representations Molecules can hold a charge and indeed may conduct electricity, but there are other potentially useful properties which could be used for storing information and ultimately providing switching capabilities. These include: nuclear spins, conformation (shape) of the molecule, energy levels, absorption frequencies, lock and key recognition properties, quasi-mechanical and topological properties, colour, voltage sensitivity, chemical specificity. Many of these properties can be altered by putting the molecule into close proximity with other molecules. The basic idea in molecular electronics is to design functionality into molecules which can be assembled into a system by self-assembly using

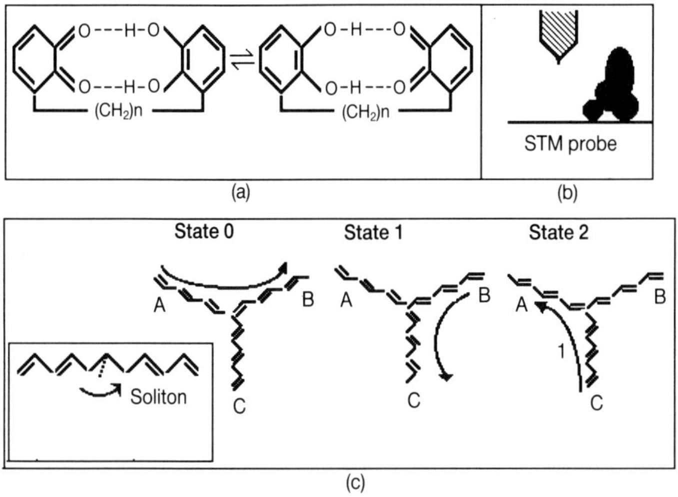

Fig. 16.9   Molecular candidates for switches.

intermolecular forces, rather than the topographic influence of a substrate, to form the desired system. Langmuir–Blodgett film technology has shown the potential for this sort of assembly concept.

The first proposals for molecular switches involved: (i) directional electron transfer in single molecules,[13] (ii) bistable systems based on proton transfer reactions,[14] (iii) storage elements based on molecular isomerization[15] and (iv) optoelectronic response of molecules.[16] Figure 16.9 shows two of the earliest proposals for molecular scale logic. In Fig. 16.9(a) the hemiquinone molecule is regarded as fixed in space and stores data 0 or 1, corresponding to being in one or the other of the two possible (degenerate states) states.[14] Proton tunnelling will convert one state to the other. Attempts to emplace such a molecule at a gold surface and drive it by a STM probe (Fig. 16.9(b)) have not succeeded. Figure 16.9(c) illustrates an idea heavily promoted by Carter[2] that solitons in polymer networks might be switched and steered. The diagram illustrates the change of state of a three-state network following the passage of a soliton. Of some interest theoretically the problem of building, interconnecting and interfacing such a system has not been attempted.

There are indeed severe problems of how to interconnect the molecule and how to read-in and read-out data at a molecular level. Although there have been exciting speculations for self-assembling, self-repairing, fault-tolerant molecular systems, none have been well-specified or fabricated. A less ambitious approach would require first the identification of a suitable high-level logic function on a molecule. Second, one must be able to emplace such molecules with precision, perform an interface to the molecules and verify both emplacement and interface. Third, it is crucial to be able to interconnect and/or isolate such molecules without destroying the functionality. Finally, an efficient assembly scheme for very large numbers of such molecules must become available.

The Granular Limit Although there have been many proposals for suitable molecules to fulfil logic functions, it is not known whether or not such systems could be 'wired up' and interconnected, or to what extent thermal and quantum fluctuations would overwhelm any device(s) as is the case for many semiconductor nanostructures. The problems are particularly acute for *granular electronic* systems,[17] in which information is carried by *individual discrete carriers*. Examples include the experimentally feasible, hemiquinone proton-tunnelling structure,[1] the shuttle molecules[18,19] (see Section 16.8), the recently proposed molecular NAND gate and molecular shift registers (Lindsey;[16] Baratan *et al.*[20]), natural systems such as the matter-transporting microtubule,[9] and hypothesized schemes such as soliton channels and valves.[2] We shall return to this question in Sections 16.7 and 16.8.

16.7 Logic Possibilities

Before examining candidates for molecular logic let us first briefly review some of the concepts required to examine non-standard logic forms.

Restoring Logic Conventional silicon circuits utilize restoring logic to prevent the physical attenuation and asynchronicity of the data flows. A simple example is afforded by a pass transistor (Fig. 16.1): a signal travelling onto the gate can be transformed into a new signal flowing between source and drain by gating the channel. It is usual not to mix data paths and control signal paths, so a more convenient way of restoring a signal which is passing along a line of pass transistors is to use an inverter gate within the line (Fig. 16.10). Without restoring logic the effective resistance-capacitance networks would lead to diffusion delays and signal attenuation. This problem would be likely to occur in molecular media also.

Neural Logic and Neural Networks Formal neurons have been extensively studied in computer science since the 1930s. Figure 16.11 gives a simple example of a neural gate. The neuron gate may be constructed in analogue or digital form. In the latter, the neuron is designed to respond to

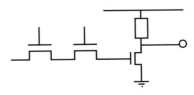

Fig. 16.10 Restoring logic.

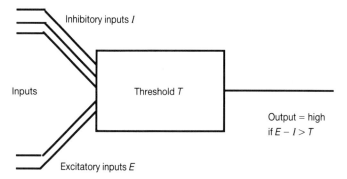

Inhibitory inputs I

Inputs

Threshold T

Output = high
if $E - I > T$

Excitatory inputs E

Fig. 16.11   Neutral logic.

the sum of its N inputs (which may be inhibitory or excitatory). Provided the sum exceeds a given threshold T the neuron will output logic 1, otherwise it outputs logic 0. More sophisticated models use a three-state version in which a period of rest follows a firing state.

By combining neurons into totally connected networks it is possible to construct adaptive learning systems, control systems and pattern recognition systems. In recent years there has been considerable interest in building solid state neural networks formed from large resistor arrays driven by sets of amplifiers. Although technically of interest there are no known proposals for building molecular equivalents. However neural networks are very similar to globally connected cellular automata which we discuss later.

Conservative Logic   The possibility of working in the granular limit—1 bit per information carrier—is familiar from the days of magnetic bubble technology. The latter was one of the first practical applications for conservative logic, where data are steered through a system such that the total number of bit zeros and bit ones in the input data is equal to the total numbers present in the output. Conservative logic is useful when the data carrier is not destroyed or created, it is ideal for discrete physical systems including single electronics (Section 16.8).

The basic building block for conservative logic[21] is usually taken as the computation-universal, invertible Fredkin gate illustrated in Fig. 16.12(a).[6,8] In the Fredkin gate there is one control line which sets the gate into one of two possible modes. Either the gate passes all data undisturbed or it swaps over the data (see Fig.16.12(b)).

There is no bit destruction/creation or signal amplification involved, but functions such as fan-out can only be emulated at the expense of introducing external constant input data streams and output garbage ('don't care') data streams. This is inevitable if standard logical functions are emulated, because replication of a signal as in fan-out requires us to physically bring in extra carriers: the power supply problem remains with us and indeed becomes more complex.

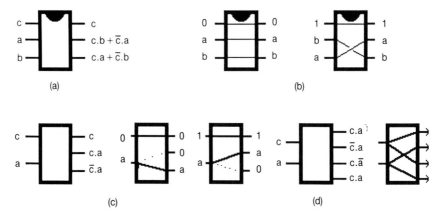

Fig. 16.12 Conservative logic.

The Fredkin gate may be built from subgates which make no distinction between the input lines being control or data lines. One method, discussed by Toffoli[21] builds the Fredkin gate from the cascading of six interaction gates, which are invertible gates that use no control-specific signals. The composite Fredkin gates may be cascaded into arbitary reversible sequential circuits. The interaction gate, a reversible computation-universal gate,[8,22–24] is shown in Fig. 16.12(d). It is related to the Priese gate,[24] Fig. 16.12(c), another computation-universal gate. The Priese gate is a fundamental route switcher using a control to route an input signal between two alternative output paths. The similarity of the soliton net in Fig. 16.9(c) is discussed in Barker.[7] Priese[22] has shown that these gates provide a powerful means to implement abstract asynchronous cellular networks, thus breaking the rigid clocking scheme inherent in many cellular automata. This may prove vital for any future molecular logic. Earlier studies of asynchronous switching networks may be found in Miller.[25]

High Level Logic   Complexity issues suggest that it is pointless to start with a 'bottom up' approach: devising a range of molecular equivalents of conventional logic gates (AND, NOR, etc.). Instead we might look for the equivalent of *high-level logic modules*. A good example is furnished by the *finite-state machine* sketched in Fig. 16.13.

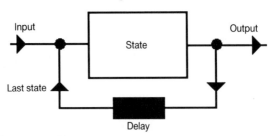

Fig. 16.13 Finite-state machine.

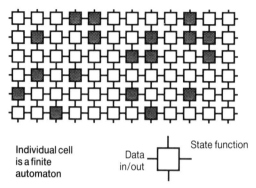

Fig. 16.14   Cellular automaton.

This structure is characterized by a state variable, which is a function of the current inputs to the module and a possible function of the previous state. The finite-state machine thus has memory. The module is also characterized by a transition rule, which describes how the state changes as a function of the inputs and previous state values. The output from the machine is either the current value of the state variable or some function of the state variables and the inputs. Finite-state machines provide elegant ways for controlling traffic lights or security systems. Such a module has important applications in advanced computer science. Represented in silicon, the finite-state machine requires a considerable number of wires and transistors. Instead such a module might be represented by a single molecule, in which the state might correspond to the colour, or conformation of the molecule and the inputs and outputs correspond to interactions with neighbouring molecules. See Fig. 16.4.

Cellular Automata   It is already obvious from VLSI studies that truly complex systems must be designed to be regular, preferably locally communicating, arrays of logic/memory modules. Cellular automata which are regular arrays of finite-state machines provide a useful architectural target, both from the standpoint of fabrication and from logical issues,[3,4,7,26] The concept of a cellular automaton is illustrated in Fig. 16.14.

A cellular automaton is a regular array of identical finite-state machines each connected to a finite number of its neighbours using the same interconnection net. It follows that the cellular automaton is characterized by: (i) the local transition function F for each finite-state machine; (ii) the lattice or cellular space-neighbourhood. Cellular automata were first investigated by von Neumann as models for completely discrete physical dynamical systems such as the brain. The early potential of cellular automata for possible highly parallel computer architectures was not achieved for practical reasons and because the mathematical theory was poorly developed. Figure 16.14 shows an example of a two-dimensional cellular automaton layout. It consists of a regular spatial array of identical

Fig. 16.15 Example of a cellular automaton: replicator.

cells, located at positions X, Y in the lattice and each characterized by a finite discrete-valued state function $f(X, Y, t)$. At successive discrete intervals of time t each cell makes a transition to a new state $f(X, Y, t+1)$ determined by a transition function (mapping) $F[f(X', Y', t)]$ which depends on the previous state values of the connected neighbours (at X', Y') of the cell (including itself). An initial state of the system is represented as a pattern of values of the various states. Iteration of the mapping F will unfold the dynamics of the cellular automaton.

Figure 16.15 shows an example of a binary state replicator[4] showing the first few time steps of its evolution. Dark cells are in state 1, light cells are in state 0. Any pattern entered into the frame pattern will reproduce at the same rate. This example has been proposed as a fault-tolerant way of shifting data from an input region to a distant part of a molecular electronic system.[4]

If each cell corresponds to a molecular assembly, the state of each cell must be a well-defined property such as charge, excitation energy, conformational setting, spin, etc. Intermolecular interactions (e.g. soliton valving, resonant electron transfer, interfacial charge transfer) are necessary for cells to communicate. There are two distinct ways in which information processing might be represented. In the first, analogous to VLSI circuits, information is represented by a transported quantity such as charge, which is routed throughout the system network using switches. In the second approach, analogous to phonons or spin waves, there is local exchange of energy and momentum but not material. A hypothetical example would have a molecule switch between different conformational states, the effect of which would be to induce a similar transition in its neighbours. Such a process, in analogy to a set of coupled nonlinear oscillators, needs a mechanism for amplifying the perturbation: an energy source, not necessarily localized, is needed. Thermodynamic considerations require that a molecular cellular automaton be kept out of thermal equilibrium. If the internal state of a cell corresponds to an occupied physical state of the molecule that state must be sufficiently long-lived for processing events to take place.

Floating Architectures   Since cellular automata support flows and interactions of spatially distributed data, they could be used to support complex computational processes in which the evolving data patterns are put into one to one correspondence with the computational process itself. This is discussed in Barker,[4,6] where the concept of a floating architecture is introduced. A floating architecture involves a dynamic data pattern on a cellular automaton which may move over the surface of the automaton state space. It might be arranged to relocate when a defect is encountered in the cellular space or to carry out computations. In a floating architecture the input data, output data and computational machine and its controls are all represented by a pattern in the cellular automaton state space.

16.8   Molecular Representations

Let us now discuss practical possibilities for molecular electronic systems.

Molecular Wires and Switches   Figure 16.16 shows two well-studied examples of molecular wires: (a) the *trans*-polyacetylene chains;[27] (b) the resonant tunnelling switch.[5] In the latter, one envisages a conducting polymer backbone with attached side arms, arranged to regularly give resonant transmission levels in the conduction. By gating a side arm it might be possible to turn the current response off or on. Synthetic schemes for producing ensembles of such systems exist. But although analogues exist for such systems in the inorganic solid state these two examples show the impracticality: how does one interface to single 'wires'? How are the wires to be gated or interconnected?

Molecular Memory and Logic   At a next level of complexity we might consider molecular memories or molecular logic but again, despite the

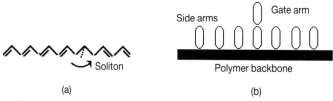

(a)   (b)

Fig. 16.16   Molecular wires.

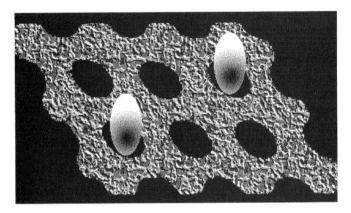

Fig. 16.17   Molecular matrices.

synthetic possibilities for designing functionality, the problem remains of how to connect up many examples of devices, and how to make the interface. This problem is partly solved by the concept of molecular matrices.

Molecular Matrices   There have been a number of attempts to produce a molecular matrix on a suitable substrate into which active molecules might be located. The principle is illustrated in Fig. 16.17 where we imagine the selective absorption of designer molecules into the host array.

The idea is to form a supported molecular film which has regular voids, preferably sensitized for specific absorption (e.g. host–guest chemistry). This approach is potentially useful for 2D cellular automaton architectures, for spectral hole burning assemblies and for spin-coupled systems. Techniques investigated range from Langmuir–Blodgett films to liquid crystal monolayers. There have been many problems with the stability of the molecular matrix films: distortion, buckling, fragmentation.

The use of patterned substrates to bind the matrix is one promising route for investigation. As an example let us quote the use of nanolithographic techniques coupled to protein engineering by Douglas *et al.*,[28] to generate a two-dimensional interacting spin system (2D magnet). The authors were able to construct a 1 nm thick metal (Ta/W) film with 15 nm holes on a 22 nm period 2D triangular grid. The process involved deposition of a two-dimensional crystalline protein monolayer onto 20 nm thick amorphous carbon films which provided a molecularly smooth surface. This was followed by metal evaporation and ion milling. Under ion milling the protein–metal heterostructure displayed differential metal removal and re-arrangement which varied on a protein molecular length scale. The technique used the crystalline proteinaceous cell wall surface layer (*S*-layer) from the thermophilic bacterium *Sulpholobus acidocalaldarius*, a sulphur oxidizing micro-organism. The *S*-layer comprises a 10 nm thick monolayer periodic array of a single glycoprotein with a molecular weight

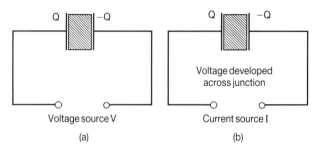

Fig. 16.18 Single electronics principle.

of 140–170 kdaltons. Purified S-layers exhibit a basis for three protein dimers arranged on a triangular 2D lattice with a 22 nm period. The 3D structure is porous where the protein occupies only 30% of the volume. It is this periodic structure which is used as a template for patterning an overlayer. By selective adsorption of individual biomolecules into the holes it was possible to produce an aligned array of horse-spleen ferritin molecules, which was used for 2D magnetic studies.

Single Electronic Systems Perhaps the biggest advance in inorganic nanoelectronics has been the fabrication of ultra-small capacitive tunnel junctions in metal-insulator and semiconductor quantum point contact structures, which display controllable *single-electron* events. (See references 29–34, reviewed in Geerligs and Harmans.[35]) The basic idea is illustrated by a single tunnel junction acting as a leaky capacitor (see Fig. 16.18).

If a bias voltage V is applied to the capacitor the electrodes will acquire a charge $Q = CV$ where C is the capacitance. The charging energy required is $W = Q^2/2C$ and this is supplied by the voltage source. After the transient the steady state will involve zero current flow provided the junction is non-tunnelling. However for a tunnel junction the applied voltage will induce a steady tunnelling current flow I between the electrodes. If G is the tunnel conductance the current-voltage characteristics will be of the form $I = GV$. Although the electrode charge Q varies quasi-continuously the tunnelling charge is discrete: equal to an integer number of electronic charges ne. The minimum transfer of one electron between the plates entails a change in charging energy of $W_e = e^2/2C$. If the applied voltage V is less than W_e/e it is impossible to supply the charging energy from the power source and the tunnelling will not proceed unless energy is acquired from thermal fluctuations (of the order of $k_B T$). It follows that if $e^2/2C > k_B T$ (the equality occurring at $T_c = e^2/2k_B C$) there is insufficient thermal energy for tunnelling. This effect is called the Coulomb blockade of the tunnel current. Below the transition temperature T_c the current-voltage characteristics are offset by $V_{offset} = e/2C$: no tunnel current flows until this voltage V_{offset} (or equivalently a charge of e) is applied across the electrodes, and then one electron at a time can cross the junction.

Under constant current bias (Fig. 16.18(b)) below the transition temperature the single junction displays periodic charge (voltage) oscillations as a function of time owing to the discrete tunnel events.

Most demonstrated systems have been relatively large, leading to transition temperatures less than 1 K. For capacitances in the attofarad regime the transition temperature can reach 300 K. Since simple parallel plate capacitances have C proportional to area, it is possible to make ultra-small capacitances by nanolithography. Coulomb blockade has been observed at room temperature for junctions composed of a molecule positioned on a conducting substrate and interfaced by a scanning tunnelling microscope probe. If we consider a hydrogen atom as the smallest practical capacitor, it becomes obvious that atomic and molecular scale capacitors will have extremely small capacitances with T_c reaching very high temperatures.

The behaviour of an array of coupled tunnel junctions under Coulomb blockade conditions is much more interesting. The need to satisfy energy conservation, when charging effects are important for the array, leads to a strong spatial and temporal correlation of electrons tunnelling through the array under an applied bias. The correlation is brought about by the extended polarization fields due to each electron and gives rise to soliton states.[29,35] Babiker et al.[36] have shown that there is a precise analogy with the way traffic flows along a road system. In a double junction (a three electrode system) it is possible to clock individual electrons along the array by arranging a suitable source-drain bias, and gating the flow by controlling the electrical potential (or charge) of the central electrode via a non-tunnelling capacitor connected to a gate voltage. Geerligs et al. used a 4-junction version of this idea to construct a single electron turnstile or shift register.[30] It is now possible to build large extended structures containing very many single electronic devices. Correlated single electron tunnelling is highly stable and fluctuations are heavily damped due to the soliton polarization field.[37]

These developments suggest a future single-electronics technology in which one bit of information might be assigned to a few electrons or even a single electron.[17,37] An equally exciting possibility is the development of sophisticated atomic scale instrumentation, such as electrometers and time-domain on-chip capacitance measuring devices, which will open up new experimental access to mesoscopic phenomena in both the inorganic and organic solid state.

The main approach to fabrication of single-electronic devices has utilized the hanging-resist technology developed for metal-insulator-metal tunnel junctions.[31-33] Although this method has allowed construction of large numbers of interconnected junctions in complex circuits, it is limited by the lithography to relatively large capacitances and consequently very low operating temperatures. Very recently a much finer resolution lithography—which has fewer processing stages and involves forming ultra-small (<5 nm radii and spacings) metal on semiconductor electrodes, Schottky islands and dot arrays—has been developed at Glasgow[34] (Figs 16.19, 16.20). These new structures involve ultra-small capacitances with equiva-

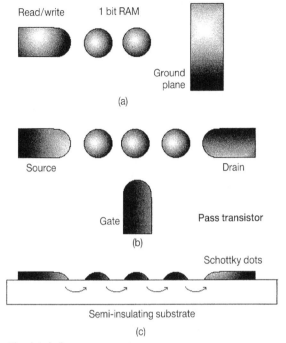

Read/write   1 bit RAM

Ground
plane

(a)

Source   Drain

Gate   Pass transistor

(b)

Schottky dots

Semi-insulating substrate

(c)

Fig. 16.19   Schottky dot devices.

lent Coulomb blockade temperatures of 60 K scaleable to much greater than room temperature. The dot structures may be arranged laterally to form capacitor chains which act as pass transistors (Fig. 16.19(b,c)), RAM cells (Fig. 16.19(a)) or more complex circuits. An essential feature of these systems is the requirement for the 'tunnelling tails' of the Schottky islands to overlap, thus permitting correlated electron tunnelling from dot to dot via the semiconductor.

A typical experimental silicon system based on the Schottky dot structure is shown in Fig. 16.20(a). The aluminium dot size is 40 nm and inter-dot spacing is 12 nm.

Most of the recent effort in the field of single electronics has been devoted to revealing and understanding the physics of phenomena such as correlated single electron tunnelling and single electron tunnelling oscillations. Progress towards a controllable single electronics technology and novel instrumentation will require a significant level of *engineering-oriented* studies of single electronic systems. One of the problems will be to overcome macroscopic quantum tunnelling, a many-body process which limits the lifetime of the Coulomb blockade, due to tunnelling of the electrode charge in charge configuration space.

A new lithographic technique developed at Glasgow has demonstrated sub-10 nm electrode structures with a present best of 3 nm tip radius and 3 nm tip separation for molybdenum electrodes (shown schematically in

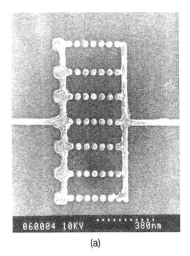

(a)

Fig. 16.20a   Experimental Schottky dot array: 40 nm dots on 12 nm spacing Al on silicon.

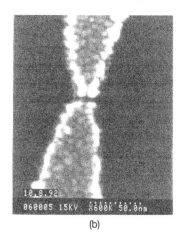

(b)

Fig. 16.20b   Experimental nanoelectrode showing 3 nm gap.

Fig. 16.20(b)). This technology should be capable of producing dot tunnel junctions with very high transition temperatures ($T_c = e^2/2Ck_B > 300$ K).

There are strong parallels to correlated electron tunnelling in molecular systems, for example macroscopic quantum tunnelling is known in both electron transfer and proton transfer chains; and solitonic correlated tunnelling is a feature of polymer conduction and redox reaction-centre chains. There are very exciting prospects for realizing single electronics technology within molecular systems.

The high resolution of the new lithographic techniques suggests that hybrid structures, such as the one sketched in Fig. 16.21, may be feasible,

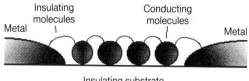

Fig. 16.21   Hypothetical molecular tunnelling capacitor array.

where the inorganic metal dots are replaced either by isolated conducting molecules or by a polymer chain comprising insulating and semiconducting organic components.

Cellular automata architectures and conservative logic schemes are ideally matched to single electronic systems.

Quantum Cellular Automata Systems   A novel form of cellular auto-maton architecture has been proposed recently by Lent *et al.*[38,39] It comprises a 2D array of quantum devices considered to be coupled quantum dots, each of which is a 2-electron cell. The charge density in each cell is very highly polarized along one of the two cell axes suggestive of a 2-state cellular automaton. The cell interactions proceed by polarization: each cell induces polarization in a neighbouring cell via the Coulomb interaction. Theoretical models of such a system suggest that they can perform useful computation. Conventional logic (AND, OR, inverters) can easily be implemented and interconnected. In this paradigm the computational process is edge-driven: input, output and power are delivered at the edge of the array. By setting the polarization state of the edge cells the cellular automaton is allowed to evolve dissipatively to a new ground state configuration. The output is sensed at the opposite edge of the array. Figure 16.22 illustrates the basic cell.

Such a system is a candidate for nanoelectronic quantum dot structures, single electronic systems, and within molecular electronics they would provide a practical target for magnetic or electronic cells in a molecular matrix.

Optoelectronic Systems   There exists an interesting and demonstrated technique—spectral hole burning (Fig. 16.23)—which has been used to

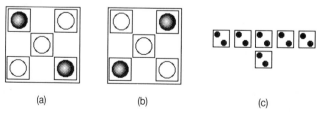

(a)   (b)   (c)

Fig. 16.22   (a, b) Bi-stable polarization states in a quantum cellular automaton cell; (c) cells are combined to form a system.

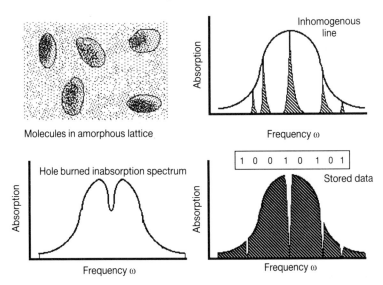

Fig. 16.23   Spectral hole burning principles.

investigate optical storage and retrieval of data in individual molecules.[40] It is based on the observation that identical molecules placed in an essentially amorphous host will be exposed to different local environments, which can lead to shifts in the absorption lines of individual molecules. The result is an inhomogeneously broadened absorption line resulting from groups of molecules. If the molecules can be bleached by exposure to light, it becomes feasible to tune into the frequency of an individual molecule and burn a hole in the inhomogeneous line at that frequency; the subsequent reading of the full broadened line by an absorption experiment will reveal the hole. For example, Haarer and co-workers have demonstrated the spectral hole burning of complex bit patterns into the dye molecules free base phthalocyanine in PMMA at 2 K. This technique is currently limited to low lattice temperatures and for a detailed account we refer to the recent review by Haarer.[40]

Spectral hole burning is in principle capable of achieving discriminating 10^3 to 10^4 holes within an individual absorption band, and could lead to ultra-thin optical storage densities which are some 10^4 times greater than current diffraction limited optical storage. This is achieved by using addressing in frequency space as well as direct space; even though we cannot locate the position of the hole in real space to better than an optical wavelength, its location in frequency space is determined with high precision.

Quasi-mechanical Systems: Molecular Networks   It has often been thought that we should approach molecular electronics as though it were a scaled down version of conventional electronics; but this loses the enormous potential offered by the properties of molecules. For example, one

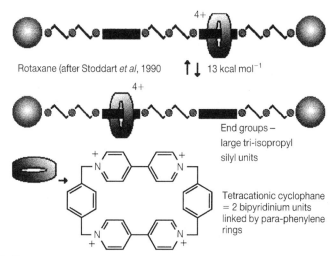

Fig. 16.24 Rotaxanes.

could imagine using the lock and key concept in chemical docking and chemical recognition to form the basis for a logic element. The 'data' however might be provided by the free drift of two or more molecules which would carry out a combinatorial process after a suitable encounter. This type of behaviour can be simulated on cellular automata models. The essence is a mechanical switch plus the unguided movement or partially guided movement of the 'data'. As an example of what is currently being achieved in this direction we refer to the recent exciting work by Stoddart and his co-workers,[18,19] who have used methods from supra-molecular chemistry[41] to synthesize and characterize a range of topologically complex 'designer molecules', such as the rotaxane or shuttle molecule sketched in Fig. 16.24.[19]

The shuttle molecule is particularly interesting because it comprises a moveable ring molecule located along the shaft of a long chain molecule with a number of energetically favoured sites. The ring shuttles back and forth thermally between the equilibrium sites. As a potential device it can be viewed as a memory, and indeed we have conjectured[42,43] that it could be used as the basis for an abacus or other classes of mechanical computational system, provided the shuttle ends could be pinned to a patterned substrate by the techniques discussed by Barker et al.[42] (see Fig. 16.25). Interestingly there has been a revival of interest in billiard ball computing particularly within the community of cellular automata theorists. It may well be that this is a good direction to pursue molecular electronics and there are interesting analogies with microtubular transport in nerve cells.

Artificial Bio-neural Networks   There is currently a strong interest in the study of *in vivo* neural networks using artificially grown neuritic

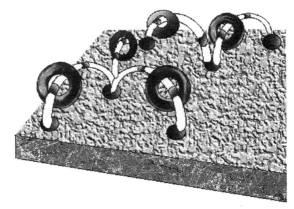

Fig. 16.25 Hypothetical quasi-mechanical molecular network.

structures with directed growth and directed synapsing. A variety of methods for the topographic control of cell behaviour have been developed,[42,44] which are based on using micron scaled grooved substrates to trap and subsequently guide the neuritic processes of different cell types (BHK, MDCK) and chick embryo cerebral neurones. By controlling the position of cells and synapses it becomes possible to make detailed cellular investigations with remote electrode sensing in previously determined buried sites in the substrate. This work provides one level at which biological information processing may be studied, and it provides a pointer to deeper methods for studying subcellular activity.

Fluid Models As pointed out in detail in Haarer,[45] nature prefers low density 'soft' fluid environments for the engineering of its informatic processes. Environments are produced which *screen* charges—ionic environments, hydrogen bonded environments, environments in which van der Waals' interactions dominate. The active sites are separated by bulky proteins. The importance of such a fluid phase in natural systems cannot be stressed too highly: geometries can be optimized and defects healed in very short periods of time. For these reasons it is pertinent to ask whether or not logic schemes can be devised for a fluid environment. In fact it is not so difficult to imagine and in a sense demonstrators already exist.[46,47]

It is already well-known that hard collision events—billiard ball collisions—may be used to represent reversible computational processes: a form of conservative logic. Figure 16.26 illustrates a collision between two spheres. We may regard the presence of a sphere on a trajectory as 'logic 1' and the absence of a sphere as 'logic 0'. The general state of a sphere (either present or not present) may be given the binary variable A, B, etc. Figure 16.26 illustrates the outcome of a collision in logical terms, and includes the special cases where 1 or no spheres collide. The basic process is identical to the interaction gate discussed in Section 16.7. Thus by arranging linear trajectories appropriately it is possible to simulate any

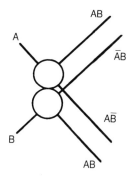

Fig. 16.26   Binary collision as a logic process.

logical operation by conservative logic processes. The trajectories plus the conservation of momentum are the analogue of confining a charge carrier to a wire. Collisional processes therefore form a basis for 'wire-less' logic, a prerequisite for fluid logic.

Collision processes may be easily studied within Margolus neighbour-hood cellular automata[26] whose state space flows are billiard-ball like flows on a discrete space-time. Figure 16.27 shows such a scheme for the steering of two bit flows by fixed reflector cells (Barker unpublished 1988).

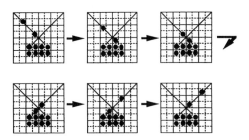

Fig. 16.27   Data steering by collision.

We may generalize the collision version of conservative logic to binary chemical reactions. The idea is illustrated in Fig. 16.28, where we sketch two different molecules colliding inelastically in a fluid environment to

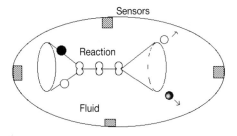

Fig. 16.28   Logical processes in a fluid.

produce a quasi-bound state. This propagates and then decays into two daughter products (not necessarily the same as the incident objects). The cones show the zones of incident trajectories which give the same result. Logic is performed by the mutual recognition of the incident states by the interacting molecules. The outcome is coded in terms of the reaction products. By building specific molecular sensors into the boundaries of such a system one might imagine the detection of the outcome of the reactions to be seen as a logic event. Actually reactions are stochastic events but this poses no problem since it is possible to deal with stochastic logic operations just as easily as deterministic ones.

It is evidently possible to use the spatial broadcasting of collision and reaction dynamic of appropriate molecular configurations to obtain versions of conservative logic; in these 'molecular wires' would not be required. Instead logic would be firmly based on molecular recognition. A demonstrator already exists in the form of spatial wave and pattern formation in chemical self-organizing mixtures, such as the Belousov–Zhabotinsky Reaction.[46,47] Such processes may be very closely modelled by cellular automata which are variants of the collisional models.

16.9   Practical Issues

Functionality and Emplacement   Recently the first practical steps have been taken towards performing electronic operations on individual molecules, whether they be alone or embedded in some large assembly (reviewed in Barker *et al.*[42]). These might eventually lead to molecular-scale electronic systems. Progress has been possible due to advances in synthetic organic chemistry,[1,2,41,48] the availability of nanometre scale inorganic technology, and the molecular scale instrumentation afforded by scanning tunnelling microscopy and atomic force microscopy.[49–51] Whereas previous excursions into molecular computing have been largely theoretical it is now clear that the way forward must involve less ambitious goals, which are forced on us by the practical problems of fabrication and verification. For example, various groups are now considering: (i) the synthesis of a molecule or molecular complex with a designed switching/ transport function; (ii) the confirmed emplacement of such a 'designer molecule' into a controlled environment; (iii) a demonstrated electrical or optical interface to the emplaced molecule; (iv) characterization of the emplaced molecule, including excite and probe measurements and/or transport measurement; (v) demonstration of a controlled molecular scale switch/gate and/or wire; (vi) demonstration of self-organization in either assembly or the dynamics of a molecular assembly which is of potential electronic or optoelectronic exploitation.

The Input-output Problem   Severe difficulties face the addressing of individual molecules. One possibility is to provide a chemical sequence of events which would amplify a single electron/photon process at a particular molecule to produce a substantial current elsewhere. This is the process used by biological photoreceptors, in which a conformational change is induced in the receptor molecule by a single photon. Surrounding molecules then chemically recognize the hitherto 'invisible' molecule and produce a chain reaction of events leading to a large photon signal elsewhere in the system.

More practically at present, the scanning tunnelling (STM) and atomic force microscopy probes have provided a direct tool for detecting, measuring and manipulating individual molecules—although there are severe limitations, problems of interpretation and lack of system wide interfacing (see Chapter 12). Spectacular pictures of molecules such as DNA have been produced, but in a haphazard way, since there has been no control of the positioning, orientation and emplacement of the molecules. Attempts to build a molecular switch using a STM as one contact have been made recently; and Likhaerev [unpublished] and Nejoh[52] have described a room-temperature Coulomb blockade structure, based on a STM probe interfaced to a molecular complex, but without more detailed model characterization it is hard to verify these results. The STM provides both spatial location and electrical, mechanical and thermal interfaces to positioned molecules. It has great potential.

Hybrid Routes   An interim hybrid approach is being developed by us, in which the problem of the electronic interface is being tackled first. Nanolithographic techniques borrowed from semiconductor electronics are routinely used (for quantum dot fabrication) to fabricate polymer dot arrays (typically polystyrene) on a solid substrate (20 nm diameter dots on 20 nm spaced arrays). We are investigating the use of such patterned substrates as sites for the growth of small numbers of organic molecules using the Merrifield–Shepherd techniques,[48] and for the attachment of presynthesized molecules.[42.] There are good possibilities for progressing to conducting polymer sites which are connected through the substrate to a solid state nanoelectronic circuit. The latter will provide a means for directly electrically addressing the polymer sites and hence the molecular array. The overall aim is to study the direct positioning of small numbers of significantly organic molecules (ultimately single molecules) on precisely defined sites and to study the electrical interface. This is currently achieved by using a scanning tunnelling microscope.

16.10   The Long-term View

Emergent Computation and Artificial Life   What will we do with all that complexity if molecular electronic systems can be constructed in the future? Let us briefly look at two possibilities. From the 'software' angle the prospect of an unlimited 'computational medium' offers exciting opportunities, not only for the floating architecture concept discussed earlier but for much more fundamental ideas to do with the evolution of computational and artificial intelligence processes. Emergent computation is the name given to a new field of computer science which deals with interacting 'programs', or rather processes/objects which may evolve to produce complex outcomes. Very high complexity levels are required.

Of more scope than emergent computation, artificial life[12] is an interdisciplinary field which considers models for artificial systems which can reproduce, evolve and carry out the functionality of living systems.

Bio-molecular Systems   What will molecular systems be made from? Perhaps the biggest challenge to molecular electronics is the relation with and interface to biological systems. From a purely fabrication point of view it has long been suggested that methods of genetic and cellular engineering might be useful in constructing molecular information processing systems, particularly by biological self-organizing processes. At present this is strictly science fiction. Practical routes forward are likely to benefit more from the detailed scientific study of bio-information processing of natural systems and from progress in realistic bioelectronic interfacing. But the construction of rudimentary model systems is not inconceivable: model ion gates in artifical membranes, artificial microtubule structures, molecular and artificial neural networks and so on are worth developing. From the point of view of biosensors and prosthetic devices the control of biological material for information engineering is very desirable.

The Silicon Factor   Is there a market for molecular electronic systems? The future of silicon technology is perhaps the biggest single factor which will determine the pace of interest in practical molecular electronic systems. Until recently silicon CMOS technology was expected to run into serious problems at the 0.2 μm design rule level (fully-scaled smallest feature size) for room temperature operation, and 0.1 μm scales if liquid nitrogen operating temperatures were to be deployed. If chip size had stayed constant this would have implied a complexity density of about 100 million transistors per chip. Because chip-to-chip communication is bandwidth limited and slow it pays to put as much complexity into single chips as possible. This is being achieved by the tremendous strides being made in the lithographic pattern transfer technology. It is now commercially feasible to consider 4 Gbit DRAM chips with over 4 billion transitions per chip. Similarly there are already prototype single chip processors with GHz clock speeds. The development of excimer (blue) lasers allowing photolithography down to shorter wavelengths foreshadows a scaleable progressive development of silicon technology well into the next century. Moreover the costs of electron beam lithography, X-ray lithography and photolithography converge dramatically as these truly complex systems are approached.

It has long been recognized that by changing some of the architecture of the basic transistor it would be possible to break through the 0.1 μm barrier for silicon. Very recently IBM and other groups have begun modelling silicon devices which should function on 30 nm scales. It has been suggested that devices might be fabricated down to the lithography limits which until recently were thought to be around the 10 nm scale. In fact, by restricting the number of lithography levels and by restricting aspect ratios and controlling the metallurgy closely it is possible to use electron beam lithography to fabricate quite complex structures with feature sizes of 3 nm set by the width of the electron beam. It is not obvious however that a commercially viable technology could exploit this limiting scale.

The inescapable conclusion is that silicon technology has the potential to achieve extremely high bit-density levels for two-dimensional layouts: typically around 10^{10} 'transistors' cm^{-2} for 30 nm devices. There are also indications that inorganic time-domain spectral hole burning crystal memory systems might be achieved with 10^{12} cm^{-2} bit densities, although the practicality is still open to question.

16.11 Conclusions

The first practical steps towards molecular electronic systems require modest targets. The most promising route is to derive logic from molecular function: the designer molecule approach. Because we are aiming at highly complex systems high level logic structures are preferred targets, as opposed to the bottom up 'wires and transistors to gates' approach. One bit to one 'carrier' suggests unitary or binary conservative logic—the need to suppress fluctuations will be a major source of problems. Cellular automata architectures satisfy many of the practical constraints. The problems of clocking and timing of molecular systems suggest asynchronous schemes, but there is a possibility of global clocking via illumination fields. A promising interim fabrication route would use separately synthesized functional molecules attached to prepared sites on a solid substrate. There are possibilities for exciting demonstrators: the molecular abacus and conservative logic schemes. Characterization of molecular circuits is limited now to atomic probe techniques. In parallel to developing practical molecular logic the same techniques could prove useful for studies of intra-cellular phenomena such as microtubule transport. Passive single-electron tunnel device arrays (Coulomb blockade devices) using molecules also look promising—there are interesting analogues of solitons in polymers. Indirect methods for local addressing of molecules with light as an extension of spectral hole burning require much more investigation, and there are possibilities here for optical parallel processing.

For molecular electronics to be competitive—in the future—with silicon it will clearly have to offer something other than speed and bit density. Three-dimensional memory/logic is often mooted as a potential advantage for molecular systems and that certainly could lead to higher bit densities than for 2D. But it is much more likely that it is the *functional* properties of molecular materials that will prove more significant. This is already obvious with liquid crystal and electroluminescent displays. One might imagine a future optoelectronic molecular technology which integrates electronic logic/memory functionality with display capability in a foldable plastic packaging. If we consider the use of molecular materials as sensors and bio-interfaces, there is a real need for complex information processing *in situ* for applications to implantable biosensors, subcellular instrumentation, prosthetic devices (e.g. a multiplexed spinal bypass), local control of artificial muscle in advanced robotics. In such instances inorganic circuit technology causes serious problems of poisoning and bio-incompatibility.

Perhaps the biggest advantage of molecular electronics has been the opportunity to examine how and why we carry out information processing in the currently favoured paradigms of binary logic based on charge packet transport and switching. The fact that information may be potentially represented in a rich variety of physical forms in molecular media, has led us to question whether the 'wires and transistors' picture of a logic system is the best and whether or not other schemes might occur. We have shown that a rich variety of possibilities exist. The science of molecular electronic systems is in its infancy, the technology is not yet born; and as for the future: it is likely to be most rewarding if the special viewpoint afforded from chemistry, biochemistry, molecular and cell-biology is fully exploited.

References

1. A. Aviram, Ed., *Molecular Electronics—Science and Technology*, (Engineering Foundation, 1989).
2. F. Carter, Ed., *Molecular Electronic Devices II*, (Marcel Dekker, New York, 1987).
3. J. R. Barker, *Prospects for Molecular Electronics*, Hybrid Circuits, **14**, 19–24 (1987).
4. J. R. Barker, *Complex Networks in Molecular Electronics and Semiconductor Systems*, Molecular Electronic Devices II, (Marcel Dekker, Inc., New York, 1987), pp. 639–74.
5. J. R. Barker, *Quantum Ballistic Transport and Tunnelling in Molecular Scale Devices*, Molecular Electronic Devices II, (Marcel Dekker, Inc., New York, 1987), pp. 675–92.
6. J. R. Barker, *System Models and Fabrication Schemes for Molecular Electronics*, Alta Frequenza, **LVIII**, 249 (1989).
7. J. R. Barker, Molecular Electronic Systems: Models and Fabrication Schemes, in *Molecular Electronics—Science and Technology*, Ed. A. Aviram, (Engineering Foundation, 1989), pp. 213–21.
8. J. R. Barker, Novel Logic and Architectures for Molecular Computing, in *Parallel Processing in Neural Systems and Computers*, Eds R. Eckmiller, G. Hartmann and G. Hauske (North-Holland, 1990), p. 519.
9. C. A. Mead and L. Conway, *Introduction to VLSI Systems*, (Addison Wesley, 1979).
10. S. Hameroff, S. Rasmussen and B. Mansson, Molecular Automata in Microtubules: Basic Computational Logic of the Living State? in *Artificial Life*, (Addison-Wesley, 1989), pp. 521–54.
11. P. Ptashne, How Gene Activators Work, *Scientific American*, **260**, 24–31 (1989).
12. C. G. Langton, Ed., *Artificial Life*, (Addison-Wesley, 1989).
13. A. Aviram and M. A. Ratner, *Chem. Phys. Lett.*, **29**, 25 (1974).
14. A. Aviram, P. E. Seiden and M. A. Ratner, in *Molecular Electronic Devices*, Ed. F. Carter, (Dekker, New York, 1982), p. 5.

15. D. Higlin and R. Sixl, *Chem. Phys.*, **77**, 391 (1983).
16. J. S. Lindsey, Synthetic Approaches to Molecular Devices, in *Molecular Electronics—Science and Technology*, Ed. A. Aviram, (Engineering Foundation, 1989), pp. 221–28.
17. J. R. Barker, in *Physics of Granular Electronic Systems*, Ed. D. K. Ferry, J. R. Barker and C. Jacoboni, (Plenum, 1991), p. 327.
18. P. Ashton, T. T. Goodnow, A. E. Kaifer *et al.*, A [2] Catenane Made to Order, *Angew. Chem. Int. Ed. Engl.*, **28**, 1396–99 (1989).
19. P. Anelli, N. Spencer and J. F. Stoddart, submitted to *J. Am. Chem. Soc.* (1990).
20. D. N. Beratan, J. N. Onuchic and J. J. Hopfield, Design of a Molecular Memory Device Based on Electron Transfer Reactions, in *Molecular Electronics—Science and Technology*, Ed. A. Aviram, (Engineering Foundation, 1989), p. 331.
21. T. Toffoli, Reversible Computing, in *Automata, Languages and Programming*, (Springer-Verlag, 1980), pp. 632–44.
22. L. J. Priese, *J. Comp. Syst. Sci.*, **17**, 237 (1978).
23. R. M. Keller, *IEEE Trans. Computers*, **C-23**, 21 (1974).
24. L. J. Priese, *J. Cybernetics*, **6**, 101 (1976).
25. R. E. Miller, *Switching Theory*, (Wiley, New York, 1965), Vols 1, 2.
26. T. Toffoli and N. Margolus, *Cellular Automata Machines*, (MIT Press, Cambridge, 1987).
27. S. Roth and H. Bleier, in *Molecular Electronics—Science and Technology*, Ed. A. Aviram, (Engineering Foundation, 1989), p. 317.
28. K. Douglas, N. A. Clark and K. J. Rothschild, *Appl. Phys. Lett.*, **48**, 676 (1986); *Appl. Phys. Lett.*, **56**, 692 (1990).
29. K. Likhaerev, B. Bakhvalov, G. S. Kazachan and S. I. Serdyukova, *IEEE Trans. Magnet.*, **25**, 1436 (1989).
30. D. V. Averin and K. K. Likhaerev, *J. Low Temp. Phys.*, **62**, 345 (1986).
31. T. A. Fulton and G. J. Dolan, *Phys. Rev. Letters*, **89**, 109 (1987).
32. L. S. Kuzmin, P. Delsing, T. Claeson and K. K. Likhaerev, *Phys. Rev. Lett.*, **60**, 309 (1989); **62**, 2539 (1989).
33. L. J. Geeligs, V. F. Anderagg, P. A. M. Holweg *et al.*, *Phys. Rev. Lett.*, **64**, 2691 (1990).
34. J. R. Barker, J. M. R. Weaver, S. Babiker and S. Roy, *Theory, Modelling and Construction of Single-electronic Systems*, Proceedings Second International Symposium on New Phenomena in Mesoscopic Structures, Hawaii (1992).
35. *The Physics of Few-electron Nanostructures*, Eds L. J. Geerligs C. J. P. M. Harmans and L. P. Kouwenhoven, (North Holland, 1993).
36. S. Babiker, J. R. Barker and A. Asenov, *Proc. International Workshop on Computational Electronics*, Ed. C. Snowden, (University of Leeds Press, 1993), p. 260.
37. J. R. Barker, S. Roy and S. Babiker, in *Science and Technology of Mesoscopic Structures*, Eds S. Namba, C. Hamaguchi and T. Ando, (Springer-Verlag, London, Tokyo, New York, 1992), Chapter 22, p. 213.
38. C. S. Lent, P. Douglas and W. Porod, *Appl. Phys. Lett.*, **62**, 714 (1993).
39. C. S. Lent, P. Douglas, W. Porod and G. H. Bernstein, *Nanotechnology*, **4**, 49 (1993).
40. D. Haarer, in *Parallel Processing in Neural Systems and Computers*, Eds R. Eckmiller, G. Hartmann and G. Hauske, (North-Holland, 1990), p. 519.
41. J-M. Lehn, Nobel Lecture, *Angew. Chem. Int. Ed. Engl.*, **27**, 89 (1988).
42. J. R. Barker, P. Connolly and G. Moores, in *Physics of Granular Electronic*

Systems, Eds D. K. Ferry, J. R. Barker and C. Jacoboni, (Plenum, 1991), pp. 425–40.

43. J. R. Barker, *Chemistry in Britain*, **27**, 728–31 (1991).
44. P. Clark, P. Connolly, A. S. G. Curtis, J. A. T. Dow and C. D. W. Wilkinson, Topographical Control of Cell Behaviour: II Multiple Grooved Substrata, *Development*, **108**, 635–44 (1990).
45. D. Haarer, *Adv. Mater.*, **101**, 362 (1989).
46. J. J. Tyson, *The Belousov–Zhabotinsky Reaction*, Lecture notes in biomath, 10 (Springer, Berlin, Heidelberg, New York, 1976).
47. H. Haken, *Advanced Synergetics*, (Springer, Berlin, Heidelberg, New York, 1983), Chapter 1.
48. C. Birr, *Aspects of the Merrifield Peptide Synthesis*, (Springer-Verlag, Berlin, Heidelberg, New York, 1978).
49. C. F. Quate and C. A. Lang, Molecular Imaging with the Tunnelling and Force Microscopes, in *Molecular Electronics—Science and Technology*, Ed. A. Aviram, (Engineering Foundation, 1989), pp. 79–86.
50. M. E. Welland, *Int. J. Biol. Macromol.*, **11**, 29–32 (1989).
51. P. Connolly, J. M. Cooper, G. Moores, J. Shen and G. Thompson, *Nanotechnology*, **2**, 160 (1991).
52. H. Nejoh, *Nature*, **353**, 640 (1991).

Indexes

Subject

α helix 69, 244, 283, 289
α optical polarizability coefficient 98, 99
ab initio calculations, NDO methods 32
Absorption *see* Optical absorption
Acceptor, electron acceptor 101, 171, 175
Acoustic immunosensors 311
Acoustic impedance 47, 48
Actinometry 126, 135
Activation energy, electrical conductivity 171
Adiabatic approximation 31
Alkyl chain 223, 224, 226, 238
Alpha helix 69, 244, 283, 289
Alternate layer trough 228
Amperometric glucose sensors 303–4
Amperometric immunosensors 308–9
    see also Biosensors
Analytical chemistry, electrochemical
    biosensors 302–10
Anisotropic liquid crystals 201
Antibody-binding 308
Aromaticity 22
Aromatics, *para*-substituted 101, 107, 110
Artificial bioneural networks, molecular
    representation 367–8
Artificial brain 349
Artificial life 371–2
Aspergillus ochraceus 309
Atomic force microscopy (AFM) 17, 253–4,
    262
Atomic orbitals 75, 146
    linear combination (LCAO) 31, 146, 148

β carotene 143

β helix 293
β optical hyperpolarizability coefficient 32–4,
    98, 99
β sheet 244, 283, 289
Bacteriorhodopsin 316–32
    bacteriorhodopsin-based materials, properties
        320
    light-adapted photocycle 318
    linear optoelectronic devices 316–31
        higher-order polarizabilities 333–4
        holographic optical recording 321–32
    nonlinear optoelectronic devices 332–41
        optical volumetric memories 337–9
        read operation 340–1
        second-order polarizability 334–6
        third-order polarizability 336–7
        two-photon properties 332–4
Band gap 107, 144, 171
Band theory 147, 170–1
    valance band 147, 171
Bathochromic shift 107, 131
BCS theory, superconductivity 172
Bessel function 54–5
Binary state replicator 358
Bioaffinity sensors 298–302
Biocatalytic sensors 299
Bioelectronic interfacing 372
Biological membranes 279–94
    channel systems 288–91
    electron transport system 290–1
    material transport 285–91
    nonchannel systems 285–8
    structure 279–84

Biological membranes (*cont.*)
   carbohydrate 283–4
      lipid bilayers 279–24
      proteins 282–3
   transmembrane signalling 284–5
Biomolecular optoelectronics 315–44
   linear optoelectronic devices 316–31
   nonlinear optoelectronic devices 332–41
   see also Bacteriorhodopsin
Biosensors 295–314
   amperometric 300–1
   amperometric glucose sensors 303–4
   amperometric immunosensors 308–9
   bioaffinity 298–302
   biocatalytic 299
   defined 295
   design principles of biosensors 296–302
   electrochemical 302–10
   immunosensors 300
   molecular information transduction in
      biological systems 296
   optical 310–12
   voltammetric 303
Bipolarons 155–9
Birefringence 201, 210
Bloch wall, magnetism 73
Blood cells 280
Blood typing 308
Boltzmann statistics 110
Bonds
   bond alternation 103, 104
   bond alternation parameter 103, 105
   covalent bond 142–3
   dangling bond 249
   σ bond 142–3
   valence bond 142–4
Born–Oppenheimer approximation 31
Bragg diffraction 232
Brain, human, bit-count 349

Carbohydrate 283–5
Cauchy integral 324
Cell(s)
   blood cells 280, 308
   cytosol 22, 280
   microtubule networks 349–50
   transport 349–50
Cellular automata 357–9
   quantum systems 365
Chain alignment 159–60
Chain length, γ coefficient 104
Channel systems, biological membrane 288–91
Charge carrier mobility 41
Charge packet model 351–2
Charge transport 7
Charge-coupled device (CCD) camera 246, 249,
   252
Charge-transfer complex 12, 85, 134–8, 170

Charge-transfer excitons 39
Chemical vapour deposition 243–5
Chemiosmosis 316
Chiral molecules 192–5
Chromophores 19, 41, 316
   see also Bacteriorhodopsin, *Materials index*
Cis–trans isomerization 119–26, 206–10
   photochrome 121
Clark cell 303, 308
Clausius–Mossotti equation 39
Collision events, logic processes 368–70
Conductive charge-transfer complexes 168–84
   basic concepts 170–1
   fullerenes 181–3
   metal-dithiolate systems 178–9
   metallomacrocycles 179–81
   TCNQ and TTF systems 171–8
Conductive polymers 3, 142–67
   basic concepts 142–6
   chain alignment 159–60
   device applications 163–5
   doping 153–5
   photoinduced absorption 160–2
   polarons, bipolarons and charged solitons
      155–9
   polyacetylene 146–51
   polyaniline 162–3
   solitons 151–3
Conjugated chain 101
Cooper pair, superconductivity 173
Coulomb blockade 362–3
Coulomb's law 37, 150
Covalent bond 142–3
Crystal classes 49
Crystal polarization, transition dipole moment
   43, 44, 101
Crystal symmetry 48
   inversion centre 48
Crystal systems 48–50
   slipped-stack form 254
Curie law 74
Curie temperature 57, 67, 74
   ferromagnetic 47, 73, 90
Curie–Weiss law 74
Cyclization 126–34
Cytosol, cell(s) 22, 280

d-orbital 75
Dangling bond 249
Data storage 114–17
Davydov components 44
Delocalization, electron 103, 105, 143, 147, 170
Delocalization parameter 105, 179
Device applications 163–5
Dielectric anisotropy 209
Dielectric displacement 50, 56
Differential thermal analysis 196
Dimerization 119

Dipole, libration frequency 66
Dipole moment 42, 44, 101, 257
  see also Polarization
Directional coupler devices 97
Director, liquid crystals 191
Discotic molecules, liquid crystals 185, 187
Dissociation 118–19
  photochromic materials 118
Donor, electron donor 101, 138, 170
Doping 153–5
  p-type, *n*-type 155

Einstein coefficient 43
Einstein convention 42, 43
Elastic stiffness constant 50, 53
Electric field 37–42, 53, 56
  local 38, 41
  macroscopic 38
  see also Field effect transistors
Electrical conductivity, activation energy 171
Electro-optic coefficient *r* 96
Electrochemical biosensors 302–10
  amperometric glucose sensors 303–4
  amperometric immunosensors 308–9
  electrochemical sensing principle 302–3
  enzyme FETs 307
  molecular interfacing for electron transfer of
    enzymes 304–6
  potentiometric enzyme sensors 306–7
  potentiometric immunosensors 307–8
Electrochromism 113
Electrocrystallization 170
Electroluminescence 312
Electromechanical coupling coefficient 53
Electron
  delocalization 103, 105, 143, 147, 170
  delocalization parameter 105, 179
  free electron models 103–4
  transition moments 99–100, 105
  ungerade, singlet state 336
  see also π electron; Single electronic sytems
Electron acceptor 101, 171, 175
Electron diffraction 232
  reflection high energy (RHEED) 246–52
Electron donor 101, 138, 170
Electron ionization energy 170
Electron mediator 305
Electron mobility 41, 170
Electron promotor 304
Electron spin resonance (ESR) 135, 158
Electron transfer, enzymes, molecular
    interfacing 304–6
Electron transport system, biological
    membranes 290–1
Electron tunnelling 304
  single electronic systems 363–5
Electron valence 142–4, 147
Electron wave-function 40, 146

Electron-beam lithography 235
Electron–photon coupling 150
Emergent computation 371–2
Endocrine system 296
Energy gap (band gap) 107, 144
Enthalpy 197
Enzymes
  electron transfer 304–6
  field effect transistors (FETs) 304, 307
  immunoassay 308, 309
  labels 301
  optoelectronics 309
  redox reactions 8, 298–9
Epitaxy
  phase-locked epitaxy 246, 252
  van der Waal's 249
  see also Organic molecular beam epitaxy
Excitons 39, 40, 44, 116, 267

Femtosecond spectroscopy
  optical absorption 255–7
  see also Optical absorption
Fermi energy 105, 149, 263
Ferroelectric liquid crystals display 215–17
Ferroelectricity 47–71, 215–17
  hysteresis 63
Ferromagnetics 12, 72–90
  hysteresis 73, 78
Ferromagnets 85–9
  see also Magnetism
Field effect transistors (FETs) 303, 304, 308
  enzymes 304, 307
  as memory cell 346
Figures-of-merit, pyroelectricity 48, 57–8, 61,
    109
Film growth - RHEED 249–53
Finite-state machine 356–7
Floating architectures 359
Fluid models, molecular representation 368–70
Förster excitation energy 116, 134
Fourier transform 38, 331
Fourier transform holographic 330–1
Frank–Condon principle 155, 156
Frank–Oseen continuum theory 201
Fredkin gate 355–6
Free electron models 103–4
Freedericksz cell 202–3
Fullerenes, conductive charge-transfer
    complexes 181–3

γ
  optical hyperpolarizability equation 334
  second nonlinear polarizability 103–9
γ coefficient, chain length 104
Gas sensor 165, 240
Glass transition temperature 110, 245
Glucose sensor 8, 303–4
Glycosylation 284

Gold *see* Noble metals
Gradient operator 37
Granular electronic systems 354
Growth factor receptor 285
Gruneissen constant 66
Guest–host complex 122, 360

H-aggregate 131, 133
Halobacterium halobium 120, 287, 315, 316
  see also Bacteriorhodopsin, *Materials index*
Hamiltonian operator 148, 263
Hanging-resist technology 362
Harmonic generation 95–6
  second harmonic 96–7, 238, 334
  third harmonic 97, 336
Hemiquinone proton tunnelling 353–4
High definition television (HDTV) 217–18
High electron mobility transistors (HEMT) 319
Highest occupied molecular orbital (HOMO)
  149, 156, 265
Holographic associative memories 329–32
Holographic Fourier transform 330
Holographic optical recording 321–32
Holography 11
HOMO (highest occupied molecular orbital)
  149, 156, 265
Host–guest chemistry 122, 360
Hückel theory 32, 86
Hünd's rule, magnetism 74
Hydrogen tautomerism 117–18
Hydrophilic group 222, 224, 279–80
Hydrophobic group 222, 224, 226, 231, 238,
  279–80
Hydrophone imaging 47, 53
Hydrostatic stress 53
Hyperpolarizability *see* Polarizability
Hypsochromic shift 131
Hysteresis
  ferroelectricity 63
  ferromagnetism 73, 78

Image reduction and collimation optics (IRCO)
  331
Imine site 162
Immunoassay, enzyme 308, 309
Information, optical writing 125
Information molecules 114, 297–9
Infrared spectroscopy, photochromic switching
  132, 135
Inorganic electronic systems 346–8
Input–output problem 370–1
Insulin receptor 287
Inversion centre, crystal symmetry 48
Ion channels 22–4, 288–90
Ion-selective field effect transistors (ISFETs)
  303
Ionization energy, electron 170
Ionophore 293

J-aggregate 131, 133

Knudsen cells 245–6, 249
Kramers–Kronig transform 323–5
Krypton-ion laser 322

Langmuir films 120, 223–6
Langmuir trough 227–9
Langmuir–Blodgett deposition 115, 224–8
Langmuir–Blodgett films 36, 41, 115, 220–42,
  282
  basic concepts 222–8
  characterization of LB assemblies 228–34
  fundamental and applied research 234–41
Laser
  excimer 372
  krypton-ion 322
  Nd-YAG 238
Layered materials 268
LB *see* Langmuir–Blodgett
Lennard–Jones potential 34
Ligand-gated channels 283, 285
Ligand–receptor interaction 298
Light-emitting diode (LED) 165
Light-to-energy transducer *see*
  Bacteriorhodopsin; Photosynthetic reaction
  centre
Linear combination of atomic orbitals (LCAO)
  31, 146, 148
Linear optoelectronic devices 316
Lipid bilayers 279–82
Liquid crystal polymers 198–200
Liquid crystals 185–219
  active matrix display 214–15
  alignment 124, 236
  anisotropic properties 201, 201–2
  applications 200–17
  basic concepts 186–8
  calamatic 185, 187
  director 191
  discotic 185, 187
  Frank–Oseen continuum theory 201
  identification of phases 196–8
  mesogen 188
  mesophases
    cholesteric phase (chiral nematic) 192–5
    nematic phase 4–5, 191–2
    smectic phase 4–5, 191
  nematogen 191
  orientational elasticity 201
  phase sequences 195–6
  pitch of molecule 192–3
  smectogen 191
Liquid crystals display
  active matrix 214–15
  double layer supertwist 213
  ferroelectric 215–17

Freedericksz cell 202–3
nematic curvilinear aligned phase (NCAP) 200
supertwisted nematic 212–14
twisted nematic 12, 196, 203–11
twisted smectic 195
Liquid phase epitaxy 243–5
Logic
cellular automata 357–9
conservative 355–6
floating architectures 359
high level 356–7
NAND gate 346–7
neural and neural networks 354–5
restoring 354
Logic processes, collision events 368–70
Lorenz correction 98
Lorenz–Lorenz equation 39, 98
Lowest occupied molecular orbital (LUMO) 149, 156, 265
Luminescence 236–7

Magnetism
antiferromagnetism 73, 74, 78, 83–4, 178
Bloch wall 73
coercivity 75, 90
diamagnetism 73, 181
ferrimagnetism 73, 75, 84
ferromagnetism 12, 72–90
ferromagnets 85–9
Hünd's rule 74, 75
interstack interaction 178–9, 183
magnetic domain 73
magnetic moment 79
paramagnetism 73
remanent 75
susceptibility 74, 177
see also Curie law and temperature;
Molecular magnets; Néel temperature
Maxwell displacement current 125
Meissner effect, superconductivity 174
Mesophases, liquid crystals 192–5
Metal oxide field effect transistors (MOFETs) 303
Metal-dithiolate systems, conductive charge-transfer complexes 178–9
Metal–insulator–metal structure (MIM) 258
Metal–insulator–semiconductor (MIS) structure 163
Metallic conductivity 146, 154, 160, 170, 173, 178–80, 183
Metallomacrocycles, conductive charge-transfer complexes 179–81
Microchannel plate spatial light modulators (MSLM) 331
Microtubule networks, cell 349–50
MIM structure 258
Mobility, charge carrier mobility 41

Molecular adsorbates under ultra-high vacuum, scanning tunnelling microscopy 269
Molecular beam epitaxy see Organic molecular beam epitaxy
Molecular conductors, classes 168
Molecular dynamics 35–7
Molecular electronic devices, categories 1–2, 243–5
Molecular electronic logic and architectures 345–74
complexity and systems specification 351
guidelines from biological and chemical systems 349–51
inorganic electronic systems 346–8
logic possibilities 354–9
long-term view 371
molecular electronic systems 348–9
molecular representations 359–70
physical representations of data and logic 351–4
practical issues 370
Molecular electronic switching 125
Molecular information, and signal transduction, biosensors 296–7
Molecular interfacing, electron transfer of enzymes 304–6
Molecular magnets 72–90
basic concepts 73
ferromagnets 85–9
transition metal complexes 73–84
Molecular materials 2–12
Molecular matrices 360
Molecular memory and logic 359–60
Molecular networks 366–7
Molecular nonlinear optics 98–9
Molecular orbital (MO) 31–3
highest occupied (HOMO) 149, 156, 265
Hückel theory (HMO) 32, 86
lowest occupied (LUMO) 149, 156
p-orbital 143
see also π electron, σ bond
Molecular rectifier 29
Molecular scale electronics 13–25
Molecular switches 138, 353–4
Molecular wires and switches 359
Molecules at solid-fluid intervaces, scanning tunnelling microscopy 269–74
Monte-Carlo technique 35–6
Multi-frequency optical memory (MFOM) 115

p-NA, NLO molecule 100–1
Nanoelectronics, single electronic systems 361–5
Nanolithographics 17, 371
protein engineering 360
Near field optical microscopy 17
Néel temperature 75
Neglect of differential overlap (NDO) 32

Nematic curvilinear aligned phase (NCAP) 200
Nernst equation 303
Neural cell, transport 349–50
Neural logic and neural networks 354–5
Neural networks, artificial 367–8
Neural synapse 296
Neural transmitter 296
m-Nitrophenol 36
Noble metals, STM 266–8
Non-polymeric materials 108–9
Nonchannel systems, biological membrane 285–8
Nonlinear optics, organics for 91–111
    linear and first nonlinear polarizabilities 99–102
        enhanced β 101–2
        p-NA, model NLO 100–1
    macroscopic assemblies 109–10
    molecular nonlinear optics 98
    second nonlinear polarizability, γ 103–9
        free electron models 103–4
        non-polymeric materials 108–9
        quantum chemical models 105–8
        semiconductor models 104–5
    volumetric memories, bacteriorhodopsin 337–9
Nonlinear optoelectronic devices 332

OMBE *see* Organic molecular beam epitaxy
One-dimensional conductor 105, 151
Onsager correction 98
Optical absorption 230–2
    femtosecond spectroscopy 255–7
    nonlinear absorption 108
    one-photon absorption 320
    photoinduced 160–2
    $\pi\pi^*$ absorption 121–2
    Q-band 254, 256
    Soret band 254, 256
    two-photon absorption 106, 138
Optical anisotropy 201, 210
Optical band gap 107
Optical biosensors 310–12
Optical energy gap 105
Optical fibre electrode 310–12
Optical fibre pH sensor 310
Optical gate 134
Optical hyperpolarizability coefficient, β 32–4, 98, 99
Optical immunosensors 310–12
Optical microscopy
    liquid crystal 197
    near field (NFOM) 17
Optical polarizability coefficient, α 98, 99
Optical second harmonic generation 238
Optical storage media, bit densities 349
Optical writing, information 125

Optoelectronic systems, biomolecular 315–44
    linear devices 316–31
        bacteriorhodopsin, function and photochemistry 316–32
        holographic recording 321–32
    nonlinear devices 332–40
    spectral hole burning 365–6
    see also Bacteriorhodopsin
Organic adsorbates, UHV-STM 269
Organic conductors, STM 265–6
Organic molecular beam epitaxy 243–60
    basic concepts 246–9
    characterization of assembled molecules - STM 253–4
    film growth - RHEED 249–53
    prospects for OMBE 254–9
    UHV-OMBE 246–9
Organics
    nonlinear optics 91–111
        basic concepts 93–8
        linear and first nonlinear polarizabilities, α and β 99–103
        macroscopic assemblies 109
        second nonlinear polarizability, τ 103–9
Oxygen
    partial pressure measurement 310
    sensor 303, 310
Oxygen electrode 301

$\pi\pi^*$ absorption 121–2
π-bridge 135
π-electrons 143
p-NA, a model NLO molecule 100–1
p-orbital 74, 143
Pariser–Parr–Pople (PPP) theory 32, 39
Pauli exclusion principle 147, 157
Peierls distortion 103, 174
Peierls transition 149, 150–1, 175
Penicillium viridicatum 309
pH electrode 306
pH sensor, optical fibre 310
Phase-locked epitaxy 246, 252
Phases, liquid crystals and devices 196–8
Phonon 115
Photobleaching 126, 128, 136
Photochemical hole burning 115–16, 365–6
Photochromic materials, dissociation 118
Photochromism 112–40
    basic concepts 112–14
    charge-transfer 134–8
    cis–trans isomerization 119–26
    classes of photochromic materials 117–39
        hydrogen tautomerism 117–18
    cyclization 126–34
    data storage and other applications 114–17
    dimerization 119
    dissociation 118–19
    fatigue 13, 126

Photochromism – *cont.*
  photoinduced optical anisotropy 123
  quantum yield 131–2
  reaction rate constant 131
  switching, infrared spectroscopy 132, 135
Photoinduced absorption 160–2
Photoinduced optical anisotropy 123
Photosynthetic reaction centre 121, 283, 290–2
Physical vapour deposition 243–5
  sputtering 244, 245
  vacuum evaporation 244
Piezoelectric and pyroelectric materials 47–70
  basic concepts 48–60
  polymers 60–70
Piezoelectric effect 50–6
Piezoelectricity 47–71
Pipeline structure, data path 347, 351
Pitch of molecule, liquid crystal 192–3
Pixel 211
Planar coupling coefficient 53–4
Poisson's ratio 54–5
Polarizability 32
  hyperpolarizability 32, 33, 334
  second nonlinear polarizability γ 103–9
  see also Nonlinear optics, organics for
Polarizability coefficient
  α 98, 99
  β 32–4, 98, 99
Polarization 64
  higher-order 333–7
  linear interaction 93
  optical spectra 44
  remanent 98
  see also Dipole moment
Polarons and bipolarons 155–9
Poled polymers 109
Polyacetylene 146–51
Polyaniline 162–3
Polymerization, plasma 245
Polymers
  liquid crystals 200
  poled 110
  see also Materials index
Poly(vinylidene fluoride)
  copolymers 67–8
  other polymers 68–9
  structural forms 60–7
Potentiometric enzyme sensors 306–7
Potentiometric immunosensors 307–8
Protein engineering 279–91
  nanolithographics 360
Proteins 282–3
  α helix 69, 244, 283, 289
  β helix 293
Proton transfer reactions 353
Proton tunnelling 353
Protonation 162–3, 336–7

  enzymes 304–6
Pyroelectric effect 56–60
Pyroelectricity 47–70
  direct measurement 57
  dynamic measurement 58
  figure-of-merit 48, 57–8, 61, 109
  in polymers 48, 60–9
  primary/secondary effects 56–7
  pyroelectric coefficient 57–9, 67, 239
  static measurement 239, 240

Q-band optical absorption 254, 256
Quadrupole moment 42, 43
Quantum cellular automata systems 365
Quantum chemical models 105–8
Quantum chemistry 39–41, 105
Quartz crystal microbalance 230
Quasi-mechanical systems 366–7
Quinoid ring 100, 102, 155–6

Read operation 340–1
Receptors 286–7
  signal transduction 296–7
Red shift 107, 131
Redox reactions 8, 298–9
Reflection high energy electron diffraction (RHEED) 246–52
Refractive index 97
Repressor molecules 350
Resonance absorption 104
Rhodopseudomonas viridis 121
Rotaxanes (shuttle molecule) 354, 367

σ bond 142–3
Scalar 43
Scanning near-field optical microscopy (SNOM) 262
Scanning tunnelling microscopy (STM) 125, 253–4, 261–78
  basic concepts 263
  layered materials 268
  molecular adsorbates under ultra-high vacuum 269
  molecules at solid–fluid interfaces 269–74
  organic conductors 264–9
  prospects for molecular manipulation 274–6
  solid surfaces 264–9
Schottky islands and dot devices 362–3
Schrödinger equation 146, 148
Second nonlinear polarizability, γ 103–9
Second-harmonic generation 96–7, 238
  optical 238, 334
Semiconductors
  metal–insulator–semiconductor (MIS) structure 163
  models 104–5
Shuttle molecule 354, 367

Signal transduction, and molecular information 296–7
Silicon factor 372
Silicon technology
    30nm scale 372
    progress 346–8
Silver *see* Noble metals
Single electronic systems 361–5
    Coulomb blockade 361–2
    electron tunnelling 363
    hanging-resist technology 362
    Schottky islands and dot devices 362–3
Singlet state 336
Solid surfaces, scanning tunnelling microscopy 264–9
Solid-fluid interfaces, STM 269–74
Solitons 151–3, 155–9
Soret band optical absorption 254, 256
Spatial light modulators (SLM) 329
    electronically addressable 331
    microchannel plate 331
Spectral hole burning 115–16, 365–6
sp^2, sp^3 hybridized orbital 143
SSH theory 148
STM *see* Scanning tunnelling microscopy
Sulpholobus acidocaldarius, S-layer 360
Superconductivity 170–83
    BCS theory 172
    Cooper pair 173
    high T_c 78, 181
    Meissner effect 174
Surface plasmon microscopy 124
Surface plasmon resonance (SPR) 310
Surface pressure/tension 227–9
Switches 353–4
Switching
    Fredkin gate 355–6
    molecular electronic 125
Switching speed 6
Synapse 296

Tautomerization 117
TCNQ and TTF systems, conductive charge-transfer complexes 171–8
Television, high definition (HDTV) 217–18
Tensor 43, 63–4, 93
    Einstein convention 42, 43
    matrix notation 52, 63
    notation 43, 52, 93
    piezoelectric 63
    rank 43, 63
Theory of molecular electronics 29–46
    material properties 41–4

molecular arrangement 33–7
molecular interactions 37–41
molecular properties 30–3
Thermal analysis, thermogram 196–7
Thermochromism 113
Thin film transistors (TFT) 207, 214–15
Third-harmonic generation 97, 336
Tight-binding approximation 39, 146, 149, 150
Transfer ratio 225
Transistors
    high electron mobility transistor devices (HEMT) 319
    ion-selective field effect transistors (ISFETs) 303
    thin film 207, 214–15
    see also Field effect transistors
Transition dipole moment 43, 44, 101
Transition metal complexes 73–84
Transition probability, electrons 99–100, 105
Transmembrane signalling 284–5
Triplet state 74
Tunnelling *see* Electron tunnelling
Two-dimensional interacting spin system 360
Two-photon absorption 106, 138
Two-photon data 332, 340

Ultrahigh vacuum OMBE 246–9
Ultrathin film 243–5

Vacuum evaporation, physical vapour deposition 244
Valence band 147, 171
Valence bond 142–4
Van der Waal's epitaxy 249
Very large scale integrated circuits (VLSI) 347, 351
    architectures used 352
Vinylidene fluoride, copolymers 68–9
Voltammetric biosensors 303

Wannier function 87
Wave mixing 96–7
Wave-function, electron 40, 146
Wavevector 146
Wilhemy plate 227
Write once, read many (WORM) device 137

Y-type deposition 226
    see also Langmuir–Blodgett films

Zwitterion compounds 19, 127, 134–9, 238, 239
Zwitterions *see also Materials index*

Materials

Aberchrome 127
Acetylcholine 285, 289
Acridizinium salts 119
Adenosine diphosphate (ADP) 287
Adenosine triphosphate (ATP) 287, 289
Alcohol 298, 306
Alcohol dehydrogenase 298, 306
Alkanes 269–71, 275
Alkanols 271
Alpha-fetoprotein (AFP) 308
Ammonium ion 307
Anaesthetic 240, 282
Anthracene 36, 39, 44, 119, 236, 237
Antibody/antigen 8, 302, 307, 308, 311, 312
Arachidic acid 134, 135
Arsenic pentafluoride 89, 145, 154
Aziridine 126
Azobenzene 116, 120–6, 127, 139, 271
Azulene 35

β-carotene 142, 143
Bacteriorhodopsin BR 11, 12, 21, 25, 120–1,
    287, 315–41
Barium titanate 57, 58, 62
BEDO-TTF (bis(ethylenedioxo)-TTF) 169,
    171–7, 183
Behenic acid 132
Benzenes
    para-disubstituted benzenes 101, 110
    tetrathiobenzyl benzene 119
Betaine 126
Bichromophores 119
Bidentate nitroxide 78
Biological membrane see Cell
BIPS 127, 129
4,4'-Bipyrydyl 304
Blood cells 280
Bromine 179
Buckminsterfullerene see Fullerene
Butadiene 86, 87

Cadmium arachidate 233
Carbohydrate 283
Carboxylic acid 224, 239
Catalase 308
Cell, eukaryotic 350
Cell membrane 279–94
Charge–transfer complex 39, 80, 81, 85–6, 117,
    168–79, 183
Chloroform 122, 222
Cholesterol 192, 281–2
Chromophores 41, 115, 119–20, 124–5, 130–2,
    137, 160
Cinnamate 119
Copper 142, 154, 160, 172, 176

Copper acetate 76, 77
Copper phthalocyanine 240–1, 250–1, 253–4
Cu[N(CN)$_2$]Br 174, 176
Cu(NCS) 174
Cyanine 104, 116, 127, 131, 134, 236–7
Cyanobiphenyl 4, 6, 271
Cyclobutane 206, 208–9
Cytochromes 290

DAN 11
Decamethyl ferrocene (DMeFc) 82, 83, 85
Deoxyribonucleic acid (DNA) 114, 371
Diacetylene 69
Diacylglycerol 286
Dichalcogenide 249
Dihydronaphthalene 118
Diphenyl-3,4,9,10-perylenebis(dicarboximide)
    138
DM-DCNQI ligand 169, 176, 177
DMET 169
Dmit 169, 175, 178, 179
Docosylamine 240
Dynein 350

Emeraldine 145, 162
Eosin 126, 310
Ethane 142
Ethanol 306
Ethylene 143
Ethylenediamine tetraacetate (EDTA) 78
bis(ethylenedioxo)-TTF (BEDO-TTF) 169
bis(ethylenedithio)-TTF (BEDT-TTF) 169,
    171–7, 183

Fatty acids 221–6, 232, 234, 236, 280–3
Ferric chloride 154
Ferritin 361
Ferrocenyl nitrophenylethylene 11
Ferroelectric materials 49–50
Fluorinated biphenyl 6
Fulgide 126, 127, 139
Fullerene 20, 89, 168, 175, 181–3, 224

G-protein 285–6, 296
Gallium arsenide 145, 165
GDP 286
Glucose 165
Glucose oxidase 8, 165, 304, 306
Glycogen synthase 287
Gold 266, 275
Gramicidin 293
Graphite see Highly orientated pyrolytic
    graphite (HOPG)

H aggregation 31, 133, 134

H-PMC 131, 134
Haem *see* Porphyrin, iron
HC18 (octadecane) 130, 132, 133
Hexakis-heptyloxytriphenylenes 271
Hfac 79
Highly orientated pyrolytic graphite (HOPG) 249, 253, 268–74
HP-PBDCI-HP 138, 138
Human chorionic gonadotropin (hCG) hormone 308
Hydrogen peroxide 165, 304, 308
1-Hydroxypyrene-3,6,8-trisulfonate (HPTS) 310

Immunoglobulin G (IgG) 312
Indium tin oxide (ITO) 164, 203, 205, 213, 258
Inositol phosphates 289
Iodine 145, 179
Ionic polymers 122

J aggregates 131–4
J-PMC 131–4

Kinesin 350

Lactate 298
Lactate dehydrogenase 298
Leucoemeraldine 162
Lipid *see* Phospholipid
Liquid crystal 3–5, 20, 35–6, 41–3, 123–4, 185–6, 260, 273, 282
Liquid crystal polymer 69, 126, 186
    main chain 186, 198
    side chain 186, 198, 199
Liquid crystal polymers 198–200
Luciferase 309, 319
Luminol 312

M(dmit) 169
MDT-TTF 169
Meldola blue 306
MeN 178, 179
Metallomacrocycle 179
Metallophthalocyanines 254
4-Methoxyphenyl 33
Methyl stearate 133
1-Methyl-1,4-dithianium (MDT) 169, 171, 172, 173
Mitochondria 350
MMONS 11
mNA 11
Molybdenum disulphide 249–52
M(PC) 169, 169
MSP 129, 133, 134

NAD and NADP 291, 306
Naphthalene 36, 39, 44, 119
Naphthoxazine 132

Nickel dithiolene complex 108
Nickel phthalocyanine 180, 258–9
Nicotinamide adenine dinucleotide (phosphate) 291, 306
Ni(dmit)$_2$ 178, 179
Ni(Pc)I 180
NITE 90
NITR 78
para-nitroaniline (*p*-nitroaniline) 100–2
m-Nitrophenol 36
4-Nitrophenyl 33
Nitroso dimers 119
N,N-dimethylindoaniline 102
Noble metals 266–8

Ochratoxin A (OTA) 309
Octadecane (HC18) 130–3
Octadecanoic acid 222, 223
Octaethylporphrin 116
[*p*-(octyloxy)cinnamylidene]acetic acid 119, 120
Organoruthenium complex 230, 240
Oxirane 126

p-NA, a model NLO molecule 100–1
Pba 78
PBPCI 138
Penicillin 310
Pentacene 36
Pernigraniline 145, 162, 163
Peroxidase 309
Perylene 168
Phenanthrene 119
Phenol red 310
Phenylalanine 133
Phenylbenzoate 123, 271
Phospholipids 131, 281–3
    phosphatidylcholine 280, 281
    phosphatidylethanolamine 280, 281
    phosphatidylserine 280, 281
Photochromic materials, dissociation 118
Photomerocyanine (PMC) 127–34
Photosynthetic reaction centre 121, 283, 290–2, 121, 283, 290–2
Phthalocyanine (Pc) 115, 179, 254–8, 366
    copper phthalocyanine 240, 241, 250–1, 253–4
    nickel phthalocyanine 180, 258–9
    vanadyl phthalocyanine 255–8
Phytochrome protein 120
Piezoelectric and pyroelectric materials 48–60
    polymers 60–70
Platinum 177, 312
PMC (photomerocyanine) 127–34
Poly-L-glutamate 124
Poly(acetylene) (PA) 3, 9–11, 14, 87–8, 104, 107, 146–64, 359
    trans chain 359

Poly(aniline) (PANi) 3, 9, 145, 159, 162–3
Poly(benzyl glutamate) 69
Poly(carbene) 88
Polychloropolyphenylenes 89
Poly(diacetylene) 9, 11, 69
Polyene 106, 107
Polyhalide 179
Poly(isothianaphthene) (PITH) 145
Poly(methyl glutamate) 69
Poly(paraphenylene) (PPP) 145, 155
Poly(paraphenylene vinylene) (PPV) 9, 106, 107, 145, 159, 163–5
Poly(phenylene) 89
Poly(*p*-phylene sulphide) (PPS) 245
Poly(pyrrole) (Ppy) 145, 158, 162, 165, 306
Poly(thienylene vinylene) (PTV) 107, 108, 145, 159, 163, 164
Poly(thiophene) (PT) 9, 145, 161, 162, 164
Poly(urea) 69
Poly(vinyl carbazole) 40
Poly(vinylidene fluoride) (PVDF) 7, 47, 48, 60–9
Poly(vinylidene fluoride-trifluoroethylene) 69
Porin 283–4
Porphyrin 115, 290
Potassium bromide 255
Potassium chloride 181, 182, 251–2
Protein 281, 282, 285, 349, 350, 365
    integral 282
    peripheral 282
Protein engineering 360
Purple membrane 120, 121
2-Pyrazolines 33
Pyrenebutyric acid 310
Pyridinium derivative 125, 126, 135
α-Pyridone 119
Pyrolitic graphite *see* Highly orientated pyrolytic graphite (HOPG)

Quinodimethane 87, 88
Quinolinium derivative 135, 137, 171
Quinone 86

R-P3CNQ, R-Q3CNQ 135, 136, 137
Retinal 120, 121
Rhodium 269
Rhodopsin 120, 286
Rotaxanes 367
Rubidium 181
Ruthenium complex 230, 240

Saccharide 284
Salicylidine aniline 117, 118, 120
Schiff base 77
Sexithienyl 164
Silica 109
Silicon 7, 10, 14, 24, 25, 149, 346, 357, 362, 372
Silver 266–8, 275, 306
Sodium 178

Solitons 151–63, 355, 359
Sphingomyelin 281
Spirooxazine 117, 127, 128, 129
Spiropyrans 120–34, 139
Stearic acid 133, 230
Stilbene 121
Strontium barium titanate 57
Styryldicyanomethanides, R-P3CNQ, R-Q3CNQ 135, 136, 137
Sugar *see* Saccharide

Terphthalyidene-bis-4-n-pentylaniline 191
Tetracationic cyclophanes 367
Tetracene 2
Tetracyano-*p*-quinodimethane (TCNQ) 114, 125, 134–5, 168–79
Tetracyanoethylene (TCNE) 82–5, 90
    V(TCNE) 90
Tetrafluoroethylene (TeFE) 67, 68
Tetramethyltetraselenafulvalene (TMTSF) 169, 171, 172, 174, 175, 179
Tetrathiafulvalene (TTF) 168–74, 178, 179, 183
    structures 169
Tetrathiobenzyl benzene 119
TFPB-viologen-TFPB 134, 135
Thioindigo derivative 121
Transition metal complex 75, 85
Transition metal dichalcogenide 249
Tri-isopropyl silyl 367
Triarylimidazole 119
Triarylmethane 119
22-Tricosenoic acid 236, 240
Trifluoroacetate 77
Trifluoroethylene copolymer 48, 67
tetrakis[3,5-bis(trifluoromethyl)phenyl]borate (TFPB) 134–5
Triglycine sulphate (TGS) 7, 57, 60
Triiodobenzoate 271
Trimethylene methane 86, 87
Tripalmitine TP 130
TTF *see* Tetrathiafulvalene
Tyrosine kinases 285

Urease 301

Valinomycin 232, 293
Vanadium (TCNE) 90
Vanadyl phthalocyanine 255–8
Vinyl acetate copolymer (VCA) 68
Vinylidene cyanide copolymer (VDCN) 60, 68, 69
Vinylidene fluoride copolymer (VDF) 48, 60, 61, 67–9
Viologen 134, 135
Vitamin A aldehyde 316
VOPc *see* Vanadyl phthalocyanine

Zwitterion compounds 19, 127, 134–9, 238, 239
    D-π-A 134–9